Graphics with Mathematica

Fractals, Julia Sets, Patterns and Natural Forms

Graphics with Mathematica

Fractals, Julia Sets, Patterns and Natural Forms

by

Chonat Getz

and

Janet Helmstedt

2004
ELSEVIER
Amsterdam – Boston – Heidelberg – London – New York – Oxford
Paris – San Diego – San Francisco – Singapore – Sydney – Tokyo

ELSEVIER B.V.
Sara Burgerhartstraat 25
P.O. Box 211, 1000 AE Amsterdam
The Netherlands

ELSEVIER Inc.
525 B Street, Suite 1900
San Diego, CA 92101-4495
USA

ELSEVIER Ltd
The Boulevard, Langford Lane
Kidlington, Oxford OX5 1GB
UK

ELSEVIER Ltd
84 Theobalds Road
London WC1X 8RR
UK

First edition 2004

Library of Congress Cataloging in Publication Data
A catalog record is available from the Library of Congress.

British Library Cataloguing in Publication Data
A catalogue record is available from the British Library.

ISBN: 0-444-51760-x

♾ The paper used in this publication meets the requirements of ANSI/NISO Z39.48-1992 (Permanence of Paper).
Printed in The Netherlands.

Preface

The aim of this book is to introduce beginners and more advanced users of *Mathematica* to some of the software's graphic capababilities in the areas of Fractals, Discrete Dynamical Systems, Patterns and some Natural Forms. For beginners, it includes an introductory chapter on the basics of *Mathematica*. Detailed explanations are given for the use of *Mathematica* commands. The programmes used to construct the more complicated images are often built up step by step.

Fractals, including Julia and Mandelbrot sets, are of interest from the aesthetic and the mathematical points of view; this book will cater for both of these interests. Many people are intrigued by the infinite complexity of fractals like the Mandelbrot set and want to construct the images themselves. Indeed, graphical illustrations are essential for the scientific study of fractals. We describe the mathematical theory underlying the generation of fractals in an intuitive way, and give references to allow the more experienced mathematician to study the theory in detail.

Because of its order and regularity, mathematics is saturated with patterns. Only recently, and thanks to the advent of computers, have some of the visual patterns been revealed. In this book we show how to construct many types of patterns. Most of the routines described are straightforward, easy to understand and readily adaptable to variation (ideas for variations are usually presented), enabling readers to use their own creativity to construct a multitude of different graphics.

Some of the natural forms we construct are fractal-like, others, such as coiled shells are constructed using sequences of 2D or 3D parametric curves, others as 3D surface plots.

Beginning *Mathematica* users will find our method a very interesting way to learn basic commands and programming techniques such as pure functions, the use of tables and other tools because the results of their efforts will often be aesthetically pleasing, sometimes unexpected, sometimes amusing, sometimes amazing. These techniques can form the basis for further study of the use of *Mathematica* in other areas of mathematics and science.

Chapter 1 provides the basic *Mathematica* commands which will be required in the rest of the book. With each copy of *Mathematica*, Wolfram Research provides an easy-to-follow booklet entitled 'Getting Started with *Mathematica*'. Beginners receive directions on the use of the booklet The ability to use *Mathematica*'s Help section is is vital for the successful use of the software. A guide is provided for the use of this section.

Chapters 2 and 3 describe techniques for constructing many patterns and natural forms both 2D and 3D. These chapters also provide practice in the use of color and in constructing certain types of graphics which will be required in the rest of the book. The techniques described can be used to construct a wealth of graphics.

Chapter 4 provides some of the basic constructions and mathematical ideas which will be needed in the later chapters on fractals.

In Chapter 5, we use Roman Maeder's packages 'AffineMaps', 'IFS' and 'ChaosGame' which are shipped with *Mathematica,* to construct affine 2D fractals, beginning with the Sierpinski triangle, moving on to Sierpinski relatives, tree-like forms and others. We illustrate the idea of an Iterated Function System by showing the successive images derived from the step-by-step construction of a fractal. Readers are encouraged to construct their own variations on the fractals introduced in this chapter.

In Chapter 6, we introduce programs for constructing non-affine 2D fractals such as Julia sets as well as some for 3D fractals, which are naturally more complicated, but detailed explanations are given.

In Chapter 7, we show how to construct certain Julia sets, the Mandelbrot set and other parameter sets, using the ideas and constructions of Chapter 4. The main ideas needed for the construction of the (filled) Julia sets of polynomials and the Julia sets of certain transcendental functions, as well as parameter sets, are the notions of bounded and unbounded orbits of points under the action of a function. The more difficult Julia sets to construct are those of rational functions. Further intuitive mathematical background is given in this section, such as a description of the Riemann sphere. We give detailed instructions on how to attempt the construction of the Julia set of a rational function. However, some Julia sets may prove intractable We also display graphically some interesting results on the use of Newton's method for solving equations. The ideas used in the section on Newton's method are fairly simple and can be employed both to find possible starting points for solving equations and to construct some beautiful graphics.

Chapter 8 has three sections: construction of Sierpinski relatives as Julia sets of piecewise defined functions; Truchet-like patterns; and construction of images of coiled shells.

Many exercises are given throughout the book.

A CD-ROM accompanies the book. It contains the text of the book.
In the printed version of the book, most of the colored images are collected together in the

color pages. On the CD-ROM they are shown in context. Further, some of the colored images in the printed version are in gray scale, while they appear in color on the CD-ROM. In the printed version of the book, many programs for generating images are given as examples for the reader to try. Some of these images are shown on the CD-ROM.

Acknowledgements

We wish to thank Wolfram Research for permission to quote extracts from *Mathematica*'s Help section in this book.

We are grateful to members of Wolfram Research Technical Support, especially Henry Kwong, for help with formatting.

We also thank Roman Maeder for permission to use his packages 'AffineMaps', 'IFS' and 'ChaosGame', which are shipped with *Mathematica* and are available to all *Mathematica* users.

We thank Robert L. Devaney for assistance with a problem on Julia sets.

Contents

Chapter 5   **Using Roman Maeder's Packages Affine**
                     Maps, Iterated Function Systems and
                     Chaos Game to Construct Affine Fractals   161

Chapter 6   **Constructing Non-affine and 3D**
                     Fractals Using the Deterministic
                     and Random Algorithms   197

Chapter 1

Basics

Introduction

In this Chapter we introduce the reader to the basic commands, calculations and their syntax which will be needed in the later chapters. Some interesting patterns are also included. There are many conventions to be learned, such as using capital letters for built-in functions, the bracketing conventions etc. Practice examples are given for these. It is important for a *Mathematica* user to know how to use the Help section of *Mathematica*, as books on the use of *Mathematica* would have to be very large indeed to cover all the capabilities of *Mathematica*. So, in this chapter, detailed instructions are given on the use of the Help section of *Mathematica*. Also, we give exercises involving the use of Help, where the reader is required to look up a certain section of Help and then try an example. We also include some quotations from Help to stress important points.

1.1 The Booklet: Getting Started with *Mathematica*

1.1.1 Getting Started with *Mathematica*

A booklet entitled 'Getting Started with *Mathematica'*, published by Wolfram Research, accompanies every copy of *Mathematica*. In this Chapter, we start by showing you how to perform basic calculations, learn the basic syntax of *Mathematica* and some basic commands using this booklet and the Help in *Mathematica*.

Start by using the booklet to install and register your copy of *Mathematica*.
(You may need to consult the appendix entitled 'If You've Never Used Windows Before'.)
Now go to the section entitled 'Starting *Mathematica'*, and learn how to start a *Mathematica* session. Do the simple calculation, noticing the input and output labels you obtain, and read the tip at the bottom of the page on a different way of implementing your input. Try a few more additions on your own. You should now be able to add two or more natural numbers with *Mathematica*.

1.1.2 Your First *Mathematica* Calculations

Go to this section of the booklet and try the first 2 examples given, noting the method of executing a *Mathematica* command. Read the section on 'Some *Mathematica* conventions', noting the bracketing conventions, the capitalisation of built-in function names and the 3 ways of expressing the product of 2 numbers or expressions. Note that the functions mentioned can have complex arguments.

Exercise:
1) Find the product of the following numbers:
a) 3 and 4;
b) $Sin[\pi / 3]$ and 24;
c) $Cos[\pi / 4]$ and $Sin[2\pi / 3]$;
d) $2^{10} + Tan[2]$ and $Cos[5]$.
2) Evaluate $Sin[2+3I]$.

1.1.3 Error Messages

If you make a syntax error you may receive an error message. For example, suppose you type:

sin[3 Pi]

After pressing Shift-Enter (or Enter on the extended key board), you receive the following message:

General::spell1 : Possible spelling error: new symbol name "sin" is similar to existing symbol "Sin".

$\sin[3\,\pi]$

The mistake, as you can see, was that the convention of using capital letters for built-in functions was not obeyed.

In the following case, no error message was received, but no calculation was done. What mistake was made here?

Cos[2] Cos[3]   Cos[2] Cos[3]

In this case, the product was incorrectly written. There should have been a space or an asterisk between the factors, or at least one factor should have been enclosed in brackets.

1.2 Using Help in *Mathematica*

We shall use the Help section of *Mathematica* a great deal in this chapter. We show you how to learn about some syntax, how to do simple calculations and how to use Help. Besides reading this section, more information can be found in the section Using Online Help in the booklet. Sections of Help can be printed for reference.

Click on Help in the tool-bar. A menu drops down. Click on Help Browser. You will see a notebook. Under the button Go, there are a number of headings including: Built-in Functions, Add-ons, The *Mathematica* Book, Getting Started, Front End ('Other Information' in Version 4), and Master Index. In this section we illustrate the use of these sections of Help, with the exception of Add-ons, which will be illustrated in 1.8.7.

Note: The Help Browser differs slightly in different versions of *Mathematica*. In this book, we shall refer to the version of the Help Browser in Version 5.

1.2.1 Using 'The *Mathematica* Book' Section of Help

We start with the use of the Contents section of 'The *Mathematica* Book'. Here you may obtain an idea of the different capabilities of *Mathematica* and of the different topics covered.

Click on The *Mathematica* Book, a menu drops down. Click on Contents. A list of contents appears. The list consists of headings in black and sub-headings in blue. Notice that the sections and sub-sections are numbered. You can click on any of the blue sub-headings to view the material in it. Scroll down to the heading: Notebook Interfaces, click on this and read this section which gives information about starting *Mathematica*, and about Input and Output. In the space at the end of the paragraph, practise a couple of addition examples, such as: $12 + 28$ and notice the blue input and output signs which appear.

Here is an important extract from the section entitled: Suggestions about Learning *Mathematica* ('Getting Started' in Version 4.2):

As with any other computer system, there are a few points that you need to get straight before you can even start using Mathematica. For example, you must know how to type your input to Mathematica..... Once you know the basics, you can begin to get a feeling for Mathematica by typing in some examples from this book. Always be sure that you type in exactly what appears in the book—do not change any capitalization, bracketing, etc. After you have tried a few examples from the book, you should start experimenting for yourself. Change the examples slightly, and see what happens. You should look at each piece of output carefully, and try to understand why it came out as it did. After you have run through some simple examples, you should be ready to take the next step: learning to go through what is needed to solve a complete problem with Mathematica.

■ Arithmetical Calculations

Now click on Back on the tool-bar at the top. You will return to Contents. There is another way to open a section of 'The *Mathematica* Book'. Underneath the Contents is a heading: A Practical Introduction to *Mathematica*. Click on this, and a list of sub-sections of this section drops down in the next column. Click on Numerical Calculations in the second column of subject headings. A third column drops down. Click on Arithmetic. You will see the statement: 'You can do arithmetic with *Mathematica* just as you would on an electronic calculator'. Read this section, and then scroll to the end of the section on Arithmetic, and you can practise your own calculations there. First practise with simple calculations, so that, if you obtain the wrong answer, or no answer, you may scroll back to find your mistake.

Exercise:
Use *Mathematica* to calculate: $12+3$, 7×5, 2^3, $\frac{12}{4}$, $\frac{6}{1+2}$.

■ The command N

In future, when requesting you to read a section of the *Mathematica* book we shall sometimes write something like: 'Go to the Help-Browser and click along the route - *Mathematica* Book - Practical Introduction - Numerical calulations - Exact and Approximate Results'. We request that you do this, and learn about the use of *Mathematica*'s command **N**.

Exercise:
Find, using 2 different methods, approximations to π^2, $375/68$.

■ Some Mathematical Functions

Another method we shall sometimes use to refer you to a section of the *Mathematica* book is to give you the number of the section. We wish you to study the section numbered 1.1.3. To access this section, proceed as follows: In the Help Browser, in the space next to Go, type the number 1.1.3, click on 'The *Mathematica* Book' and then on Go. The section 1.1.3, 'Some Mathematical Functions', is opened. For the moment, we shall only be concerned with the exponential, logarithmic, trigonometric and square-root functions. So notice the syntax for these. Scroll down past the list of common functions and study the rest of the section, noting particularly the paragraphs on bracketing and on capital letters. We shall use *Mathematica*'s notation for the numbers e and i namely E and I. Notice that π can also be written as Pi.

Exercise:
Find the exact values of $\sqrt{16129}$, $\text{Arcsin}[1/2]$, $\text{Log}_E[1/E]$, $\text{Cos}[\pi/3] + \text{Sin}[\pi/6]$.

■ The Transformation Rule /.

This is a very useful technique for finding values of an expression involving one or more variables for different values of the variable(s). Go to the section: Applying Transformation Rules, in the Contents of the *Mathematica* Book. Read the first part of this section and find out about the use of the replacement command /.

Exercise:
Use /. to find the value of:
$\text{Sin}[x] + \text{Cos}[x]$ when $x = \frac{\pi}{8}$; yE^x when $x = 10$ and $y = 0.003$.

1.2.2 Using the Master Index

The Master Index is very useful if you wish to find out about a general topic, such as equations or polynomials, but do not know any specific terms used by *Mathematica* in these areas. We use the Master Index to find out about brackets and complex numbers.

■ Brackets

Suppose you wish to learn about the different types of brackets and their uses in *Mathematica*. In the Help Browser, next to Go, type Brackets, click on Master Index and then on Go. A list of sections of 'The *Mathematica* Book' containing information on brackets drops down. Click on 1.2.5 and the section 1.2.5 entitled 'The Four Kinds of Bracketing in *Mathematica*' drops down. Scroll upwards to the start of the section. We shall not be concerned with double square brackets for now. Having studied the section, click on Back again, and then on 1.1.6. Scroll to the top of this section which is entitled 'Getting Used to *Mathematica*'. This section gives reasons for the syntax and notation used in *Mathematica*.

Exercise:

Correct the following commands:

$$\textbf{Sin (2)}; \ \textbf{log[e]}; \ \ \textbf{[3 x + 1] [2 x − 4]}$$

■ Complex Numbers

In the Help Browser, next to Go, type Complex Numbers, click on Master Index and then on Go. A list of sections of the *Mathematica* book containing information on Complex numbers drops down. Click on 1.1.5. and read this section, noting carefully the syntax for the complex number I and the list of complex number operations.

Exercise:

Calculate the following:

1) The product of $1 + I$ and $2 − I$.

2) $1 − I$ divided by $1 + I$.

3) $\sqrt{2.3 − I}$, and check your result.

4) The conjugate of $Sin[2 + I \frac{\pi}{4}]$.

5) $|3 − 2 I|$.

6) The imaginary part of $Sin[1.0 + 2 I]$.

1.2.3 Built-in Functions

To find out about some of the many built-in functions, proceed as follows: In the Help-Browser, click on Built-in Functions. A menu drops down, giving a list of basic categories of Built-in Functions. Click on one of these categories, Algebraic Computation, say. A menu of sub-categories drops down. Click on one of these, Basic Algebra, say. A list of Built-in Functions for Basic Algebra appears.

We now give an example on the use of Built-in Functions. Suppose you wish to find out about the Built-in Function **Mod**. In the Help Browser, next to Go, type Mod, then click on Built-in Functions and then Go. Read the definition of **Mod[m, n]**. Click on Further Examples to see an application of this command.

Exercise:

Find the remainder when 9768321 is divided by 193.

Another method of finding out about a built-in function is to type ? followed by the name of the built-in function and then press shift-enter. For example:

> **? Factorial**

> n! gives the factorial of n.

? D

D[f, x] gives the partial derivative of f with respect to x. D[f, {x, n}] gives the
    nth partial derivative of f with respect to x. D[f, x1, x2, ...] gives a mixed derivative.

Exercise:

1) Look up **Im** in the Master Index and find the imaginary part of:

a) $[3 + 2\,I]^3$;

b) $7\,E^{\frac{I\pi}{8}}$.

2) Find the derivative with respect to x of Arctan[Sin[x]].

3) Find the value of 19!.

We suggest you browse through the section Built-in Functions.

A third method of finding out about a *Mathematica* Built-in Function is to type the name of the function, and then press F1. The relevant section in Help will appear.

Exercise:

Type N, press key F1, read the section on this command, and then find the numerical value of:

1) E^2 .

2) $|\operatorname{Sin}[\pi/15] - 23/51|$.

1.2.4 Front End (**Other Information** in Version 4.2)

We use this section to learn about Menu Commands. In Help Browser, click on Front End (or Other Information) then Menu Commands. If you wish to know how to start a new file or open, close or save an existing file, click on File Menu, and then the relevant section thereof. If you wish to undo your most recent command or copy or paste a selection from a notebook, click on Edit Menu and then the relevant section thereof.

Keyboard shortcuts for all of these commands are obtained by clicking on Keyboard Shortcuts and then choosing the relevant operating system i.e. Microsoft Windows, Macintosh, Next or X.

1.2.5 Getting Started

We use the Getting Started section in the Help Browser to learn about Cells. In Version 5, click along the route: Getting Started - Working with Notebooks - Working with Cells and Notebooks. In Version 4.2, click along the route: Getting Started/Demos - Getting Started - Working with Notebooks - Working with Notebooks and Cells.

Study the first part on Creating a new cell. If interested, look at the section on creating a text cell. A cell may be selected by clicking on the bracket at the right of the cell. The cell contents may then be modified using the drop down menus on the menu bar.

1.3 Using Previous Results

The symbol **%** denotes the last result generated, and the symbol **%%**, the last but one generated. Here is an example of their use:

Sin$\left[3\frac{\pi}{5}\right]$ // N   0.951057

2 Tan$\left[\frac{\pi}{3}\right]$ $2\sqrt{3}$

% + %%   4.41516

Another example:

Sin$\left[\frac{\pi}{3}\right]$ $\frac{\sqrt{3}}{2}$

N[%]   0.866025

Exercise:
1) Find the derivative with respect to x of Sin[x] / x.
2) Use your result from 1) above to find the value of the derivative at x = $\frac{\pi}{3}$.
3) Click along the route - *Mathematica* Book - Practical Introduction - Algebraic Calculations - Transforming Algebraic Expressions. Read the notes on the commands **Expand** and **Factor**, and then expand and re-factor the expression: $(x - 2y)^3 (1 + y^2)$, using **%**.
4) Look up Partial Fractions in the Master Index, and then:
a) Find partial fractions for $(x + 1)/(x - 1)^2 (x^2 + 1)$, and, using **%**, check your answer;
b) Express as a single rational function: $1/(x + 2)^2 - 2/(x^2 + 1)$, and check your result.

From Help - The *Mathematica* Book 1.2.1:

If you use a text-based interface to Mathematica, then successive input and output lines will always appear in order, as they do in the dialogs in this book. However, if you use a notebook interface to Mathematica, as discussed in Section 1.0.1, then successive input and output lines need not appear in order. You can for example "scroll back" and insert your next calculation wherever you want in the notebook. You should realize that % is always defined to be the last result that Mathematica generated. This may or may not be the result that appears immediately above your present position in the notebook. With a notebook interface, the only way to tell when a particular result was generated is to look at the Out[n] *label that it has. Because you can insert and delete anywhere in a notebook, the textual ordering of results in a notebook need have no relation to the order in which the results were generated.*

1.4 Some Type-setting

In this section, we shall learn how to enter expressions such as $\frac{x}{y+z}$, π, $\in$, $\leq$, 2^r, $\sum_{r=1}^{n} a_r$. This can either be done using palettes or using key-board shortcuts.

1.4.1 Using Palettes

In Version 5, click along the route: Getting Started - Working with Notebooks - Using Palettes. In Version 4.2, click along the route: Getting Started/Demos - Getting Started - Working with Notebooks - Using Palettes.

Read the opening paragraph, click on Graphic to see the Basic Input palette, scroll down and learn how to enter a Greek letter and how to enter and evaluate an indefinite integral. The 7 different types of palettes may be accessed by clicking on File on the tool-bar and then on Palettes and then on the particular palette you wish to use. Explore these palettes to find the expressions available.

Exercise:
Open the Basic Input palette, and enter and evaluate:

a) $\sum_{r=1}^{n} 2^r$;

b) $\sqrt[3]{59}$;

c) $\frac{1+2\,\mathrm{Log}[15]}{7}$.

1.4.2 Using Keyboard Shortcuts

It is very convenient to use keyboard shortcuts instead of palettes to enter certain frequently used expressions. We suggest you learn or make a copy of key-board short cuts for entering what *Mathematica* calls '2D Expressions'; these are fractions, exponents, subscripts etc. Also of importance are Greek letters, the double-struck letters: $\mathbb{N}$, $\mathbb{R}$, $\mathbb{C}$; the symbols ∞, $\rightarrow$, $\leq$ which will be used often. Most of these are easy to remember.

1.4.3 Entering 2D Expressions

In the Help Browser click on Front End (Other Information in Version 4.2) - 2D Expressions Input - Entering 2D Expressions - Overview.
Study methods of entering $\frac{x}{y}$, x^m, $\sqrt{x}$, x_i. Then go to the sections: Fractions (in this section, note particularly how to enter $\frac{x+y}{z}$), Powers, Square Roots, Subscripts, Superscripts. Here is a summary of methods of entering some of the 2D expressions:

Action	Keystrokes	Result
power	x CTRL ^ 3	x^3
fraction	x CTRL / y	$\frac{x}{y}$
square root	CTRL 2 x	$\sqrt{x}$
subscript	x CTRL _ 3	x_3
move cursor out of 2 D structure	CTRL - SPACE	□

To enter $\frac{x+y}{z}$:   CTRL [/] x+y Tab z

Exercise:

Enter the following expressions:

E^x, a_n, $\sqrt{3}$, $\frac{5}{2}$, 2^{3+y}, $\frac{5}{2^x}$, $\sqrt{x^2 + y^2}$, $\frac{2+y^3}{15}$, $\sum_{r=1}^{n} a_r$.

1.4.4 Entering Special Characters

The character ∞ is entered in the following way: Press the Escape key and you will see the symbol ⋮ then type inf, you will see ⋮inf. Now press the Escape key again. You will see ∞. All special characters can be entered in this way: Escape key - expression - Escape key. You need only remember what expression to use for each character. It is a good idea to make a list of ways of entering the special characters you will need. To find out about entering Greek letters, in the Help Browser, go to section 3.10.3, Letters and Letter-like Forms, of the *Mathematica* Book. Then scroll down to the heading: Variants of English Letters to find how to enter the double struck letters. Keyboard shortcuts for entering many other special characters can be found in the section 3.10.4 . Here are some of the ones we shall need:

Form	Keyboard Shortcut
$\leq$	ESC < = ESC
$\rightarrow$	ESC − > ESC
$\in$	ESC elem ESC
$30\,°$	30 ESC deg ESC
$\mathbb{R}$	ESC dsR ESC
∞	ESC inf ESC
α	ESC a ESC
π	ESC p ESC

Exercise:

1) Enter the following expressions: $Sin[\theta]$, $E^{I\phi}$, $\mathbb{R}^2$, If $x \leq 0$, $E^x \leq 1$, $a_n \rightarrow 1$ as $n \rightarrow \infty$, $Tan[60\,°]$;

2) Go to 3.10.4 and find out how to enter $\neq$ and then enter $2 \neq 3$.

Exercise:

In this exercise, we give you practice in the use of type-setting and the use of Help.

1) Use the Master Index to find out about Partial Fractions, and then use *Mathematica* to resolve $\frac{(4\,x^2-5\,x+3)}{(x^2-1)\,(x-1)}$ into partial fractions.

2) Find out about the built-in function **Factor**, and then factorise $8 + 20\,x + 6\,x^2 - 5\,x^3 - 2\,x^4$.

3) Find the remainder when $x^5 - 2\,x^2 + 3\,x + 1$ is divided by $x^2 - 2$.

4) Find out about the command **D** using ?D and then find the derivative with respect to x of $\mathrm{Tan}[x^2]\,/\,(x^3 + 3)$.

5) Evaluate 3^5, followed, with a minimum amount of typing, by $3^5 \times 17$.

6) In the Help Browser, click along the route Built-in Functions - Mathematical Functions - Elementary Functions and then ask *Mathematica* to evaluate the numerical value of Arctan[3].

7) Find the remainder when the polynomial $x^5 + 5\,x^4 - 3$ is divided by $2\,x - 1$.

8) Use /. to find the numerical value of $\sqrt{x^4 + x}$ when $x = 2.78$.

1.5 Naming Expressions

Suppose you want to carry out a few operations on the expression: $\frac{x^3+2\,x^2-3\,x+5}{(x-)\,(x+1)}$. We assign the name 'a' to the expression, and follow the assignment with a semi-colon and then press shift-enter. The semi-colon ensures that *Mathematica* records the instruction but does not show any output, which can be a useful space-saver.

$$a = \frac{x^3 + 2\,x^2 - 3\,x + 5}{(x - 10)\,(x + 1)};$$

From now on, during the present session, unless instructed otherwise, *Mathematica* will replace 'a' by the above expression.

We find the derivative of a, find the value of a when $x = 2$, and resolve a into partial fractions.

D[a, x]

$$\frac{-3 + 4\,x + 3\,x^2}{(-1 + x)\,(1 + x)} - \frac{5 - 3\,x + 2\,x^2 + x^3}{(-1 + x)\,(1 + x)^2} - \frac{5 - 3\,x + 2\,x^2 + x^3}{(-1 + x)^2\,(1 + x)}$$

Apart[a, x] $11 + \frac{1175}{11\,(-10+x)} + x - \frac{9}{11\,(1+x)}$

a /. x -> 2 $-\frac{5}{8}$

Throughout the current session, *Mathematica* will replace a by the above expression (even if you open a new file and do not exit *Mathematica*). If you wish to use the symbol a to represent other expressions later on in the same session type: **a =.** or **Clear[a]** and press Shift-enter.

a =.

You can now give a new definition for a.

a = 5   5

a²⁰   95367431640625

From Help - The *Mathematica* Book 1.2.2:

It is very important to realize that values you assign to variables are permanent. Once you have assigned a value to a particular variable, the value will be kept until you explicitly remove it. The value will, of course, disappear if you start a whole new Mathematica session.
Forgetting about definitions you made earlier is the single most common cause of mistakes when using Mathematica. If you set x = 5, Mathematica assumes that you always want x to have the value 5, until or unless you explicitly tell it otherwise. To avoid mistakes, you should remove values you have defined as soon as you have finished using them.
*The variables you define can have almost any names. There is no limit on the length of their names. One constraint, however, is that variable names can never start with numbers. For example, x2 could be a variable, but 2x means 2 * x.*
Mathematica uses both upper- and lower-case letters. There is a convention that built-in Mathematica objects always have names starting with upper-case [capital] letters. To avoid confusion, you should always choose names for your own variables that start with lower-case letters.

Mathematica does not remember anything defined in a previous *Mathematica* session, although it keeps screen records. If you want to use again the name given to an expression or command used in a previous session you must 'reactivate' it by placing the cursor after the expression and pressing shift-enter.

Exercise:
Assign a name to E^{2x+3Iy} then find its its absolute value and the value of $5E^{2x+3Iy}+2I$ when $x = 2$ and $y = -2$.

1.6 Lists

1.6.1 Making Lists of Objects

We shall frequently use lists in this book. We shall plot lists of points, curves, polygons etc.

From Help - The *Mathematica* Book 1.2.3:
In doing calculations, it is often convenient to collect together several objects, and treat them

as a single entity. Lists give you a way to make collections of objects in Mathematica. As you will see later, lists are very important and general structures in Mathematica.

A list such as {3, 5, 1} is a collection of three objects. But in many ways, you can treat the whole list as a single object. You can, for example, do arithmetic on the whole list at once, or assign the whole list to be the value of a variable.

list1 = {0.4, 2, 5, 1}; list2 = {−3, 4, 0.7, 4};

6 list1 + list2   {−0.6, 16, 30.7, 10}

list2 + 2   {−1, 6, 2.7, 6}

One of the most useful properties of lists is that built-in functions can be applied to the elements of a list. Here is an example:

N[Log[list1]]   {−0.916291, 0.693147, 1.60944, 0.}

Exercise:

1) A{2, -3, 4} and B{0, 5, -1} are 2 points in $\mathbb{R}^3$. Find vector AC = $\frac{2}{3}$ vector AB.

2) Find the numerical value of the Cosine of each number in the list: {3 °, 6 °, 9 °, 12 °}.

1.6.2 Constructing Lists using the Command Table

Suppose you wish to construct a list whose entries are of the form $\frac{\text{Sin}[n\,\theta]}{n}$ for $1 \le n \le 10$. It would be tedious to type out all 10 terms. The command **Table** can be used to do this:

? Table

Table[expr, {imax}] generates a list of imax copies of expr. Table[expr, {i, imax}] generates a list of the values of expr when i runs from 1 to imax. Table[expr, {i, imin, imax}] starts with i = imin. Table[expr, {i, imin, imax, di}] uses steps di. Table[expr, {i, imin, imax}, {j, jmin, jmax}, ...] gives a nested list. The list associated with i is outermost.

Here is an example:

$$\textbf{Table}\left[\frac{\textbf{Sin}[\textbf{n}\,\theta]}{\textbf{n}}, \{\textbf{n, 2, 10}\}\right]$$

$$\left\{\frac{1}{2}\,\text{Sin}[2\,\theta], \frac{1}{3}\,\text{Sin}[3\,\theta], \frac{1}{4}\,\text{Sin}[4\,\theta], \frac{1}{5}\,\text{Sin}[5\,\theta],\right.$$
$$\left.\frac{1}{6}\,\text{Sin}[6\,\theta], \frac{1}{7}\,\text{Sin}[7\,\theta], \frac{1}{8}\,\text{Sin}[8\,\theta], \frac{1}{9}\,\text{Sin}[9\,\theta], \frac{1}{10}\,\text{Sin}[10\,\theta]\right\}$$

We now find the values of the elements of the above list when $\theta = \frac{\pi}{6}$.

% /. $\theta \to \frac{\pi}{6}$ $\left\{\frac{\sqrt{3}}{4}, \frac{1}{3}, \frac{\sqrt{3}}{8}, \frac{1}{10}, 0, -\frac{1}{14}, -\frac{\sqrt{3}}{16}, -\frac{1}{9}, -\frac{\sqrt{3}}{20}\right\}$

We shall use the command **Table** frequently.

1.6.3 Applying Built-in Functions to Lists

Here is an example of a built-in function applied successively to the elements of a list defined with the command **Table**:

$$\text{Sin}\left[\text{Table}\left[n\,\frac{\pi}{12},\,\{n,\,0,\,12\}\right]\right]$$

$$\left\{0,\,\frac{-1+\sqrt{3}}{2\sqrt{2}},\,\frac{1}{2},\,\frac{1}{\sqrt{2}},\,\frac{\sqrt{3}}{2},\,\frac{1+\sqrt{3}}{2\sqrt{2}},\,1,\,\frac{1+\sqrt{3}}{2\sqrt{2}},\,\frac{\sqrt{3}}{2},\,\frac{1}{\sqrt{2}},\,\frac{1}{2},\,\frac{-1+\sqrt{3}}{2\sqrt{2}},\,0\right\}$$

Exercise:
Calculate the numerical values of $\text{Exp}[\frac{n}{20}]$ for n from 0 to 20.

1.6.4 Some Operations on Lists

■ The command Flatten

The elements of a list may themselves be lists. Such a list is called a nested list by *Mathematica*. For example: {{a,b},{c,d},{{e}}}. The command **Flatten** may be used to eliminate some of the interior brackets:

? Flatten

Flatten[list] flattens out nested lists. Flatten[list, n]
    flattens to level n. Flatten[list, n, h] flattens subexpressions with head h.

Here **Flatten** is applied to a nested list without any qualification, and it removes all brackets except the outermost ones.

Flatten[{{a, b}, {c, d}, {{e}}}]   {a, b, c, d, e}

The 'levels' of a nested list are measured from the outside inwards, with the outermost brackets having level 0. In the following example the level 1 brackets are removed.

Flatten[{{a, b}, {c, d}, {{e}}}, 1]   {a, b, c, d, {e}}

Exercise:
Experiment with the command **Flatten**. Examples of the use of this command are given in 1.8.4.

■ The commands Join and Union

? Union

Union[list1, list2, ...] gives a sorted list of all the distinct elements that appear in any of the listi.
Union[list] gives a sorted version of a list, in which all duplicated elements have been dropped.

Here is an example:

Union[{2, 1, −3}, {4, 2, 1, 0}]   {−3, 0, 1, 2, 4}

? Join

Join[list1, list2, ...] concatenates lists together.
Join can be used on any set of expressions that have the same head.

The command **Join** applied to the above example. Notice that repeated elements are not omitted and the elements are not sorted.

Join[{2, 1, −3}, {4, 2, 1, 0}]   {2, 1, −3, 4, 2, 1, 0}

Exercise:
1) Experiment with the above 2 commands.
2) Using **Table**, make lists of the first 20 even and the first 20 odd natural numbers, and then apply the commands **Join** and **Union** to the pair of lists.

There are many operations that can be performed on lists. Look up Lists in the Master Index and inspect some of the sections of the *Mathematica* Book which apply to lists.

1.7 Mathematical Functions

1.7.1 Standard Built-in Functions

In 'The *Mathematica* Book', 1.1.3, as we have seen, there is a list of the more common built-in functions. Section 3.2 of the *Mathematica* Book contains a complete list of all standard mathematical functions used in *Mathematica*.

From Help - The *Mathematica* Book 3.2: Mathematical Functions:
Mathematical functions in Mathematica are given names according to definite rules. As with most Mathematica functions the names are usually complete English words, fully spelled out. For a few very common functions, Mathematica uses the traditional abbreviations. Thus the modulo function, for example, is Mod, *not Modulo.*

▪ Numerical Functions

A real number x can be expressed uniquely in the form $x = x_i + x_f$, where x_i is an integer and $0 \leq x_f < 1$. In such a case:

a) **Floor[x]** $= x_i =$ **IntegerPart[x]**, if $x \geq 0$;

b) **IntegerPart[x]** $= x_i$ if $x < 0$;

c) **FractionalPart[x]** $= x -$ **IntegerPart[x]**.

Numerical functions that we need in this book can be found in Section 3.2.2 of 'The *Mathematica* Book'.

IntegerPart[−2.3] $\qquad\qquad$ −2

Floor[−2.3] $\qquad\qquad$ −3

Another numerical function we shall some-times use is **Sign:**

? Sign

Sign[x] gives −1, 0 or 1 depending on whether x is negative, zero, or positive.

Exercise:

1) Find the **Floor** and **IntegerPart** of Sin[3.95 n], for n from 1 to 12.

2) Find the **Sign** of Cos[3].

▪ Pseudorandom Numbers

We shall use the command **Random** quite often, in adding perturbations to graphic shapes and in coloring graphics, and in generating graphics which have a great deal of randomness.

? Random

Random[ ] gives a uniformly distributed pseudorandom Real in the range 0 to 1. Random[type, range] gives a pseudorandom number of the specified type, lying in the specified range. Possible types are: Integer, Real and Complex. The default range is 0 to 1. You can give the range {min, max} explicitly; a range specification of max is equivalent to {0, max}.

Here is an example of a table of 16 pseudorandom real numbers between -5 and 5.

Table[Random[Real, {−5, 5}], {n, 1, 16}]

{−4.17883, −4.2987, 3.83673, 3.33992, 1.12981, 3.09228, 1.26555, −3.63101, 3.03258, 0.199143, 2.86381, −1.35706, 1.51259, −1.87778, 2.43393, 3.16453}

Here is a list of 5 random real numbers between 0 and 1:

Table[Random[ ], {n, 1, 5}]

{0.843793, 0.357468, 0.724706, 0.358927, 0.498827}

Exercise:

Look up **Random** in Built-in Functions, and make a randomly chosen list of 20 zeros and ones.

Read the following extract from Help - 3.2.7: Functions that do not have Unique Values, if you are not familiar with principal values of real or complex functions:

There are many mathematical functions which, like roots, essentially give solutions to equations. The logarithm function and the inverse trigonometric functions are examples. In almost all cases, there are many possible solutions to the equations. Unique "principal" values nevertheless have to be chosen for the functions. The choices cannot be made continuous over the whole complex plane. Instead, lines of discontinuity, or branch cuts, must occur. The positions of these branch cuts are often quite arbitrary. Mathematica makes the most standard mathematical choices for them.

1.7.2 User-defined Functions

In this section, we show how you may define your own functions.

From Help - The *Mathematica* Book 1.7.1:
In this part of the book, we have seen many examples of functions that are built into Mathematica. In this section, we discuss how you can add your own simple functions to Mathematica. Part 2 will describe in much greater detail the mechanisms for adding functions to Mathematica.

As a first example, consider adding a function called f *which squares its argument. The Mathematica command to define this function is* f[x_] := x^2. *The* _ *[referred to as "blank"] on the left-hand side is very important; what it means will be discussed below. For now, just remember to put a* _ *on the left-hand side, but not on the right-hand side, of your definition.*

In order to define the function f [x] = x^2, we use the command:

f[x_] := x^2

We now can evaluate f [x] for different numerical or symbolic values of values of x:

f[−2]   4

f[2 + 3 I] $-5 + 12\,i$

We can make a list of values of f [x] :

Table$\left[f\left[\frac{n}{2}\right], \{n, 1, 10\}\right]$ $\{\frac{1}{4}, 1, \frac{9}{4}, 4, \frac{25}{4}, 9, \frac{49}{4}, 16, \frac{81}{4}, 25\}$

From Help 1.7.4:

Probably the most powerful aspect of transformation rules in Mathematica is that they can involve not only literal expressions, but also patterns. A pattern is an expression such as f[t_] *which contains a blank [underscore]. The blank can stand for any expression. Thus, a transformation rule for* f[t_] *specifies how the function* f *with any argument should be transformed. Notice that, in contrast, a transformation rule for* f[x] *without a blank, specifies only how the literal expression* f[x] *should be transformed, and does not, for example, say anything about the transformation of* f[y].

When you give a function definition such as f[t_] := t^2, *all you are doing is telling Mathematica to automatically apply the transformation rule* f[t_] -> t^2 *whenever possible.*

From Help 1.7.1:

The names like f that you use for functions in Mathematica are just symbols. Because of this you should make sure to avoid using names that begin with capital letters, to prevent confusion with Built-in Mathematica Functions. You should also make sure that you have not used the names for anything else earlier in your session.

*When you have finished with a particular function it is always a good idea to clear definitions you have made for it. If you do not do this then you will run into trouble if you try to use the same function for a different purpose later in your Mathematica session. You can clear all definitions you have made for a function or symbol f by using **Clear[f]**.*

A function you define can have more than one argument. For example:

$$g[x\_, y\_] := \frac{x^2 + y}{x + 1};$$

We shall later learn how to plot the graph of such a function.

The argument of a function may be a set:

h[{x_, y_}] := x + y; j[{x_, y_}] := {x + y, x − y};

Note that the underscore makes the argument a pattern. Suppose we define a function and leave out the underscore:

g[x] := Sin[x]

g[2 π]   g[2 π]

Here *Mathematica* knows the value of g only for x. *Mathematica* reads g[x] as a single symbol. Thus it is very important to write the underscore when defining a function.

Exercise:
1) Define a function which maps z onto z Sin[z], find the approximate and exact function value at $z = \frac{\pi}{3} + 2\,I$.
2) Find the derivative of the function z Sin[z].

■ **Applying a User-Defined Function to a List using Map or /@**

In 1.6, we applied Built-in Functions directly to lists. User-defined functions may be applied to lists, using the command **Map** or **/@**. The examples make the syntax clear:

? Map

> Map[f, expr] or f /@ expr applies f to each element on the first level
>   in expr. Map[f, expr, levelspec] applies f to parts of expr specified by levelspec.

f[x_] := x²;

f /@ {1, 2, 3}   {1, 4, 9}

Map[f, {1, 2, 3}]   {1, 4, 9}

Exercise:
1) Define a function which maps x onto $1 + 2\,x + 3\,x^2 + ... + 21\,x^{20}$, and then find the values of the function at $\frac{1}{n}$, for $2 \le n \le 6$;
2) Define a function f, which maps x onto the integer part of E^x, and find f [n], for n from 1 to 10.

1.7.3 Pure Functions

When we define a function which does not have a standard name such as Sin or Log, we usually give it a name such as f or g. For example, we might write 'Let $f[x] = x^2 + x, x \in \mathbb{R}$'. However we sometimes write: 'Consider the function $x \to x^2 + x$'. This time, we have not given the function a name, but have defined it by its effect on the variable x. In *Mathematica*, we can use the above idea to great effect. *Mathematica* has the following method of defining an 'anonymous' or 'pure' function. The function we defined above can be defined in *Mathematica* as:

$\#^2 + \# \&$

The symbol # stands for the argument of the function, and the ampersand, &, informs *Mathematica* that we are concerned with a 'pure' function. Think of such a function as 'the function which maps x onto $x^2 + x$'. We apply the pure function so defined to an argument:

#^{2} + # &[3] 12

Mathematica has another notation for a pure function, which can be found in the Master Index. We shall not use this notation. A pure function can be applied to the entries of a list using /@.

N[Sin[2 #]] & /@ $\{-2.84724 - 2.37067\,i, -1.05122 - 3.4824\,i, 0. + 3.62686\,i\}$
{2 + I, 3 − I, I}

Exercise:
1) Define the pure function which maps z onto E^{-z^2}, and find its values when $z = n + n\,I$, for n from 1 to 3;
2) Use a pure function to find Log[n] / n for n from 1 to 10.

1.7.4 Compiling Functions

Consider the following:

f[x_] := x^2;

f[2] 4

f[{3.2, 1.9, 7, 0.1}] {10.24, 3.61, 49, 0.01}

f[a + 2 b] $[a + 2\,b]^2$

Having defined the function f as above, *Mathematica* prepares code for evaluating f [x] for cases where the argument x could be a number or a matrix or a list or an algebraic expression or... Having to take account of all these possibilities makes the evaluation process slower. If *Mathematica* could assume that all the arguments of a function were numbers, then the process of evaluation could be speeded up.

■ The Command Compile

? Compile

Compile[{x1, x2, ... }, expr] creates a compiled function which evaluates expr assuming numerical values of the xi. Compile[{{x1, t1}, ... }, expr] assumes that xi is of a type which matches ti. Compile[{{x1, t1, n1}, ... }, expr] assumes that xi is a rank ni array of objects each of a type which matches ti. Compile[vars, expr, {{p1, pt1}, ... }] assumes that subexpressions in expr which match pi are of types which match pti.

To see how **Compile** speeds up calculations we define a function g and calculate g [x] for a particular value of x, and time the calculation:

$$g[x\_] := \sum_{r=1}^{1000} x^r;$$

g[0.25679] // Timing   {0.11 Second, 0.345515}

We now define a function h which is the compiled version of g, and calculate h [x] for the same value of x, timing the result:

$$h = \text{Compile}\left[\{x\}, \sum_{r=1}^{1000} x^r\right];$$

h[0.25679] // Timing   {0. Second, 0.345515}

Notice the syntax in the definition of h: **Compile[{x}, $\sum_{r=1}^{1000} x^r$]** is a function which maps the number x onto $\sum_{r=1}^{1000} x^r$ so the argument can be given numerical values. It would be incorrect to write **h[x_] := Compile[{x}, $\sum_{r=1}^{1000} x^r$];**

Exercise:
Let f [x] = $\prod_{r=1}^{n} x^r$. Use the command **Compile** to calculate f [1.3].

When we use the command **Compile** it is understood that all arguments are numbers or logical variables. We can speed up the process of calculation further if we specify the type of argument as integer, real, complex or logical.

Here is an example:

j = Compile[{x, {n, _Integer}}, x^n];

j[2.3, 4]   27.9841

        j[4, 3.1]

        CompiledFunction::cfsa : Argument 3.1` at position 2 should be a machine−size integer.

        73.5167

In the above, the compiled code could not be used, so *Mathematica* printed a warning and evaluated the expression with normal code.

Exercise:
Let q [z, n] = $\sum_{r=1}^{n} \text{Sin}[z]$. Use the command **Compile** to calculate, with maximum speed:

a) $q[1+I, 3]$;

b) the numerical value of $q[\frac{\pi}{5}, 7]$;

c) the sequence $\{q[1, 1], q[1, 2], \ldots q[1, 12]\}$.

1.7.5 Functions as Procedures

In 1.3, you were asked to look up the commands **Expand** and **Factor**, and apply these commands to polynomials. These are examples of procedures. Other examples of procedures which we will encounter later are **Plot** (for plotting 2D graphs) and **Solve** (for solving certain types of equations). Now suppose you wish to expand the polynomial $(1 + x)^n$ for various values of n. *Mathematica* uses functional notation to define procedures which are to be carried out for values of the argument(s) of the function. The rules stated in 1.7.2 apply. In the following example, a function **binomialExpand** with argument n defines a procedure for expanding $(1 + x)^n$.

> **binomialExpand[n_] := Expand[(1 + x)^n];**

Having entered the above command, we can use it calculate the expansion for different values of n:

> **binomialExpand[8]**
>
> $1 + 8x + 28x^2 + 56x^3 + 70x^4 + 56x^5 + 28x^6 + 8x^7 + x^8$

Another example:

> **myProd[n_] := Expand$\left[\prod_{r=1}^{n}(1 + x^r)\right]$;**
>
> **myProd[4]**
>
> $1 + x + x^2 + 2x^3 + 2x^4 + 2x^5 + 2x^6 + 2x^7 + x^8 + x^9 + x^{10}$

Exercise:

1) Look up the command **Apart** in the Master Index and then define a procedure which for each n, resolves $\frac{x^n}{(1+x)^n}$ into partial fractions, and use it to resolve $\frac{x^{10}}{(1+x)^{10}}$ into partial fractions

2) Look up the command **TrigExpand** in the Master Index, and then define a function cosExpand which, for each n, expresses $Cos[a_1 + a_2 + \ldots + a_n]$ in terms of Sines and Cosines of a_1, a_2, ..., a_n. Calculate cosExpand[n] for various values of n.

3) Look up the command **Log** in the Master Index and then define a function logbase[x, y], which, for each x and y finds the logarithm of x to the base y. Calculate logbase[-3+I, 4]

1.7.6 Logical Operators and Conditionals

■ The Logical Operators && (and) and || (or).

? &&

e1 && e2 && ... is the logical AND function. It evaluates its arguments in
    order, giving False immediately if any of them are False, and True if they are all True.

Examples:

2 < 4 && 5 > 1 && 2 == 2   `True`

2 < 4 && 5 < 1   False

? ||

e1 || e2 || ... is the logical OR function. It evaluates its arguments in order,
    giving True immediately if any of them are True, and False if they are all False.

Examples:

2 < 4 || 5 > 1   True

4 < 2 || 5 < 1   False

■ The Conditionals If, Which and /;

? If

If[condition, t, f] gives t if condition evaluates to True, and f if it evaluates
    to False. If[condition, t, f, u] gives u if condition evaluates to neither True nor False.

Examples:

If[7 > 10, x, y]   y

If[7 < 10, x, y]   x

? Which

Which[test1, value1, test2, value2, ...] evaluates each of the testi in
    turn, returning the value of the valuei corresponding to the first one that yields True.

Example:

Which[4 > 5, a, 4 > 6, b, 4 ≤ 4, c, 4 ≤ 6, d]   c

? /;

> patt /; test is a pattern which matches only if the evaluation of test
> yields True. lhs :> rhs /; test represents a rule which applies only if the evaluation
> of test yields True. lhs := rhs /; test is a definition to be used only if test yields True.

We can think of the command **/;** as meaning 'provided that'. We can use this command to restrict the domain of a function:

f[x_] := $\sqrt{x}$ /; x ≥ 0;

f[9]   3

f[−2]   f[−2]

Now consider:

g[x_] := $\sqrt{x}$;

g[−2.0]   1.41421 i

▪ Piece-wise Defined Functions

The above logical operations and conditionals can be used to define functions which are defined in different ways on different parts of their domains. We give some examples. Graphs of such functions will be plotted in the sections on graphics.

f[x_, y_] := x + y /; x > 0 && y < 0;
f[x_, y_] := x − y /; x < 0 && y > 0;

f[2, −3]   −1

f[−3, 4]   −7

r[x_] := If$\left[x > \dfrac{\pi}{6}, x^2, 1 - \text{Sin}[x]\right]$;

r[0.5 π]   2.4674

r[π] π^2

g[x_] := Which[x < 0, 0, x ≥ 0 || x ≤ 1, 1, x > 1, 2];

g[−2]   0

g[0.5]   1

1.8 2D Graphics

1.8.1 Options

The commands for graphics have options associated with them. For example, one of the commands we shall use is the command **Plot**, which is used to plot the graph of an equation of the form y = f [x]. To find the options for this command, enter:

Options[Plot]

$\left\{ \text{AspectRatio} \rightarrow \dfrac{1}{\text{GoldenRatio}}, \text{Axes} \rightarrow \text{Automatic}, \text{AxesLabel} \rightarrow \text{None}, \right.$
   AxesOrigin → Automatic, AxesStyle → Automatic, Background → Automatic,
   ColorOutput → Automatic, Compiled → True, DefaultColor → Automatic,
   Epilog → {}, Frame → False, FrameLabel → None, FrameStyle → Automatic,
   FrameTicks → Automatic, GridLines → None, ImageSize → Automatic,
   MaxBend → 10., PlotDivision → 30., PlotLabel → None, PlotPoints → 25,
   PlotRange → Automatic, PlotRegion → Automatic, PlotStyle → Automatic,
   Prolog → {}, RotateLabel → True, Ticks → Automatic, DefaultFont :→ $DefaultFont,
   $\left. \text{DisplayFunction} :\rightarrow \$\text{DisplayFunction}, \text{FormatType} :\rightarrow \$\text{FormatType}, \text{TextStyle} :\rightarrow \$\text{TextStyle} \right\}$

The default values for the options are specified in the above list. We shall discuss some of the options for the commands we use, and how to implement them.
A useful option for 2D graphics is **AspectRatio**.

? AspectRatio

AspectRatio is an option for Show and related functions which specifies the ratio of height to width for a plot.

From Help - Built-in Functions:
AspectRatio → Automatic determines the ratio of height to width from the actual coordinate values in the plot.
The default value AspectRatio → 1/GoldenRatio is used for two-dimensional plots.AspectRatio → Automatic is used for three-dimensional plots.

The plots of 2D graphics can also be modified with graphics directives such as **Thickness, GrayLevel, PointSize**. These are grouped with the definition of the graphic primitive. This will be demonstrated in 1.8.2.

1.8.2 Plotting a Sequence of Points Using the Command ListPlot

? ListPlot

ListPlot[{y1, y2, ... }] plots a list of values. The x coordinates for each point are taken to be 1, 2, ... ListPlot[{{x1, y1}, {x2, y2}, ... }] plots a list of values with specified x and y coordinates.

In the following example, we use the first method of plotting a list of points, by giving their y-co-ordinates. This is a good method for plotting a sequence of numbers. Options can be added after the list of points or numbers. A useful option is **PlotJoined→True.**

ListPlot[{1, 4, 3, 0, 1}, PlotJoined → True]

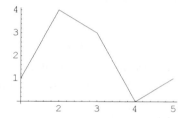

In the following example, we plot a table of points. As the points form the vertices of a regular hexagon, we use the option **AspectRatio→Automatic.**

ListPlot[Table[{Cos[n 60°], Sin[n 60°]}, {n, 0, 6}], PlotJoined → True,
AspectRatio → Automatic, Ticks → {{−1, 1}, Automatic}]

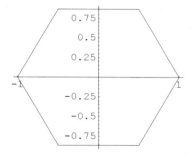

Exercise:

1) Use **ListPlot** to plot the sequence $(\frac{n}{n+1})$, $1 \le n \le 20$.

2) Look up **Axes** in Built-in Functions, and construct a square without showing the axes.

■ Plotting a Sequence of Points Representing Complex Numbers

Mathematica does not have a built-in command for plotting a sequence of complex numbers. In order to plot a complex number z in the plane we must first calculate the point with co-ordinates {Re [z], Im [z]}. We can use an anonymous function to do this:

{Re[#], Im[#]} &[1 + I]   {1, 1}

The above function can be applied to a list of complex numbers using the command **/@**:

$$\{Re[\#], \ Im[\#]\} \ \& \ /@ \ Table\left[E^{n \ I \ \frac{\pi}{6}}, \ \{n, \ 1, \ 12\}\right]$$

$$\left\{\left\{\frac{\sqrt{3}}{2}, \frac{1}{2}\right\}, \left\{\frac{1}{2}, \frac{\sqrt{3}}{2}\right\}, \{0, 1\}, \left\{-\frac{1}{2}, \frac{\sqrt{3}}{2}\right\}, \left\{-\frac{\sqrt{3}}{2}, \frac{1}{2}\right\}, \{-1, 0\},\right.$$

$$\left.\left\{-\frac{\sqrt{3}}{2}, -\frac{1}{2}\right\}, \left\{-\frac{1}{2}, -\frac{\sqrt{3}}{2}\right\}, \{0, -1\}, \left\{\frac{1}{2}, -\frac{\sqrt{3}}{2}\right\}, \left\{\frac{\sqrt{3}}{2}, -\frac{1}{2}\right\}, \{1, 0\}\right\}$$

We now plot the above list of points. We shall use the command **PlotStyle** in order to obtain larger points, as we shall not use the command **PlotJoined**.

From Help - Built-in Functions:

PlotStyle is an option for Plot and ListPlot that specifies the style of lines or points to be plotted. PlotStyle -> style specifies that all lines or points are to be generated with the specified graphics directive, or list of graphics directives.

$$ListPlot\left[\{Re[\#], \ Im[\#]\} \ \& \ /@ \ Table\left[E^{n \ I \ \frac{\pi}{6}}, \ \{n, \ 1, \ 12\}\right],\right.$$
$$\left.PlotStyle \rightarrow PointSize[0.03], \ AspectRatio \rightarrow Automatic\right]$$

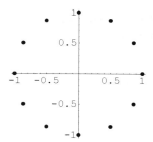

Exercise:
Plot the first 30 points in the sequence $Sin[\frac{n}{20} + I \ \frac{n}{20}]$.

1.8.3 2D Graphics Elements

The 2D graphics elements: **Point, Line, Rectangle, Polygon, Circle** and **Disk** can be generated using the commands **Show** and **Graphics**. The command **Graphics** has a list of options such as **AspectRatio, Axes,** etc. which can be used with any of the above elements. Each graphics element can also be paired with a Graphics Directive. For example: **Point** and **Point-Size**; **Line, Thickness** and **GrayLevel**. The general form of a command to construct a graphics primitive is of the form:

Show[Graphics[Graphics Element], Options] or
    Show[Graphics[{Graphics Directive, Graphics Element}], Options]

Notice that the **Graphics Directive** and **Graphics Element** are enclosed in braces.

From Help: The *Mathematica* Book, Two-Dimensional Graphics Elements:
You can combine different graphical elements simply by giving them in a list. In two-dimensional graphics, Mathematica will render the elements in exactly the order you give them. Later elements are therefore effectively drawn on top of earlier ones.

■ Line

? Line

Line[{pt1, pt2, ... }] is a graphics primitive which represents a line joining a sequence of points.

Show[Graphics[Line[{{0, 0}, {2, 1}, {−2, 1}, {0, 0}}]]],
    AspectRatio → Automatic, Axes → True, Ticks → False]

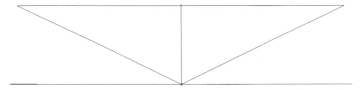

Exercise:

1) Construct a regular hexagon using the command **Line**.

2) In Built-in Functions, look up **Rectangle**, and construct a square.

■ Polygon

? Polygon

Polygon[{pt1, pt2, ... }] is a graphics primitive that represents a filled polygon.

In the following example, the directive **GrayLevel** is used to color a polygon.

? GrayLevel

GrayLevel[level] is a graphics directive which specifies the
    gray−level intensity with which graphical objects that follow should be displayed.

From Help: Built-in Functions-GrayLevel:
The gray level must be a number between 0 and 1.0 represents black and 1 represents white.

Show[Graphics[{GrayLevel[0.6], Polygon[{{0, 0}, {2, 1}, {−2, 1}}]}],
   AspectRatio → Automatic]

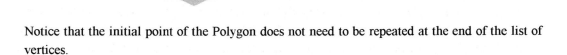

Notice that the initial point of the Polygon does not need to be repeated at the end of the list of vertices.

Exercise:
Construct a non-rectangular light gray parallelogram.

▪ Circle

? Circle

Circle[{x, y}, r] is a two−dimensional graphics primitive that represents
   a circle of radius r centered at the point x, y. Circle[{x, y}, {rx, ry}] yields an ellipse
   with semi−axes rx and ry. Circle[{x, y}, r, {theta1, theta2}] represents a circular arc.

Two or more graphics elements can be constructed in the same diagram. The list of graphics elements must be enclosed in braces. Each graphic may be grouped with one or more graphics directives. In the following example, we use the option **Background**, choosing **GrayLevel [0.6]**, which is medium gray. The first graphic primitive has 2 directives and the second 1.

Show[Graphics[{{Thickness[0.05], GrayLevel[1], Line[{{0, 0}, {2, 1}, {−2, 1}, {0, 0}}]},
   {Thickness[0.02], Circle[{0, 0}, {1, 2}]}}],
   AspectRatio → Automatic, Background −> GrayLevel[0.6]]

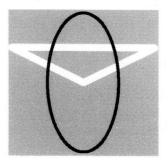

Exercise:
1) Construct a leaf-shaped form with 2 circular arcs.
2) Construct 3 concentric circles of different radii in different shades of gray.

1.8.4 Constructing a Sequence of Graphics Primitives

A sequence of graphics elements can be constructed using the command **Table**. Each can be paired with a graphics directive. In the following example, a sequence of disks with random centers and decreasing radii is constructed. The colors black or white are assigned to the disks at random, using the directive **GrayLevel[Random[Integer]]**. The white disks are not visible, but conceal parts of the black disks, and so interesting patterns can sometimes be constructed. (Recall that **Random[Integer]** returns 0 or 1, and **Random[ ]** returns a number between 0 and 1.)

?Disk

Disk[{x, y}, r] is a two−dimensional graphics primitive that represents a filled
    disk of radius r centered at the point x, y. Disk[{x, y}, {rx, ry}] yields an elliptical disk
    with semi−axes rx and ry. Disk[{x, y}, r, {theta1, theta2}] represents a segment of a disk.

Show[

   Graphics[Table[$\left\{$GrayLevel[Random[Integer]], Disk[$\{$Random[], Random[]$\}$, $1 - \dfrac{n}{20}$ $]\right\}$,

   {n, 0, 20}], AspectRatio → Automatic, Axes → False]]

In the following example we use the command **Background**:

?Background

Background is an option which specifies the background color to use.

Show[

   Graphics[Table[$\left\{$GrayLevel[Random[Integer]], Disk[$\{$Random[], Random[]$\}$, $1 - \dfrac{n}{20}$ $]\right\}$,

   {n, 1, 20}], AspectRatio → Automatic, Axes → False, Background –> GrayLevel[0]]]

Using a larger number of disks:

$$\textbf{Show}\Big[\textbf{Graphics}\Big[$$
$$\textbf{Table}\Big[\Big\{\textbf{GrayLevel[Random[Integer]], Disk}\Big[\{\textbf{Random[ ], Random[ ]}\}, 1 - \frac{n}{10000}\Big]\Big\},$$
$$\{\textbf{n, 0, 10000}\}\Big], \textbf{AspectRatio} \rightarrow \textbf{Automatic, Axes} \rightarrow \textbf{False}\Big]\Big]$$

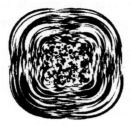

Every time the above program is implemented, a different image is obtained.

Exercise:

1) Try the above program with different circle centers, using **Random[Real, {a, b}]**, or with different values for n (remember to change the circle radii accordingly so as to avoid negative radii).

2) Try the above program using ellipses instead of circles.

3) Try the above program using thick circles, or other graphics primitives.

4) Look up **Random** in Built-in Functions, and adapt the above program to obtain disks in GrayLevel[0], GrayLevel[0.5] or GrayLevel[1], randomly chosen.

More regular patterns can also be constructed. In the following example, a sequence of lines is constructed joining points on the y-axis to points alternately on the left, right of the y-axis. This is achieved by choosing the x-co-ordinate of the second end-point to be $(-1)^n \sin[\frac{n\pi}{30}]$. The directive **Thickness$[0.02\,(1.5 - \frac{n}{30})]$** ensures that the thickness of the lines decreases as n increases.

$$\textbf{Show}\Big[\textbf{Graphics}\Big[\textbf{Table}\Big[\Big\{\textbf{Thickness}\Big[0.02\Big(1.5 - \frac{n}{30}\Big)\Big],$$
$$\textbf{Line}\Big[\Big\{\Big\{0, \frac{n}{30}\Big\}, \Big\{(-1)^n \sin\Big[n\,\frac{\pi}{30}\Big], \frac{2\,n}{30}\Big\}\Big\}\Big]\Big\}, \{\textbf{n, 1, 29}\}\Big]\Big], \textbf{AspectRatio} \rightarrow \textbf{1.1}\Big]$$

Exercise:

Try an example similar to the above with thin circle sectors.

The following example illustrates the use of the commands **Transpose** and **Flatten**. A pattern is formed by a sequence consisting of black ellipses and white circles of decreasing size. Recall that **Disk[{x, y}, {r, s}]** represents an elliptical disk with semi-axes r and s.

Suppose we have 2 lists of equal length, say {a, b, c} and {d, e, f}, and we wish to 'interlace' them with *Mathematica* to form the list {a, d, b, e, c, f}. We use the commands **Transpose** and **Flatten** to do this:

Transpose[{{a, b, c}, {d, e, f}}]   {{a, d}, {b, e}, {c, f}}

Flatten[%, 1]   {a, d, b, e, c, f}

Here is a list of black filled ellipses with semi-axes of decreasing size:

$$\textbf{Table}\left[\textbf{Disk}\left[\{0, 0\}, \left\{\left(\frac{3}{4}\right)^{n} 3, \left(\frac{3}{4}\right)^{n} 4\right\}\right], \{n, 0, 5\}\right];$$

Here is a list of white disks with decreasing radii. The nth disk is inscribed inside the nth ellipse.

$$\textbf{Table}\left[\left\{\textbf{GrayLevel[1], Disk}\left[\{0, 0\}, \left(\frac{3}{4}\right)^{n}\right]\right\}, \{n, 1, 5\}\right];$$

We wish to plot the ellipses and disks in order of decreasing size, as *Mathematica* plots the graphics elements in the order given, and so we need to 'interlace' the ellipses and disks. We have also decreased the disks radii slightly for visual clarity.

$$\textbf{Flatten}\left[\textbf{Transpose}\left[\left\{\textbf{Table}\left[\textbf{Disk}\left[\{0, 0\}, \left\{\left(\frac{3}{4}\right)^{n} 3, \left(\frac{3}{4}\right)^{n} 4\right\}\right], \{n, 0, 6\}\right],\right.\right.\right.$$

$$\left.\left.\left.\textbf{Table}\left[\left\{\textbf{GrayLevel[1], Disk}\left[\{0, 0\}, \left(\frac{3}{4}\right)^{n} 3 - .025\right]\right\}, \{n, 0, 6\}\right]\right\}\right], 1\right];$$

Show[Graphics[%], AspectRatio → Automatic, PlotRange → All]

We now use the options **AspectRatio** and **PlotRange** to obtain:

Show[Graphics[%%], PlotRange → {Automatic, {−4, 0}}, AspectRatio → 0.4]

Exercise:

Construct a decreasing squence of filled black squares alternating with white inscribed discs.

In the following example, 4 sets of circles are plotted. The command **Join** is used to make a single list, so that the circles can be plotted using **Show** and **Graphics.** The circles are plotted in white (**GrayLevel[1]**), on a black disk (the default color). The disk must be plotted before the circles. This can be achieved by using the command **Prolog**.

? Prolog

Prolog is an option for graphics functions which gives a list of
    graphics primitives to be rendered before the main part of the graphics is rendered.

$$\textbf{Show}\Big[\textbf{Graphics}\Big[$$

$$\textbf{Join}\Big[\Big\{\textbf{Table}\Big[\Big\{\textbf{Thickness[0.01], GrayLevel[1], Circle}\Big[\Big\{0, 1 - \frac{n}{10}\Big\}, 1 - \frac{n}{10}\Big]\Big\}, \{n, 0, 10\}\Big],$$

$$\textbf{Table}\Big[\Big\{\textbf{Thickness[0.01], GrayLevel[1], Circle}\Big[\Big\{\frac{n}{10}, 1\Big\}, 1 - \frac{n}{10}\Big]\Big\}, \{n, 0, 10\}\Big],$$

$$\textbf{Table}\Big[\Big\{\textbf{Thickness[0.01], GrayLevel[1], Circle}\Big[\Big\{-\frac{n}{10}, 1\Big\}, 1 - \frac{n}{10}\Big]\Big\}, \{n, 0, 10\}\Big],$$

$$\textbf{Table}\Big[\Big\{\textbf{Thickness[0.01], GrayLevel[1], Circle}\Big[\Big\{0, 2 - \frac{n}{10}\Big\}, \frac{n}{10}\Big]\Big\}, \{n, 0, 10\}\Big]\Big\}\Big],$$

AspectRatio → Automatic, Prolog → {Disk[{0, 1}, 1.05]}, PlotRange → All$\Big]$

Exercise:

1) Experiment with the above techniques applied to various graphics primitives such as semi-

disks, disk sectors, arcs of circles etc.

2) Construct a sequence of 16 half-disks with centers at the points $\{n \, Cos[\frac{n\pi}{6}], n\}$ with colors alternating between black and white (GrayLevel[$\frac{1+(-1)^n}{2}$] is white if n is even and black if n is odd) and radii equal to n for n going from 1 to m for various values of m.

1.8.5 Graphs of Equations of the Form y = f [x]

? Plot

Plot[f, {x, xmin, xmax}] generates a plot of f as a function of x
    from xmin to xmax. Plot[{f1, f2, ... }, {x, xmin, xmax}] plots several functions fi.

Here is an example:

Plot[3 x^6 − 10 x^3 + x, {x, −2, 2}, Ticks → {Automatic, {0, 10, 20}}]

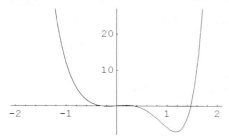

As the range of the function plotted is large in comparison to the domain, it is inadvisable to use the option **AspectRatio→Automatic**. Notice that *Mathematica* has dropped some of the points on the curve with very large y-co-ordinates, in order to display more detail elsewhere. One can include more points using the option **PlotRange**. The option **PlotRange →All**, includes all points. (This option should not be used if the graph has a vertical asymptote.)

Plot[3 x^6 − 10 x^3 + x, {x, −2, 2}, PlotRange → All]

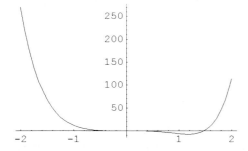

Here is the plot of a piece-wise defined function:

Plot[Which[x < −1, 1, x >= −1 && x < 1, x, True, x ≥ 1], {x, −2, 2}]

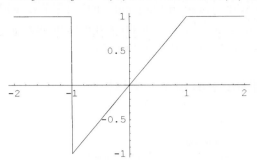

In the section below, we show how to plot piece-wise defined functions without vertical lines.

One can also use the option **PlotRange→{min, max}** to reduce the range of the plot to lie between min and max.

Another useful option is **PlotStyle**, which we used in the section on **ListPlot.** The list of styles that can be generated for plots of this type include: **GrayLevel, Thickness** and **Dashing.** Graphics directives, were used to modify the plots of graphics primitives in 1.8.2, by grouping them with the definition of the primitive. Graphics directives cannot be used with other 2D plots. To modify other 2D graphics, the command **PlotStyle** can be used, as is demonstrated below.

? Thickness

Thickness[r] is a graphics directive which specifies that lines which follow are to be drawn
    with a thickness r. The thickness r is given as a fraction of the total width of the graph.

Plot[Abs[Gamma[1 + I x]], {x, −3, 3}, PlotStyle → Thickness[0.01]]

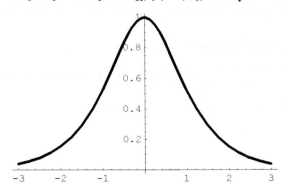

Two or more styles can be assigned to a plot, and the styles should be enclosed in braces.

$$\text{Plot}\left[\frac{\text{Sin}[x]}{x}, \{x, 0, 6\pi\}, \text{PlotRange} \rightarrow \{-1, 1\},\right.$$

$$\left.\text{PlotStyle} \rightarrow \{\text{Thickness}[0.0075], \text{GrayLevel}[0.2], \text{Dashing}[\{0.01\}]\}\right]$$

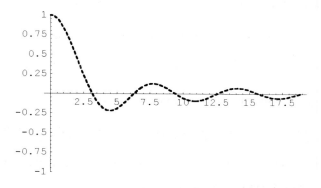

Exercise:

Experiment with the above commands applied to various functions.

Two or more plots can be included in the same diagram, each with its own **PlotStyle**. The **PlotStyle** for the function must be given in the same order as the listing of the functions to be plotted. In the following example, 2 plots are displayed, the first with dashed lines and the second in gray with lines of thickness 0.008.

$$\text{Plot}[\{3\,x^3, \text{IntegerPart}[3\,x^3]\}, \{x, -1.2, 1.2\},$$

$$\text{PlotStyle} \rightarrow \{\text{Dashing}[\{0.01\}], \{\text{Thickness}[0.008], \text{GrayLevel}[0.3]\}\}]$$

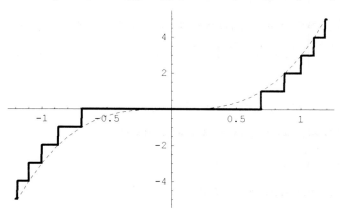

Exercise:

1) Replace **IntegerPart** with **Floor** in the above example.

2) Construct on the same diagram the graphs of the following pair of functions, using different directives for the members of the pair:

$\text{Sin}[x], \quad x - \frac{x^3}{3!} + \frac{x^5}{5!}, \quad -\pi \le x \le \pi.$

The graphs of 2 or more functions with different domains can be plotted on the same diagram, using the command **Show**. In the following example we firstly assign names to the commands for plotting the graphs of 3 functions. In each case, we use the option **DisplayFunction→Identity** , so that *Mathematica* calculates the points on the graphs, but does not display them.

$$g1 = Plot[E^x, \{x, -1, 1\}, AspectRatio \rightarrow Automatic, DisplayFunction \rightarrow Identity];$$

$$g2 = Plot[x, \{x, -1, E\}, AspectRatio \rightarrow Automatic, DisplayFunction \rightarrow Identity];$$

$$g3 = Plot\left[Log[x], \left\{x, \frac{1}{E}, E\right\}, AspectRatio \rightarrow Automatic, DisplayFunction \rightarrow Identity\right];$$

The following command includes the option **DisplayFunction → $DisplayFunction**, in order to display the 3 graphics:

Show[g1, g2, g3, DisplayFunction → $DisplayFunction]

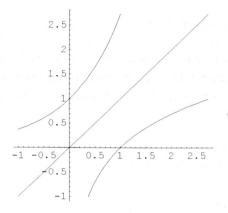

We use the above method to display the graph of a piece-wise defined function. Here is an example:

$$g3 = Plot[x^2, \{x, -2, -1\}, DisplayFunction \rightarrow Identity];$$

$$g4 = Plot[-1 - x, \{x, -1, 1\}, DisplayFunction \rightarrow Identity];$$

Show[g3, g4, DisplayFunction → $DisplayFunction]

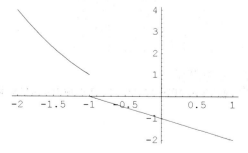

■ Constructing a Sequence of Plots

One can also plot a sequence of graphs, using the command **Table**. In this case the command **Evaluate** is needed. Here is an extract from Help - The *Mathematica* Book 1.9.1:

To get smooth curves, Mathematica has to evaluate functions you plot at a large number of points. As a result, it is important that you set things up so that each function evaluation is as quick as possible.

When you ask Mathematica to plot an object, say f, as a function of x, there are two possible approaches it can take. One approach is first to try and evaluate f, presumably getting a symbolic expression in terms of x, and then subsequently evaluate this expression numerically for the specific values of x needed in the plot. The second approach is first to work out what values of x are needed, and only subsequently to evaluate f with those values of x.

If you type Plot[f, x, xmin, xmax}] *it is the second of these approaches that is used. This has the advantage that Mathematica only tries to evaluate f for specific numerical values of x; it does not matter whether sensible values are defined for f when x is symbolic.*

There are, however, some cases in which it is much better to have Mathematica evaluate f before it starts to make the plot. A typical case is when f is actually a command that generates a table of functions. You want to have Mathematica first produce the table, and then evaluate the functions, rather than trying to produce the table afresh for each value of x. You can do this by typing Plot[Evaluate[f], x, xmin, xmax}].

Here is an example, showing the plot of some members of the family of curves with equation of the form $y = Cos[\frac{n}{2} + Sin[x^2]]$:

$$Plot\left[Evaluate\left[Table\left[Cos\left[\frac{n}{2} + Sin[x^2]\right], \{n, 1, 30\}\right], \{x, -\pi, \pi\}\right],\right.$$
$$\left.AspectRatio \rightarrow 0.4, Axes \rightarrow False\right]$$

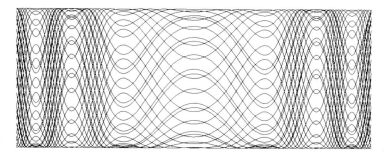

In the next example, a family of parallel curves is plotted:

$$\text{Plot}\left[\text{Evaluate}\left[\text{Table}\left[\text{Abs}[\text{Zeta}[0.5 + I\,x]] + \frac{n}{10}, \{n, 1, 40\}\right], \{x, 0, 30\}\right],\right.$$

$$\left.\text{AspectRatio} \to 0.4, \text{PlotRange} \to \text{All}, \text{Axes} \to \text{False}\right]$$

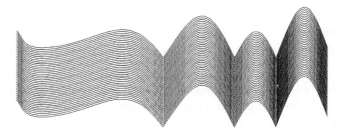

Exercise:

1) Plot a sequence of members of each of the following families of curves:

a) LegendreP[n, 2, x] for $x \in (-1, 1)$;

b) ChebyshevT[n, x] for $x \in (-1, 1)$;

c) $\text{Cos}[\frac{nx}{2}]$;

d) $\text{Cos}[\frac{5}{2} + 5\,\text{Sin}[x^2]] + \frac{n}{5}$.

2) Go to 3.2.9: Orthogonal Polynomials, in the *Mathematica* Book, to find some other interesting functions.

■ User-defined Functions which are Procedures for Plotting

Suppose we wish to plot the graph of the equation $z = \text{Sin}[y\,x]$ for various values of the parameter z. We define a function **sinGraph** as follows:

$$\text{sinGraph}[y\_] := \text{Plot}[\text{Sin}[y\,x], \{x, -\pi, \pi\}];$$

$$\text{sinGraph}\left[\frac{5}{2}\right]$$

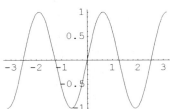

In the following example, we define a function of 2 variables, one of which is a **Thickness** option

$$\text{graph}[n\_, t\_] := \text{Plot}\left[\frac{1}{1 + x^n}, \{x, -1, 1\}, \text{PlotStyle} \to \text{Thickness}[t]\right];$$

graph[3, 0.01]

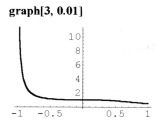

We can define procedures which have default values as well as optional values for one of the arguments. In the following example, the default value for the argument g is 0.01. The symbol **:** after g tells *Mathematica* that g has an optional value, and the 0.01 following, that the default value is 0.01. More than one variable can have a default value, but we shall not consider this case. A single argument with a default value must appear last in the list of arguments.

$$\textbf{tanGraph[n\_, g\_: 0.01] := Plot}\left[\textbf{Tan[x}^n\textbf{]}, \left\{\textbf{x}, \frac{-\pi}{3}, \frac{\pi}{3}\right\}, \textbf{PlotStyle} \rightarrow \textbf{Thickness[g]}\right];$$

In this example, the variable g is omitted, so the default graylevel is used and the graph is plotted with **Thickness**[0.01].

tanGraph[3]

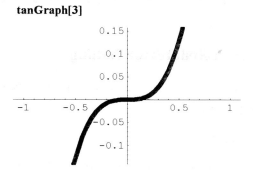

Here the variable g is given the value 0.006:

tanGraph[3, 0.006]

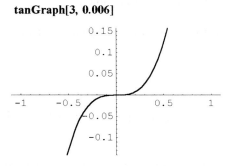

Exercise:

1) Write a procedure for plotting the graph of the equation $y = Cos[n\,x] + n\,Sin[x]$ for

$0 \leq x \leq 2\pi$ and try out your procedure for various values of n.

2) Do the same for the equation $y = E^{zx}$, and include a thickness option, t, with default value 0.005. Try out your procedure for various values of z and t.

The argument of a procedure can be a function. Here is an example:

fPlot[f_] := Plot[f[x], {x, −1, 1}];

In the example below, we replace the argument f by a pure function. It could also be replaced by a user-defined or built-in function.

fPlot[3 #² − 2 # − 1 &]

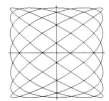

Exercise:

1) Try the above program with various functions.

2) Write a procedure plot[f, a, b] for plotting the graph of the equation $y = f[x]$ for $a \leq x \leq b$, and try out your procedure.

1.8.6 Constructing 2D Parametric Plots

? ParametricPlot

ParametricPlot[{fx, fy}, {t, tmin, tmax}] produces a parametric plot with x and y coordinates fx and fy generated as a function of t. ParametricPlot[{{fx, fy}, {gx, gy}, ... }, {t, tmin, tmax}] plots several parametric curves.

ParametricPlot has almost the same options as **Plot**. Here is an example:

ParametricPlot[{ Cos[7 t], Sin[5 t]}, {t, 0, 2 π}, AspectRatio → Automatic, Ticks → False]

Two or more parametric plots can be shown on the same diagram, if they have the same domains. The definitions must be enclosed in braces. A list of Options for the plots can be included, using **PlotStyle**. The first option will be assigned to the first plot and the second to the second plot etc. If there is only one option, it will be assigned to all plots. Two or more options must be enclosed in braces.

> **ParametricPlot[{{ E$^{0.4\,t}$ Cos[t], E$^{0.4\,t}$ Sin[t]}, {E$^{0.3\,t}$ Cos[t], E$^{0.3\,t}$ Sin[t]}},**
> **{t, 0, 4 π}, AspectRatio → Automatic, PlotRange → All,**
> **PlotStyle → {Thickness[0.015], Thickness[0.008]}, Ticks → False]**

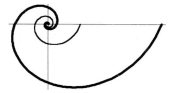

Exercise:

Write a procedure for plotting the ellipses with parametric equations $x = \frac{Cos[t]}{a}$, $y = \frac{Sin[t]}{b}$; $0 \le t \le 2\pi$, and try out your procedure for various values of a and b. Be sure to include the option **AspectRatio→Automatic**.

Plots generated by the commands **Plot** and **PolarPlot** can also be generated by **ParametricPlot** as follows:

The curve with Cartesian equation y = f [x] has parametric equations x = t, y = f [t]; while the curve with polar equation r = f [θ] has parametric equations x = f [θ] Cos[θ], y = f [θ] Sin[θ].

The following is the definition of a 4-parameter family of curves:

> **family1[a_, b_, c_, n_] :=**
> $$\left\{ n\left(2 + \frac{Sin[a\,t]}{2}\right) Cos\left[t + \frac{Sin[b\,t]}{c}\right], \; n\left(2 + \frac{Sin[a\,t]}{2}\right) Sin\left[t + \frac{Sin[b\,t]}{c}\right] \right\};$$

Remember to enter the definition of the family before using it.

Varying the parameter n changes the size, but not the shape of the curve. Here is a plot of one member of the family. We use the command **Evaluate**:

> **ParametricPlot[Evaluate[family1[6, 24, 4, 1]],**
> **{t, 0, 2 π}, AspectRatio → Automatic, Axes → False]**

Sequences of parametrically defined curves can also be plotted. In the following example, we choose particular values for the parameters a, b and c , and let the parameter n vary from 1 to 12 in the table, to obtain a sequence of similar curves of varying sizes.

ParametricPlot[Evaluate[Table[family1[7, 14, 6, n], {n, 1, 12}], {t, 0, 2π}],
AspectRatio → Automatic, Axes → False]

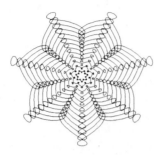

In the following example, the parameters a, c and n are chosen constants, and b varies from 1 to 20:

ParametricPlot[Evaluate[Table[family1[6, b, 5, 1], {b, 1, 20}], {t, 0, 2π}],
AspectRatio → Automatic, Axes → False]

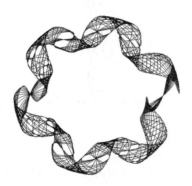

Exercise:

We suggest that you experiment with members of the following families of curves:

1) family2[a_,b_,n_]:={n Sin[a t] Cos[b t], n Cos[a t] Cos[b t]}.
2) family3[a_,b_,c_,d_, f_,g_,n_]:={ n(a Cos[b t]+Cos[c t]), n(d Sin[f t]+Sin[g t])}.
3) family4[a_,b_,n_]:={n(Sin[a t]+ Sin[b t]), n(Cos[a t]+ Cos[b t])}.
4) family5[a_,n_]:={n Sin[a t] Cos[t], n Sin[a t] Sin[t]}; (the n-leaved 'roses').
5) family6[a_,n_]:={n Cos[Cos[a t]] Cos[t], n Cos[Sin[a t]] Sin[t]}.
6) family7[a_,n_]:={ n Cos[Cos[a t]] Cos[t], n Sin[Cos[a t]] Sin[t]}.
7) family8[a_,n_]:={n t Sin[a t] Cos[t], n t Sin[a t] Sin[t]}.
8) family9[a_,b_,n_]:={n Sin[a t] Cos[t], n Sin[b t] Sin[t]}.

9) family10[a_,b_,n_]:={n Sin[a t] Sin[b t], n Cos[a t] Sin[b t]}.

10) family11[a_,b_,n_]:={n Sin[a t] Cos[t], n Sin[b t] Sin[t]}.

11) family12[a_,n_]:={n Cos[Cos[a t]] Cos[t], n Sin[Sin[a t]] Sin[t]}.

12) family13[a_,b_,n_]:={n (Cos[a t] - Cos[b t]), n(Cos[a t] - Sin[b t])}.

In Chapter 2 we shall discuss methods of coloring sequences of 2D parametrically defined curves, and if you wish to construct your own images, it will be useful to have some knowledge of interesting families. If you use a fraction for one or more of the parameters, you may need to increase the t-value range in order to obtain a complete picture. Try constructing a sequence of curves.

One can experiment with plotting one's own parametric curves. One method to try is to take a particular member of one of the above families, and modify it by, say, adding an extra term to, or multiplying by an extra factor, one or both co-ordinate functions. Here is an example:

ParametricPlot[{(8 Cos[θ] − 2 Cos[4 θ]), (8 Sin[θ] − 2 Sin[θ])},
    {θ, 0, 2 π}, AspectRatio → Automatic]

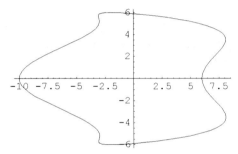

Here is a modification, which can be used as the cross-section of the trunk of a tree in an exercise in Chapter 7.

ParametricPlot[{(8 Cos[θ] − 2 Cos[4 θ]), (8 Sin[θ] − 2 Sin[θ] Sin[9 θ])},
    {θ, 0, 2 π}, AspectRatio → Automatic]

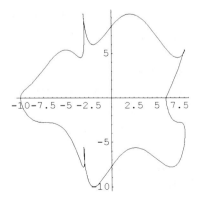

1.8.7 Add-ons, ComplexMap

From Help - *Mathematica* book 1.3.10:

One of the most important features of Mathematica is that it is an extensible system. There is a certain amount of mathematical and other functionality that is built into Mathematica. But by using the Mathematica language, it is always possible to add more functionality.

For many kinds of calculations, what is built into the standard version of Mathematica will be quite sufficient. However, if you work in a particular specialized area, you may find that you often need to use certain functions that are not built into Mathematica.

In such cases, you may well be able to find a Mathematica package that contains the functions you need. Mathematica packages are files written in the Mathematica language. They consist of collections of Mathematica definitions which "teach" Mathematica about particular application areas.

If you want to use a function from a particular package, you must first read the package into Mathematica. After you have loaded the package, you can use the function in the same way you use a built-in function.

*There are a number of subtleties associated with such issues as conflicts between the names of functions in different packages. These are briefly discussed below and in more detail in the Mathematica Book. One important point to note is that you must not refer to a function that you will read from a package before actually reading in the package. If you do this by mistake you will have to execute the command: **Remove["name"]** to get rid of the function before you read in the package that defines it. If you do not call **Remove** Mathematica will use "your" version of the function, rather than the one from the package.*

In Help, we click along the route: Add-ons - Standard Packages - Graphics - ComplexMap and find the following explanatory paragraph:

To plot the graph of a complex-valued function of a complex variable, four dimensions are required: two for the complex variable and two for the complex function value. One method to circumvent the need for four-dimensional graphics is to show how the function transforms sets of lines that lie in the complex plane. Each line will be mapped into some curve in the complex plane and these can be represented in two dimensions.

The functions CartesianMap *and* PolarMap *defined in this package make pictures of this form.* CartesianMap *shows the image of Cartesian coordinate lines while* PolarMap *shows the effect on polar coordinate lines.*

We load the package:

    <<Graphics`ComplexMap`

We find out about **CartesianMap**:

? CartesianMap

CartesianMap[f, {x0, x1, (dx)}, {y0, y1, (dy)}] plots the image of
   the cartesian coordinate lines under the function f. The default values of dx and
   dy are chosen so that the number of lines is equal to the value of the option Lines.

If f is replaced by **Identity**, we obtain a grid of straight lines parallel to the axes:

CartesianMap[Identity, { -1.1, 1.1}, { -1.1, 1.1}]

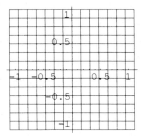

We now find the images of the above lines under the function **Tan**:

CartesianMap[Tan, {−1.1, 1.1}, {−1.1, 1.1}, Frame → False]

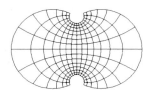

$$\text{CartesianMap}\left[\frac{\#^3}{\#^3 - 1} \ \&, \ \{-1.0, 1.0\}, \ \{-1.0, 1.0\}, \ \text{Frame} \to \text{False}, \ \text{Axes} \to \text{False}\right]$$

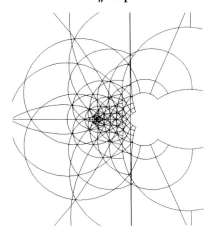

We find out about **PolarMap**:

? PolarMap

PolarMap[f, {r0:0, r1, (dr)}, {phi0, phi1, (dphi)}] plots the image of the polar coordinate
  lines under the function f. The default for the phi range is {0, 2Pi}. The default values of
  dr and dphi are chosen so that the number of lines is equal to the value of the option Lines.

PolarMap[Identity, { 0, 3}, { 0, 2 Pi}]

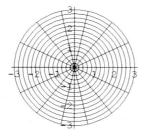

The following diagram shows the images of the above lines and circles under the function
$x \to I \, Sin[0.5 \, x]$.

PolarMap[(I Sin[0.5 #]) &, {0, 3}, {0, 2 π},
       Frame → False, Axes → False, PlotRange –> All];

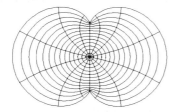

Exercise:

Experiment with the above packages. For example try the following:

CartesianMap$\left[\frac{\#^2}{\#^7-1}\right.$ &, {−1, 0, 5}, {−0.5, 0.5}, Frame → False, Axes → False$\left.\right]$

PolarMap[((4 + 2 I) Sin[#]) &, {2, 3}, {0, 2 π}, Frame → False,
    Axes → True, PlotRange → All, AspectRatio → Automatic];

We recommend that you browse through the Add-ons section of Help.

1.8.8 Polar and Implicit Plots

■ PolarPlot

Click along the route: Add-ons - Standard Packages - Graphics - Graphics. Scroll down a little,
and you will see that you need to load a package: Load the package, and scroll further to find
information on the package and study the example given then plot the graph of the polar
equation $r = E^{\frac{t}{5}}$, $0 \le t \le 4\pi$.

■ **ImplicitPlot**

Click along the route: Add-ons - Standard Packages - Graphics - **ImplicitPlot**. Study the section, load the required package and plot the graph of the equation: $\text{Sin}[x^3] = \text{Cos}[y^3]$, $-2\pi \le x, y \le 2\pi$. Use the option: **Plotpoints→180**.

1.9 3D Graphics

1.9.1 3D Plots

In contrast to 2D plots, all 3D plots are, by default, colored by *Mathematica*. In Chapter 1 of this book all graphics will be shown in gray scale. However the reader can view the images in *Mathematica's* default coloring on the computer screen. This coloring can be changed by various methods which will be discussed in Chapter 2. An option available for 3D plots is **ViewPoint**. The use of this option will be explained in 1.9.3 and used in some of the other sections. We will discuss the following options which are common to all graphics in 1.9: **Axes, Boxed, BoxRatios, ColorOutput, PlotRange, ViewPoint** and **DisplayFunction**. Other options specific to certain commands will be discussed in context. Stereograms may be constructed for any of the images in 1.9.

1.9.2 3D Graphics Elements

■ **Point, Line, Polygon**

The syntax for plotting the 3D graphics elements **Point, Line** and **Polygon** is similar to that for plotting the corresponding 2D elements, except that the command **Graphics** is replaced by **Graphics3D**. The option for 3D graphics corresponding to the option **AspectRatio** for 2D graphics is **BoxRatios**.

> **? BoxRatios**
>
> BoxRatios is an option for Graphics3D and SurfaceGraphics which
> gives the ratios of side lengths for the bounding box of the three–dimensional picture.

The default option for BoxRatios is determined from the actual ranges of the co-ordinates. Graphics directives are grouped with the name of the graphic element, as with 2D graphics elements. Two or more graphics elements may be plotted on the same diagram. Here is an example:

Show[Graphics3D[{Polygon[{{0, 0, 0}, {−1, −3, 1}, {2, 2, 3}}],
        {PointSize[0.05], Point[{0, 0, 2}]}, {Thickness[0.02], Line[{{0, 0, 0}, {2, 1, − 0.5}}]}}]]

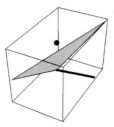

▪ Cuboid

? Cuboid

Cuboid[{xmin, ymin, zmin}] is a three−dimensional graphics primitive that
    represents a unit cuboid, oriented parallel to the axes. Cuboid[{xmin, ymin, zmin},
    {xmax, ymax, zmax}] specifies a cuboid by giving the coordinates of opposite corners.

Below is a representation of a set of 10 cuboids with bases on the x-y plane, and some co-ordinates of opposite vertices randomly chosen. The command **Boxed→False** removes the bounding box of the graphic.

Show[
    Graphics3D[Table[Cuboid[{Random[Integer, {−5, 5}], Random[Integer, {−5, 5}], 0},
            {Random[Integer, {−5, 5}], Random[Integer, {−5, 5}], Random[Integer, {0, 10}]}],
            {n, 1, 10}]], Boxed → False]

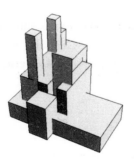

▪ Cylinder, Cone, Torus, Sphere, MoebiusStrip, Helix, DoubleHelix

In order to display the above shapes, we need to load a package:

<< Graphics`Shapes`

As with the **Cuboid**, the commands **Show** and **Graphics** are used to display these shapes.

? Cone

Cone[(r:1, h:1, (n:20r))] is a list of n polygons approximating
a cone centered around the z−axis with radius r and extending from −h to h.

In the following example of the plot of a cone, we have chosen the option **Axes → True** to emphasize that the base of the cone is not on the x-y plane.

Show[Graphics3D[Cone[2, 3, 10]], Axes → True]

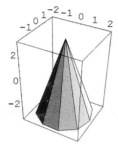

Two or more Shapes can be displayed in the same diagram. In order to do this, we need the command **TranslateShape**.

? Sphere

Sphere[(r:1, (n:20r, m:15r))] is a list of n*(m−2)+2 polygons approximating a sphere with radius r.

? TranslateShape

TranslateShape[graphics3D, {x, y, z}] translates the three−dimensional graphics object by the specified vector.

In the following example, we show a sphere center the origin and a copy thereof with center translated to the point {2, 1, 1}.

Show[Graphics3D[{Sphere[1, 20, 15], TranslateShape[Sphere[1, 20, 15], {2, 1, 1}]}]]

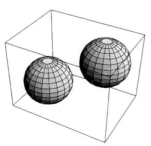

One can build up 3D objects with these shapes. Here is an example. We have included the option: **ColorOutput→CMYKColor.** This command results in a lighter coloring on the screen, and a lighter coloring for the gray level print-out.

Show[Graphics3D[{Cone[0.8, 0.4, 10], TranslateShape[Sphere[0.07, 10, 10], {0, 0, 1}],
TranslateShape[Cylinder[0.02, 0.1, 5], {0, 0, 1.15}], Cone[0.4, 1, 10]}],
ColorOutput –> CMYKColor, Axes → False, Boxed → False, PlotRange → All,
ViewPoint –> {1.085, −2.786, −1.585}, PlotRegion → {{0, 1}, {−0.45, 1.55}}]]

Note: *Mathematica* encloses each graphic output in a rectangular box, called the **PlotRegion**, which can be seen by clicking on the graphic. Sometimes the box is quite large. The size of the box can be reduced using the command: **PlotRegion** as was done in the above graphic. Here is an extract from the Master Index of Help:

■*PlotRegion is an option for graphics functions that specifies what region of the final display area a plot should fill.*

■ *PlotRegion ->   sxmin, sxmax},   symin, symax}} specifies the region in scaled coordinates that the plot should fill in the final display area.*

■ *The scaled coordinates run from 0 to 1 in each direction.*

■ *The default setting PlotRegion → {{0, 1}, {0, 1}} specifies that the plot should fill the whole display area.*

If a, b, c, d are non-negative (non-positive) and one of them is positive (negative), then the setting **PlotRegion → {{0-a, 1+b}, {0-c, 1+d}}** reduces (increases) the size of the plot region relative to the size of the plot.

■ Polyhedra

In the Help-Browser, click along the route: Add-ons - Standard Packages - Graphics - Polyhedra, to find out how to construct polyhedra.

1.9.3 Plotting Surfaces Using the Command Plot3D

? Plot3D

Plot3D[f, {x, xmin, xmax}, {y, ymin, ymax}] generates a three–dimensional plot of f as a function of x and y. Plot3D[{f, s}, {x, xmin, xmax}, {y, ymin, ymax}] generates a three–dimensional plot in which the height of the surface is specified by f, and the shading is specified by s.

Here is a plot of a surface:

Plot3D[$E^{-x^2} + E^{-4y^2}$, {x, −2, 2}, {y, −4, 4}, Ticks → {Automatic, Automatic, {0, 1, 2}}]

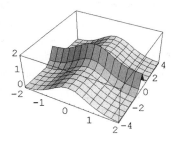

Look up the option **Ticks** which we have used in the above command. Notice that the lengths of the ranges of x, y and z are 4, 8 and 4. If we wish the surrounding box to represent these ranges, we can use the option **BoxRatios**, discussed below.

■ BoxRatios

? BoxRatios

BoxRatios is an option for Graphics3D and SurfaceGraphics which
    gives the ratios of side lengths for the bounding box of the three−dimensional picture.

Here is a plot of the surface above, with different box ratios:

Plot3D[$E^{-x^2} + E^{-4y^2}$, {x, −2, 2}, {y, −4, 4},
    BoxRatios –> {1, 2, 1}, Ticks –> {Automatic, Automatic, {0, 1, 1.5}}]

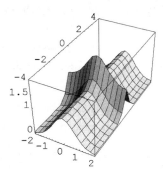

■ PlotPoints

? PlotPoints

PlotPoints is an option for plotting functions that specifies how many sample points to use.

From Help - The *Mathematica* Book - Options:
It is important to realize that since Mathematica can only sample your function at a limited number of points, it can always miss features of the function. By increasing PlotPoints,

you can make Mathematica sample your function at a larger number of points. Of course, the larger you set PlotPoints *to be, the longer it will take Mathematica to plot any function, even a smooth one.*

We use the command **GraphicsArray** to display 2 plots of a complicated surface, the first with the default number of plot points and the second with an increased number of plot points. This command is discussed in 1.9.5, the section on Two Image Stereograms. The command **Display-Function→Identity** is used to suppress the separate display of the 2 graphics.

$$b = \text{Plot3D}\Big[\frac{1}{y - x^2} + \frac{-2}{y + 15 - (2\,x - 16)^2} + \frac{-1}{x - 13 - (y + 10)^2},$$

$$\{x, -6, 20\}, \{y, -20, 20\}, \text{Ticks} \to \text{None}, \text{DisplayFunction} \to \text{Identity}\Big]$$

$$c = \text{Plot3D}\Big[\frac{1}{y - x^2} + \frac{-2}{y + 15 - (2\,x - 16)^2} + \frac{-1}{x - 13 - (y + 10)^2}, \{x, -6, 20\},$$

$$\{y, -20, 20\}, \text{PlotPoints} \to 40, \text{Ticks} \to \text{None}, \text{DisplayFunction} \to \text{Identity}\Big]$$

Show[GraphicsArray[{b, c}, GraphicsSpacing → 0.1]]

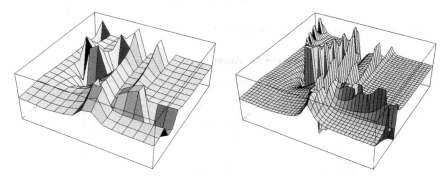

One can choose different numbers of plot points in the x- and y- directions. This is particularly useful if the lengths of the x- and y- ranges are significantly different. We show two plots of a surface, the second with plot points chosen according to the x- and y- ranges. The syntax for different numbers of plot points in the x- and y- directions is made clear in the second command:

$$d = \text{Plot3D}[\text{Sin}[x^2\,y], \{x, -2, 2\}, \{y, -4, 4\},$$
$$\text{BoxRatios} \to \{2, 4, 1\}, \text{Ticks} \to \text{None}, \text{DisplayFunction} \to \text{Identity}]$$

$$e = \text{Plot3D}[\text{Sin}[x^2\,y], \{x, -2, 2\}, \{y, -4, 4\}, \text{BoxRatios} \to \{2, 4, 1\},$$
$$\text{PlotPoints} -> \{25, 50\}, \text{Ticks} \to \text{None}, \text{DisplayFunction} \to \text{Identity}]$$

Show[GraphicsArray[{d, e}, GraphicsSpacing → 0.1]]

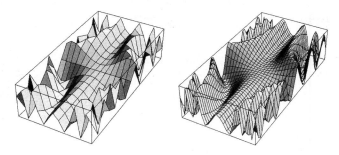

▪ Mesh, Boxed, Axes

We show two plots of a surface. In the second plot, the mesh lines, box and axes have been removed. The syntax is clear.

$$f = \text{Plot3D}\left[10\,\text{Exp}[(\text{Sin}[4\,x^2] * \text{Sin}[4\,y^2])],\ \left\{x,\ \frac{-\pi}{2},\ \frac{\pi}{2}\right\},\ \left\{y,\ \frac{-\pi}{2},\ \frac{\pi}{2}\right\},\right.$$

$$\left.\text{PlotPoints} \to 50,\ \text{AspectRatio} \to \text{Automatic},\ \text{DisplayFunction} \to \text{Identity}\right]$$

$$g = \text{Plot3D}\left[10\,\text{Exp}[(\text{Sin}[4\,x^2] * \text{Sin}[4\,y^2])],\ \left\{x,\ \frac{-\pi}{2},\ \frac{\pi}{2}\right\},\right.$$

$$\left\{y,\ \frac{-\pi}{2},\ \frac{\pi}{2}\right\},\ \text{PlotPoints} \to 50,\ \text{AspectRatio} \to \text{Automatic},\ \text{Mesh} \to \text{False},$$

$$\left.\text{Axes} \to \text{False},\ \text{Boxed} \to \text{False},\ \text{DisplayFunction} \to \text{Identity}\right]$$

Show[GraphicsArray[{f, g}, GraphicsSpacing → 0.1]]

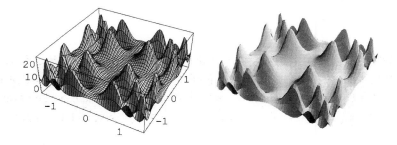

Interesting plots may sometimes be made by making small adjustments to existing plots. Here is the program for an example:

$$\text{Plot3D}\left[10\,\text{Exp}\left[\left(\text{Sin}[4\,x^2] * \text{Sin}\left[2\,y^2 + \frac{\pi}{3}\right]\right)\right],\ \left\{x,\ \frac{-\pi}{2},\ \frac{\pi}{2}\right\},\right.$$

$$\left.\left\{y,\ \frac{-\pi}{2},\ \frac{\pi}{2}\right\},\ \text{PlotPoints} \to 50,\ \text{AspectRatio} \to \text{Automatic}\right]$$

■ PlotRange

From Help - The *Mathematica* Book Co-ordinate Systems for Three-Dimensional Graphics:
Whenever Mathematica draws a three-dimensional object, it always effectively puts a cuboidal box around the object. With the default option setting `Boxed -> True`*, Mathematica in fact draws the edges of this box explicitly. But in general, Mathematica automatically "clips" any parts of your object that extend outside of the cuboidal box.*
The option `PlotRange` *specifies the range of x, y and z coordinates that Mathematica should include in the box. As in two dimensions the default setting is* `PlotRange -> Auto matic`*, which makes Mathematica use an internal algorithm to try and include the "interesting parts" of a plot, but drop outlying parts. With* `PlotRange -> All`*, Mathematica will include all parts.*

We show 2 plots of a surface, the first with the default setting for the **PlotRange** and the second with a different setting:

$$h = \textbf{Plot3D}\left[\textbf{Cos[x] Cos[y] E}^{-\sqrt{x^2+\frac{y^2}{4}}}, \{x, -2\pi, 2\pi\},\right.$$
$$\left.\{y, -2\pi, 2\pi\}, \textbf{PlotPoints} \rightarrow \textbf{40, DisplayFunction} \rightarrow \textbf{Identity}\right]$$

$$j = \textbf{Plot3D}\left[\textbf{Cos[x] Cos[y] E}^{-\sqrt{x^2+\frac{y^2}{4}}}, \{x, -2\pi, 2\pi\}, \{y, -2\pi, 2\pi\},\right.$$
$$\left.\textbf{PlotPoints} \rightarrow \textbf{40, PlotRange} \rightarrow \{-0.2, 0.4\}, \textbf{DisplayFunction} \rightarrow \textbf{Identity}\right]$$

Show[GraphicsArray[{h, j}, GraphicsSpacing → 0.1]]

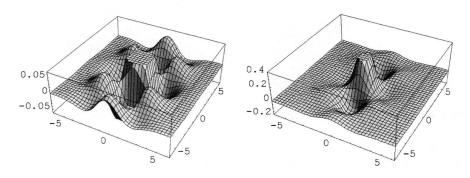

■ ViewPoint

? ViewPoint

ViewPoint is an option for Graphics3D and SurfaceGraphics
   which gives the point in space from which the objects plotted are to be viewed.

From Help - Built-in Functions - ViewPoint:

ViewPoint -> {x, y, z}gives the position of the view point relative to the center of the three-dimensional box that contains the object being plotted.

One way to change the viewpoint of a plot is as follows: insert a comma in the options section of the command for the plot, just before the closing bracket and click after the comma. Now click on Input on the toolbar and then 3D ViewPoint Selector. You will see a box representing the bounding box of a plot. Click on the box, and, holding the mouse button down, move the mouse around. The box rotates in response to your movements. Once you have chosen a viewpoint, release the mouse button. You will see the co-ordinates of the chosen viewpoint. Now click on Paste, and the command **ViewPoint→{a, b, c}** will appear next to the comma you typed earlier, with a, b, c the co-ordinates of the viewpoint you chose.

k = Plot3D[Exp[2 (Sin[$x^2 + y^2$] − Cos[x − y])], {x, −π, π}, {y, −π, π}, PlotPoints → 30, AspectRatio → Automatic, Ticks → None, DisplayFunction → Identity]

**l = Plot3D[Exp[2 (Sin[$x^2 + y^2$] − Cos[x − y])],
{x, −π, π}, {y, −π, π}, PlotPoints → 30, AspectRatio → Automatic,
ViewPoint −> {−0.006, −1.618, 2.972}, Ticks → None, DisplayFunction → Identity]**

Show[GraphicsArray[{k, l}, GraphicsSpacing → 0.1]]

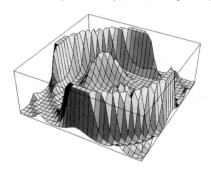 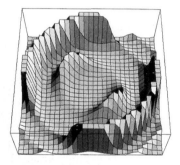

Here are programs for some other examples:

Plot3D[1 + Cos[$x^2 + y^2$], {x, −4, 4}, {y, −4, 4}, PlotPoints → 50, Mesh → False]

**Plot3D[15 (Sin[Log[$x^2 + y^6$]] − Cos[Log[$x^6 + y^2$]]),
{x, −π, π}, {y, −π, π}, PlotPoints → 30, AspectRatio → Automatic]**

The view from a point directly above or below a surface (given by **ViewPoint->{0, 0, a}**) is sometimes interesting Here is the program for such an example:

**Plot3D[Sin[x*y],{x,-4,4},{y,-4,4},PlotPoints→150,Mesh→False,Axes→False,Boxed→
False,ViewPoint→{0,0,2}]**

Exercise:

1) Try some of the above plots with different options.

2) In Help - Getting Started/Demos - Demos - Multipole Fields, the following definition is made:

$$\text{Multipole[n\_]} := \sum_{i=1}^{n} \frac{(-1)^i}{\sqrt{\left(x - \text{Cos}\left[\frac{2\pi i}{n}\right]\right)^2 + \left(y - \text{Sin}\left[\frac{2\pi i}{n}\right]\right)^2}}$$

Below is a command for plotting **Multipole[5]**, try it and for other values of n.

Plot3D[Evaluate[Multipole[5]], {x, −2, 2}, {y, −2, 2}, PlotPoints → 40]

■ Plotting the Graph of a $\mathbb{C} \to \mathbb{R}$ Function

Let $z \to f[z]$ be a function with domain, D, and range, G, subsets of $\mathbb{C}$. Then the graphs of the functions defined by $x + Iy \to \text{Abs}[f[x + Iy]]$, $\text{Im}[f[x + Iy]]$, $\text{Re}[f[x + Iy]]$ can be plotted as shown in the following example:

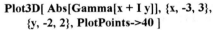

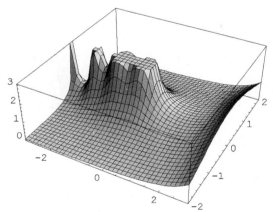

As x tends to ∞, Log[x] tends to ∞ more slowly. Log has a 'damping' effect on f as shown in the next 2 examples:

**m = Plot3D[1 + Abs[Sin[Sin[x^2 + I y]]], {x, −2, 2}, {y, −2, 2},
    PlotPoints → 40, PlotRange → All, DisplayFunction → Identity]**

**n = Plot3D[Log[1 + Abs[Sin[Sin[x^2 + I y]]]], {x, −2, 2}, {y, −2, 2},
    PlotPoints → 40, PlotRange → All, DisplayFunction → Identity]**

Show[GraphicsArray[{m, n}, GraphicsSpacing → 0.1]]

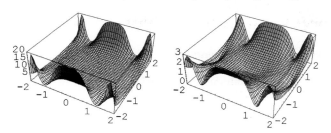

Another example:

Plot3D[Abs[Log[Sin[Log[Log[Cos[(x + I y)^6]]]]]], {y, −0.6, 1.2},
{x, 0.1, 1.9}, PlotPoints → 250, Mesh −> False, Axes → False,
Boxed → False, ViewPoint → {2, 0, 0.5}, BoxRatios → {1.6, 2.3, 0.6}]

In the following example, the view is taken from a point directly above the surface:

Plot3D[−Re[Log[Log[Sin[Cos[x^2 − I y] + Cos[y^2 − I x]]]]], {x, −4, −1}, {y, 1, 4.},
Mesh → False, PlotPoints → 250, Boxed −> False, Axes → False, ViewPoint → {0, 0, 2}]

Go to Special Functions in the Master Index, and you will find still more examples to try.

■ Displaying Two 3D Plots in the Same Diagram

The graphs of 2 or more functions from $\mathbb{R}^2$ to $\mathbb{R}$ can be shown in the same diagram, using **Show**. The functions do not need to have the same domain. Intersections of the surfaces are shown clearly. The individual plots are given names a, b, c,... say, and then executed. The command **Show[a, b, c, ...]** will display the plots on one diagram. The command **DisplayFunction->Identity** can be used.

Here is an example:

$$o = \textbf{Plot3D}\left[0.4\left((x^2 + y^2)^{\frac{3}{2}} - 4\,(x^2 + y^2)\right) + 1, \{x, -3.5, 3.5\}, \{y, -3.5, 3.5\},\right.$$

Boxed → False, Axes → True, PlotRange → All, DisplayFunction → Identity

$$p = \mathbf{Plot3D}\Big[0.4\left(-(x^2 + y^2)^{\frac{3}{2}} + 4\,(x^2 + y^2)\right) - 1,\; \{x,\, -3,\, 3\},$$

$$\{y,\, -3,\, 3\},\; \mathbf{Boxed} \to \mathbf{False},\; \mathbf{Axes} \to \mathbf{True},\; \mathbf{DisplayFunction} \to \mathbf{Identity}\Big]$$

$$q = \mathbf{Show[\%,\, \%\%,\, DisplayFunction} \to \mathbf{Identity]}$$

$$\mathbf{Show[GraphicsArray[\{o,\, p,\, q\},\; GraphicsSpacing} \to \mathbf{0.05]]}$$

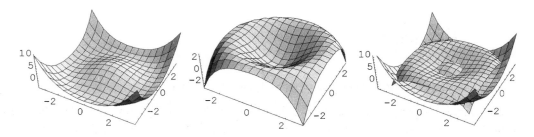

Exercise:

Display in one diagram:

1) the graphs of the paraboloids with equations $z = x^2 + y^2 - 1$, $z = -2\,x^2 - y^2 + 2$, for x and y between -2 and 2;

2) the surfaces with equations $z = \mathrm{Cos}[xy]$ and $z = -\mathrm{Cos}[xy]$, for $\mathrm{Abs}[x] < 2$ and $\mathrm{Abs}[y] < 4$.

1.9.4 3D Parametric Curve Plots

From Help - Built-in Functions - ParametricPlot3D:

"ParametricPlot3D[{fx, fy, fz}, {t, tmin, tmax}] produces a three-dimensional space curve parametrized by a variable t which runs from tmin to tmax."

Note that when using the command **ParametricPlot**, properties of the plot such as **GrayLevel** are defined using the option **PlotStyle**. This option is not available for **ParametricPlot3D**, instead plot properties are included in the definition of the plot, as graphics directives. Among the directives available are **GrayLevel**, **Thickness**, **Dashing**. These directives may be chosen to depend on the parameter t. All the options which we used for **Plot3D**, except **Mesh** are available in this case. In the following example, the graphics directives **Thickness** and **Gray-Level** are enclosed in braces and grouped with the co-ordinates of the point with the parameter t. We have chosen constant **Thickness** and **GrayLevel** a function of t, making sure that the argument of **GrayLevel** lies between 0 and 1 for all values of concern. The option **PlotPoints** has been added in order to obtain a smoother curve. The curve was derived from a member of family13 defined in 1.8.6 by adding a function of t as a z-co-ordinate.

$$\textbf{ParametricPlot3D}\Big[\Big\{\ \textbf{Cos[2 t]} - \textbf{Cos[9 t]},\ \ \textbf{Cos[2 t]} - \textbf{Sin[9 t]},$$

$$\textbf{2 Sin[2 t]},\ \Big\{\textbf{Thickness[0.018], GrayLevel}\Big[\textbf{Abs}\Big[\frac{1}{2} - \frac{t}{2\pi}\Big]\Big]\Big\}\Big\},$$

$$\textbf{\{t, 0, 2}\pi\textbf{\}, PlotPoints} \rightarrow \textbf{200, Ticks} \rightarrow \textbf{False}\Big]$$

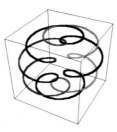

Two or more 3D parametric curves can be shown in the same diagram, provided they are defined for the same values of the parameter. Graphics directives for each curve must be grouped with the co-ordinates of the point with parameter t of the curve. The list of these curve definitions must then be enclosed in braces. The image below shows 2 parametric curves, the first with 2 and the second with 1 graphics directive.

$$\textbf{ParametricPlot3D}\Big[\Big\{\Big\{\ \textbf{Cos[2 t]},\ \ \textbf{Sin[2 t]},\ \frac{t}{2},\ \textbf{\{Thickness[0.05], GrayLevel[0.6]\}}\Big\},$$

$$\Big\{\ \textbf{Cos[2 t]},\ \ \textbf{Sin[2 t]},\ \frac{t}{2} + \textbf{0.5, Thickness[0.02]}\Big\}\Big\},\ \textbf{\{t, 0, 4}\pi\textbf{\}, Axes} \rightarrow \textbf{False}\Big]$$

Exercise:
Experiment with plotting 3D parametric curves. Try examples with parametric equations given by {Sin[a t], Sin[b t], Sin[c t]}. Try choosing a member of one of the families listed in 1.8.6 and assigning a z-co-ordinate.

1.9.5 3D Parametric Surface Plots

From Help - Built-in Functions - ParametricPlot3D:
ParametricPlot3D[{fx, fy, fz}, {t, tmin, tmax},{u,umin,umax}] produces a three-dimensional surface parametrized by t and u.

Here is the plot of a torus:

ParametricPlot3D[{Cos[θ] (2 + Cos[y]), Sin[θ] (2 + Cos[y]), Sin[y]},
{y, 0, 2π}, {θ, π, 3π}, Axes → False, ViewPoint –> {−0.010, −2.689, 2.054}]

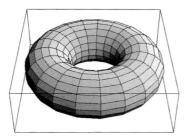

Here is a plot which will be used in Chapter 6. It was obtained from the plot of the torus by adding 2θ to the z-co-ordinate:

ParametricPlot3D[{(2 + Cos[y]) Cos[θ], (2 + Cos[y]) Sin[θ], 2θ + Sin[y]}, {y, 0, 2π},
{θ, π, 3π}, PlotRange → All, ViewPoint –> {−3.200, 1.099, 0.059}, Ticks → False]

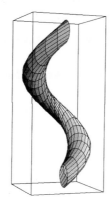

Exercise:

1) Limit the domain of y (θ) to show a horizontal (vertical) cross-section through the torus.

2) Plot the surface with parametric equations given by:

$\{(\text{Sin}[s])^2 \text{ Sin}[t], (\text{Sin}[s])^2 \text{ Cos}[t], \text{Sin}[s] \text{ Cos}[s]\}, 0 \le s \le \pi, 0 \le t \le 5 \frac{\pi}{2}.$

3) Write a procedure for constructing the ellipsoids with parametric equations given by $\{x, y, z\} = \{a \text{ Sin}[s] \text{ Cos}[t], b \text{ Sin}[s] \text{ Sin}[t], c \text{ Cos}[s]\}, 0 \le s \le \pi, 0 \le t \le 2\pi.$

Experiment with your procedure for different values of the parameters a, b and c. Display a pair of ellipsoids in the same diagram.

▪ Two Image Stereograms, the Command GraphicsArray

Here is the plot of a pair of cylinders. We have named the plot r.

r = **ParametricPlot3D**[{{Cos[t], Sin[t], s}, {Cos[t], s, Sin[t]}},
    {t, 0, 2π}, {s, −2.5, 2.5}, **Axes** → **False**, **Boxed** → **False**, **PlotPoints** → 25,
    ViewPoint −> {−1.722, −1.835, 2.263}, **ColorOutput** → **CMYKColor**]

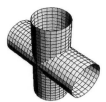

Here is the above plot with a different viewpoint obtained by slightly increasing the x- and y-co-ordinates of the previous viewpoint, while leaving the z-co-ordinate unchanged. We have called the resulting plot s.

s = **ParametricPlot3D**[{{Cos[t], Sin[t], s}, {Cos[t], s, Sin[t]}},
    {t, 0, 2π}, {s, −2.5, 2.5}, **Axes** → **False**, **Boxed** → **False**, **PlotPoints** → 25,
    ViewPoint −> {−1.622, −1.735, 2.263}, **ColorOutput** → **CMYKColor**]

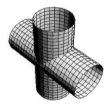

We now show the plots r and s on the same diagram, using the command **GraphicsArray** and the option **GraphicsSpacing**.

? GraphicsArray

GraphicsArray[{g1, g2, ... }] represents a row of graphics objects. GraphicsArray[
    {{g11, g12, ... }, ... }] represents a two−dimensional array of graphics objects.

? GraphicsSpacing

GraphicsSpacing is an option for GraphicsArray which specifies the spacing between elements in the array.

The h in the option **GraphicsSpacing→h**, determines the horizontal spacing between the plots as a fraction of their widths. A small negative value for h is usually needed for 2-image stereo-grams. You may need to try various values for h to obtain a good 3D image.

Show[GraphicsArray[{r, s}, GraphicsSpacing → −0.2]]

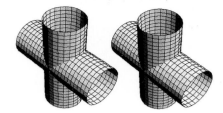

Exercise:

1) Construct a stereogram of a vertical cross-section of the above.

2) Construct a stereogram of a pair of tori intersecting at right angles.

3) Go to 1.9.9 (1.9.10 in Version 4.2) in The *Mathematica* Book section of Help to read about how parametric surfaces are generated. Read the last paragraph of the section, which explains the choice of ranges for the variables. Copy and paste the routine for generating a torus, and, by changing some of the constants, generate different tori. By limiting one or more of the ranges of the variables, generate sections of a torus.

1.9.6 Constructing Surfaces from a 2D Parametric Plot

Let C be the curve with parametric equations given by:

{9Cos[θ] - Cos[9θ], 9Sin[θ] - Sin[9θ]}, {θ, 0, 2π}.

Here is the plot of C. We wish to plot surfaces with cross-sections parallel to the x-y plane similar to C.

ParametricPlot[{9 Cos[θ] − Cos[9 θ], 9 Sin[θ] − Sin[9 θ]},
{θ, 0, 2 π}, AspectRatio → Automatic, Ticks → {{−8, 0, 8}, {−8, 0, 8}}]

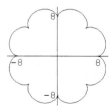

For each fixed t, the curve with parametric equations:

x = t(9 Cos[θ] − Cos[9 θ]), y = t(9 Sin[θ] − Sin[9 θ])

is similar to C. The parameter t is called the multiplying factor.

We plot some members of the family:

ParametricPlot$\Big[$

$\quad$ **Evaluate**$\Big[$**Table**$\Big[\Big\{\dfrac{\mathbf{n}}{\mathbf{n+1}}\ (9\ \text{Cos}[\theta] - \text{Cos}[9\ \theta]),\ \dfrac{\mathbf{n}}{\mathbf{n+1}}\ (9\ \text{Sin}[\theta] - \text{Sin}[9\ \theta])\Big\},\ \{\mathbf{n},\ 1,\ 4\}\Big]\Big]$,

$\quad$ $\{\theta,\ 0,\ 2\pi\}$, **AspectRatio** $\rightarrow$ **Automatic, Ticks** $\rightarrow$ **None**$\Big|$

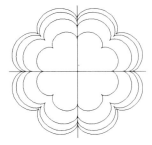

We can think of the above diagram as a contour-plot of a surface.

The 3D parametric plot:
$\{t\ (9\text{Cos}[\theta] - \text{Cos}[9\theta]),\ t(9\text{Sin}[\theta] - \text{Sin}[9\theta]),\ f[t]\}$
raises (lowers) C, f [t] units up (down) if f [t] > 0 (< 0).
Choosing f [t] = -8 t we get:

$\quad$ **ParametricPlot3D**$[\{(9\ \text{Cos}[\theta] - \text{Cos}[9\ \theta])\ t,\ (9\ \text{Sin}[\theta] - \text{Sin}[9\ \theta])\ t,\ -8\ t\},\ \{t,\ 0,\ 2\},$
$\quad\quad$ $\{\theta,\ 0,\ 2\pi\}$, **AspectRatio** $\rightarrow$ **Automatic, ViewPoint** $->$ $\{-0.010,\ -2.881,\ 1.775\}$,
$\quad\quad$ **PlotPoints** $\rightarrow$ **50, ColorOutput** $->$ **CMYKColor, Axes** $\rightarrow$ **False, Boxed** $\rightarrow$ **False]**

This time vertical cross-sections are parabolic:

$\quad$ **ParametricPlot3D**$[\{(9\ \text{Cos}[\theta] - \text{Cos}[9\ \theta])\ t,\ (9\ \text{Sin}[\theta] - \text{Sin}[9\ \theta])\ t,\ -5\ t^2\},\ \{t,\ 0,\ 2\},$
$\quad\quad$ $\{\theta,\ 0,\ 2\pi\}$, **AspectRatio** $\rightarrow$ **Automatic, ColorOutput** $->$ **CMYKColor, PlotPoints** $\rightarrow$ **50,**
$\quad\quad$ **Axes** $\rightarrow$ **False, Boxed** $\rightarrow$ **False, ViewPoint** $->$ $\{-0.010,\ -2.881,\ 1.775\}]$

ParametricPlot3D[{(9 Cos[θ] − Cos[9 θ]) t, (9 Sin[θ] − Sin[9 θ]) t, 7 Cos[π t]}, {t, 0, 2}, {θ, 0, 2 π}, AspectRatio → Automatic, PlotPoints → 50, ColorOutput –> CMYKColor, Axes → False, Boxed → False, ViewPoint –> {−0.010, −2.881, 1.775}]

One can also use a function of t instead of t itself as the multiplying factor for the co-ordinates of C. In the following example, the multiplying factor is Cos[t], which varies at a non-uniform rate from 0 to 1 as t varies from $\frac{-\pi}{2}$ to 0, and from 1 to 0 as t varies from 0 to $\frac{\pi}{2}$.

ParametricPlot3D[{(9 Cos[θ] − Cos[9 θ]) Cos[t], (9 Sin[θ] − Sin[9 θ]) Cos[t], 7 t}, $\left\{t, \dfrac{-\pi}{2}, \dfrac{\pi}{2}\right\}$, {θ, 0, 2 π}, AspectRatio → Automatic, PlotPoints → 50, ViewPoint –> {2.870, 1.385, 1.137}, Axes → False, Boxed → False

In the above example, horizontal cross-sections of the surface if projected onto the x-y plane are closed curves with the same centre. One can also add a translation term to each member of the original family of curves, to obtain a different family, in this case the horizontal cross-sections do not have the same center. For example:

ParametricPlot3D[{(9 Cos[θ] − Cos[9 θ]) Cos[t] − 5 t, (9 Sin[θ] − Sin[9 θ]) Cos[t] − 7 t, 7 t}, $\left\{t, \dfrac{-\pi}{2}, \dfrac{\pi}{2}\right\}$, {θ, 0, 2 π}, AspectRatio → Automatic, PlotPoints → 50, ViewPoint –> {2.870, 1.385, 1.137}, Axes → False, Boxed → False

Exercise:

1) Experiment with various parametric plots.

2) Use the following parametric plots to construct shell-like forms:

ParametricPlot[{1.2 (7 Cos[θ] + Cos[6 θ]), 0.7 (6 Sin[θ] − Sin[6 θ])},
 {θ, 0, 2π}, AspectRatio → Automatic, Ticks → False]

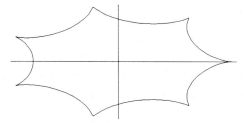

$$\textbf{ParametricPlot}\left[\left\{\textbf{Sin}\left[3\left(t+\frac{\pi}{2}\right)\right]\textbf{Cos}\left[t+\frac{\pi}{2}\right](1+0.3\,\textbf{Sign[t]}),\,\textbf{Sin}\left[3\left(t+\frac{\pi}{2}\right)\right]\textbf{Sin}\left[t+\frac{\pi}{2}\right]\right\},\right.$$
$$\left.\left\{t,\,-\frac{\pi}{6},\,\frac{\pi}{6}\right\},\,\textbf{AspectRatio -> Automatic, Ticks} \rightarrow \{\{-0.13,\,0,\,0.2\},\,\textbf{Automatic}\}\right]$$

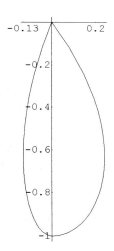

In the above example, we used a modified form of part of the plot of a member of family5 defined in 1.8.6. A single 'petal' has been plotted, the parameter t has been replaced by $t+\frac{\pi}{2}$ and the range of t is from $\frac{-\pi}{6}$ to $\frac{\pi}{6}$, to ensure that the left-hand side of the petal has negative and the right-hand side positive t values. The x-co-ordinate of the point with parameter t has been multiplied by the factor **1 + 0.3 Sign[t]**, which has the values 1.3 for t positive and 0.7 for t negative.

1.10 2D Graphics Derived from 3D Graphics

1.10.1 Density Plots

In *Mathematica*, the 3D plot of a function $f \colon \mathbb{R}^2 \to \mathbb{R}$ has a maximum and a minimum z value Maxf and Minf, say, over a rectangle R in the x-y plane. The range, r, of f in $\mathbb{R}$ is defined to be Maxf − Minf.

In the command **DensityPlot**, a number of plot points can be chosen. Suppose we choose 25 plot points. *Mathematica* divides the rectangle R into 25×25 sub-rectangles. For each n, a point P_n is chosen in the nth sub-rectangle, R_n. The rectangle R_n is assigned the number: t_n = (z-coordinate of P_n − Min f)/r, so each point in the rectangle is assigned a number between 0 and 1 representing its height above the lowest point on the plot of f. This number is called the height number of the point. The default coloring for **DensityPlot** is **GrayLevel**, so, in the command **DensityPlot**[f, {x, xmin, xmax}, {y, ymin, ymax}], each point in the rectangle R_n is assigned the color **GrayLevel**[t_n].

> **? DensityPlot**
>
> DensityPlot[f, {x, xmin, xmax}, {y, ymin, ymax}] makes a density plot of f as a function of x and y.

Consider the 3D plot below:

$$\textbf{Plot3D}\Big[\, \textbf{Sin}\big[(x^2 + y^2)^{-1}\big], \{x, -0.4, 0.4\}, \{y, -0.4, 0.4\},$$
$$\textbf{PlotPoints} \to \textbf{30, AspectRatio} \to \textbf{Automatic, Axes} \to \textbf{False}\Big]$$

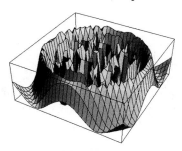

A useful option for **DensityPlot** is **Mesh**. By default, mesh lines are included. If one wishes to exclude them, one can use the option **Mesh→False.** If a large number of plot points are used, then the mesh lines should be omitted.

We show two Density plots of the above, one with 25 plot points and the mesh lines included, the other with 250 plot points and the mesh lines excluded.

$$\textbf{t} = \textbf{DensityPlot}\Big[\, \textbf{Sin}\big[(x^2 + y^2)^{-1}\big], \{y, -0.4, 0.4\}, \{y, -0.4, 0.4\},$$
$$\textbf{PlotPoints} \to \textbf{25, FrameTicks} \to \textbf{False, DisplayFunction} \to \textbf{Identity}\Big]$$

$$u = \textbf{DensityPlot}\Big[\textbf{Sin}\big[(x^2 + y^2)^{-1}\big] \Big], \{x, -0.4, 0.4\}, \{y, -0.4, 0.4\}, \textbf{PlotPoints} \rightarrow 250,$$
$$\textbf{Mesh} \mathrel{-}\mathrel{>} \textbf{False}, \textbf{FrameTicks} \rightarrow \textbf{False}, \textbf{DisplayFunction} \rightarrow \textbf{Identity}\Big]$$

Show[GraphicsArray[{t, u}, GraphicsSpacing −> 0.2]]

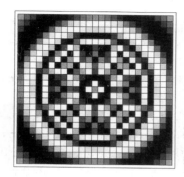 

Another example:

$$\textbf{DensityPlot}[-\textbf{Im}[\textbf{Log}[\textbf{Log}[\textbf{Sin}[\textbf{Sin}[(x^3 + I\, y^3)] + \textbf{Sin}[(y + I\, x)^3]]]]]],$$
$$\{x, -1.9, 1.2\}, \{y, -1.1, 1.1\}, \textbf{AspectRatio} \rightarrow \textbf{Automatic},$$
$$\textbf{PlotPoints} \rightarrow 250, \textbf{Mesh} \rightarrow \textbf{False}, \textbf{Frame} \rightarrow \textbf{False}]$$

Here are 2 programmes to try:

$$\textbf{DensityPlot}[-\textbf{Im}[\textbf{Log}[\textbf{Log}[\textbf{Sin}[\textbf{Cos}[x^2 + I\, y^2] + \textbf{Cos}[y + I\, x]]]]]], \{x, -3, 1\}, \{y, -0, 2\},$$
$$\textbf{AspectRatio} \rightarrow \textbf{Automatic}, \textbf{PlotPoints} \rightarrow 200, \textbf{Mesh} \rightarrow \textbf{False}, \textbf{Frame} \rightarrow \textbf{False}]$$

$$\textbf{DensityPlot}[-\textbf{Im}[\textbf{Log}[\textbf{Log}[\textbf{Sin}[(x - I\, y)^{13}]]]]], \{x, -1.3, 1.3\}, \{y, -1.3, 1.3\},$$
$$\textbf{AspectRatio} \rightarrow \textbf{Automatic}, \textbf{PlotPoints} \rightarrow 160, \textbf{Mesh} \rightarrow \textbf{False}, \textbf{Frame} \rightarrow \textbf{False}]$$

Some interesting patterns can be obtained by applying the built-in functions **Floor** or **Integer-Part** and **Mod** to a function f and then **DensityPlot**.

We show below: the plot, v, of a real function, f, of 2 variables; the plot, w, of **Floor** f and the plot, x, of **Mod[Floor[f, 2]**:

$$v = \textbf{Plot3D}[4 - x^2 - y^2, \{x, -2, 2\}, \{y, -2, 2\}, \textbf{BoxRatios} \rightarrow \{1, 1, 1.5\},$$
$$\textbf{PlotPoints} \rightarrow 30, \textbf{Ticks} \rightarrow \textbf{False}, \textbf{DisplayFunction} \rightarrow \textbf{Identity}];$$

w = Plot3D[Floor[4 − x² − y²], {x, −2, 2}, {y, −2, 2}, BoxRatios → {1, 1, 1.5},
      PlotPoints → 30, Ticks → False, DisplayFunction → Identity];

x = Plot3D[Mod[Floor[4 − x² − y²], 2], {x, −2, 2}, {y, −2, 2}, BoxRatios → {1, 1, 0.2},
      PlotPoints → 30, Ticks → False, DisplayFunction → Identity];

Show[GraphicsArray[{v, w, x}, GraphicsSpacing −> 0.15]]

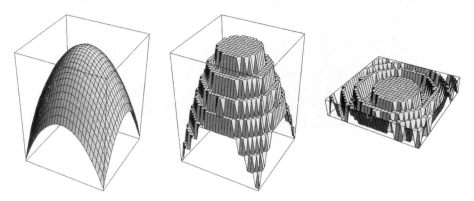

We now show Density plots of the above surfaces:

v1 = DensityPlot[4 − x² − y², {x, −2, 2},
      {y, −2, 2}, PlotPoints → 30, DisplayFunction → Identity];

w1 = DensityPlot[Floor[4 − x² − y²], {x, −2, 2},
      {y, −2, 2}, PlotPoints → 30, DisplayFunction → Identity];

x1 = DensityPlot[Mod[Floor[4 − x² − y²], 2], {x, −2, 2},
      {y, −2, 2}, PlotPoints → 30, DisplayFunction → Identity];

Show[GraphicsArray[{v1, w1, x1},
      DisplayFunction → $DisplayFunction, GraphicsSpacing → 0.01]]

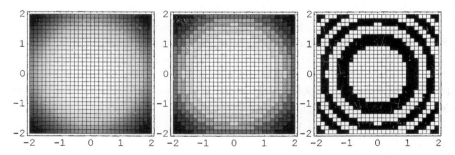

The **DensityPlot** of **Mod[IntegerPart[k f[x, y], n]** or **Mod[Floor[k f[x, y], n]** is sometimes interesting. We call this the 'Mod-Floor Process'. If n = 2, a black and white pattern is generated, while if n > 2, a gray level plot is generated. Here is an example with k = 3 and n = 2. The option: **Background→GrayLevel[g]**, where 0 ≤ g ≤ 1, provides a frame for the image.

ContourPlot[Mod[Floor[3 Abs[Log[Cos[Sin[x^2 + I y] − Sin[y + I x]]]]], 2],
{x, −2.4, 2.5}, {y, −2.3, 2.4}, PlotPoints → 200, ContourLines → False,
AspectRatio → Automatic, Frame → False, Contours → 2, Background → GrayLevel[0]]

Here are programs for other examples:

DensityPlot[Mod[Floor[Re[Log[Cos[Log[Cos[1 / (x + y + I (x − y))^2]]]]]], 2],
{x, −0.45, 0.45}, {y, −0.45, 0.45}, Background −> GrayLevel[0],
PlotPoints → 200, Mesh → False, AspectRatio → Automatic, Frame → False]

DensityPlot[Mod[Floor[−4 Re[Log[Cos[Log[Sin[(x + I y)^4]]]]]], 2],
{x, −1.2, 1.2}, {y, −1.2, 1.2}, PlotPoints → 250, Mesh → False,
AspectRatio → Automatic, Frame → False, Background → GrayLevel[0]]

DensityPlot[Mod[Floor[−4 Abs[Log[Log[Sin[Log[Sin[(x + I y)^4]]]]]]], 2],
{y, −1.3, 1.3}, {x, −1.3, 1.3}, Frame → False,
Background −> GrayLevel[0.0], Mesh → False, PlotPoints → 280]

An example of a gray scale **DensityPlot**:

DensityPlot[Mod[Floor[−2 Im[Log[Cos[Log[Sin[(x + I y)^6]]]]]], 10],
{x, −1.2, 1.2}, {y, −1.2, 1.2}, PlotPoints → 250, Mesh → False,
AspectRatio → Automatic, Frame → False, Background → GrayLevel[.9]]

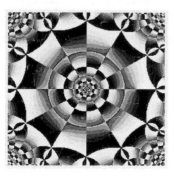

Further programs to try:

> **DensityPlot[Mod[Floor[Im[Log[Cos[Log[Cos[1 / (x + I y)]]]]]], 6],**
> **{y, −0.55, 0.55}, {x, −0.55, 0.55}, AspectRatio → Automatic, PlotPoints → 300,**
> **Frame → False, Axes → False, Mesh → False, Background → GrayLevel[0]]**

> **DensityPlot[Mod[Floor[−28 Abs[Log[Log[Cos[Log[Sin[(x + I y)⁴]]]]]]], 12],**
> **{y, −1.3, 1.3}, {x, −1.3, 1.3}, Frame → False,**
> **Background −> GrayLevel[.0], Mesh → False, PlotPoints → 80]**

Exercise:

For each example given, try one of the other techniques described in 1.10.1. Try the techniques on other functions.

1.10.2 Contour Plots

Let f be a real valued function with domain a rectangle K. Contour lines on the surface with equation $z = f[x, y]$ are lines of equal height. In the command **ContourPlot**, a number of contour lines can be chosen. Suppose we choose 25 contour lines. The mth contour line consists of all points on the graph of f with z-co-ordinate, $h_m = Min_f + m \frac{range\ of\ f}{25}$. The mth contour line is assigned the height number $\frac{m}{25}$ for coloring purposes. The default coloring is **GrayLevel**, so the region between successive contour lines is colored according to the height number.

> **? ContourPlot**
>
> ContourPlot[f, {x, xmin, xmax}, {y, ymin, ymax}] generates a contour plot of f as a function of x and y.

■ Some Options for ContourPlot

One can choose the number of contour lines in a contour plot by choosing the option **Contours→m**, where m is a chosen integer. One can also choose the actual contours one wants to be displayed. The option **Contours→{z1, z2, ...}** specifies the z values of contours to use. A useful option for **ContourPlot** is **ContourShading**. By default, regions between contour lines are shaded in **GrayLevel** according to their height. If one wishes to see the contour lines without shading one can use the option **ContourShading→False**. Contour lines can be omitted with the option **ContourLines→False**.

Here is the plot of a surface:

$$\textbf{Plot3D}[(x^2 + 3\,y^2)\,\textbf{Exp}[1 - x^2 - y^2], \{x, -2, 2\}, \{y, -2, 2\},$$
$$\textbf{PlotPoints} \rightarrow \textbf{30, ViewPoint} \rightarrow \{-0.005, -1.465, 3.050\}, \textbf{Ticks} \rightarrow \textbf{False}]$$

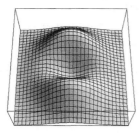

We show below contour plots α, β and γ of the above surface, α with 10 contours and no shading; β with contours 1 and 2 and no shading and γ with 10 contours and contour shading:

$$\alpha = \textbf{ContourPlot}[(x^2 + 3\,y^2)\,\textbf{Exp}[1 - x^2 - y^2], \{x, -2, 2\}, \{y, -2, 2\},$$
$$\textbf{Contours} \rightarrow \textbf{10, ContourShading} \rightarrow \textbf{False, DisplayFunction} \rightarrow \textbf{Identity}]$$

$$\beta = \textbf{ContourPlot}[(x^2 + 3\,y^2)\,\textbf{Exp}[1 - x^2 - y^2], \{x, -2, 2\}, \{y, -2, 2\},$$
$$\textbf{Contours} \rightarrow \{1, 2\}, \textbf{ContourShading} \rightarrow \textbf{False, DisplayFunction} \rightarrow \textbf{Identity}]$$

$$\gamma = \textbf{ContourPlot}[(x^2 + 3\,y^2)\,\textbf{Exp}[1 - x^2 - y^2],$$
$$\{x, -2, 2\}, \{y, -2, 2\}, \textbf{Contours} \rightarrow \textbf{10, DisplayFunction} \rightarrow \textbf{Identity}]$$

$$\textbf{Show[GraphicsArray}[\{\alpha, \beta, \gamma\}, \textbf{ GraphicsSpacing} \rightarrow \textbf{0.15}]]$$

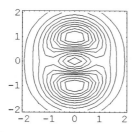 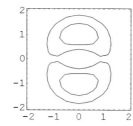 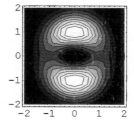

Another example:

$$\textbf{ContourPlot[Abs[Log[Sin}[1 / (x + I\,y)]]], \{x, -0.65, 0.3\},$$
$$\{y, -0.4, 0.4\}, \textbf{Contours} \rightarrow \textbf{8, Frame} \rightarrow \textbf{False, ContourLines} \rightarrow \textbf{True},$$
$$\textbf{Background} \rightarrow \textbf{RGBColor}[0, 0, 0], \textbf{PlotPoints} \rightarrow \textbf{200}]$$

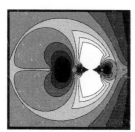

A program to try:

> ContourPlot[−Abs[Log[Sin[+Cos[(x + I (y))] − Sin[(y + I (x))]]]], {x, −3, 3},
> {y, −3, 3}, AspectRatio → Automatic, ContourLines → True, Contours → 12,
> PlotPoints → 150, Frame → False, Background → RGBColor[0, 0, 0]]

We use the 'Mod-Floor Process' described in 1.10.1 to construct Contour Plots. Here is an example of a black and white plot. The option **ContourLines→False** is used.

$$\text{ContourPlot}\left[\text{Mod}\left[\text{Floor}\left[-3\,\text{Abs}\left[\text{Log}\left[\text{Log}\left[\text{Sin}\left[\text{Sin}\left[\frac{1}{4}\,(x+y+I\,(x-y))^4\right]\right]\right]\right]\right],\,2\right],\right.$$

> {x, 0.65, 1.4}, {y, 0.65, 1.4}, AspectRatio → Automatic, ContourLines → False,
> Contours → 2, PlotPoints → 250, Frame → False, Background → GrayLevel[0]]

The program for another example:

> ContourPlot[Mod[Floor[8 Abs[Log[Log[Cos[Log[Sin[(x + I y)⁴]]]]]]], 2],
> {y, −1.4, 1.4}, {x, −1.4, 1.4}, PlotPoints → 220, ContourLines → False,
> AspectRatio → Automatic, Frame → False, Contours → 2, Background → GrayLevel[0]]

An example of a gray scale plot and the view from above of the 3D surface plot of the same function:

> ContourPlot[−Mod[Floor[10 Arg[Log[Sin[Cos[(x + I (y))] − Sin[(y + I (x))]]]]], 30],
> {x, −2, 5}, {y, 0, 4.5}, AspectRatio → Automatic, ContourLines → False,
> Contours → 52, PlotPoints → 150, Frame → False, Background → RGBColor[0, 0, 0]]

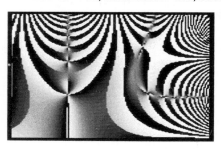

Plot3D[−Mod[Floor[10 Arg[Log[Sin[Cos[(x + I (y))] − Sin[(y + I (x))]]]]], 30],
  {x, −2., 5}, {y, 0, 4.5}, PlotPoints → 250, Mesh → False, Axes → False,
  BoxRatios → {7, 4.5, 2}, Boxed → False, ViewPoint → {0, 0, 4}]

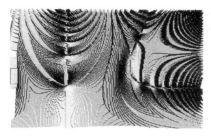

Programs for further examples of the above type: (in the first two surface plot examples, the plots are colored using **ColorOutput**)

ContourPlot[−Mod[Floor[16 Abs[Log[Log[Cos[Log[Cos[0.5 (x + y + I (x − y))^2]]]]]]], 20],
  {y, −2.1, 2.1}, {x, −2.1, 2.1}, ContourLines → False, Contours → 52,
  PlotPoints → 150, Frame → False, Background → RGBColor[0, 0, 0]]

Plot3D[−Mod[Floor[16 Abs[Log[Log[Cos[Log[Cos[0.5 (x + y + I (x − y))^2]]]]]]], 20],
  {y, −2.1, 2.1}, {x, −2.1, 2.1}, PlotPoints → 200, Mesh → False, Boxed → False,
  Axes → False, ViewPoint → {0, 0, −16}, ColorOutput → CMYKColor]

ContourPlot[Mod[Floor[5 Abs[(Log[Log[Sin[1 / (x + I (y))]]])^6]], 50], {x, −0.8, 2.8},
  {y, −1.2, 1.2}, AspectRatio → 1, ContourLines → False, Contours → 12,
  PlotPoints → 250, Frame → False, Background → RGBColor[0, 0, 0]]

Plot3D[Mod[Floor[5 Abs[(Log[Log[Sin[1 / (x + I (y))]]])^6]], 50], {x, −0.8, 2.8},
  {y, −1.2, 1.2}, BoxRatios → {1, 1, 0.3}, PlotPoints → 150, Mesh → False,
  Boxed → False, Axes → False, ViewPoint → {0, 0, −18}, ColorOutput → CMYKColor]

ContourPlot[Mod[Floor[41 Re[Log[Log[3 Sin[1 / (x + I (y))^3]]]]], 40],
  {x, −1.6, 1.99}, {y, −1.95, 1.77}, ContourLines → False,
  Contours → 52, PlotPoints → 150, Frame → False]

Plot3D[Mod[Floor[41 Re[Log[Log[3 Sin[1 / (x + I (y))^3]]]]], 40], {x, −1.6, 1.99},
  {y, −1.95, 1.8}, BoxRatios → {1, 1, 0.3}, PlotPoints → 250, Mesh → False,
  Boxed → False, Axes → False, Boxed → False, ViewPoint → {0, 0, 2}]

In the following example, the option **ContourShading→False** is used, so that only the contour lines are shown. The option **Background→GrayLevel[0.95]** provides a light gray background which serves to frame the plot.

ContourPlot[3 Abs[Log[Cos[Sin[x^2 + I y] − Sin[y + I x]]]], {x, −2.4, 2.5},
  {y, −2.3, 2.4}, ContourLines → True, Contours → 60, PlotPoints → 250,
  Frame → False, Background → GrayLevel[0.95], ContourShading → False]

A program to try:

> ContourPlot[−3 Re[Log[Cos[Sin[x² + I y] − 3 Sin[y + I x]]]], {x, −2.8, 2.7}, {y, −2.3, 2.9},
> AspectRatio → Automatic, ContourLines → True, Contours → 60, PlotPoints → 250,
> Frame → False, Background → GrayLevel[0.95], ContourShading → False]

The options **ContourStyle→GrayLevel[1]** and **Background→GrayLevel[0]** can be used to produce an image consisting of white lines on a black background:

> ContourPlot[Abs[Log[Log[Cos[Log[Sin[(x + I y)⁴]]]]]], {y, −1.4, 1.4},
> {x, −1.4, 1.4}, Background −> GrayLevel[0], PlotPoints → 250,
> ContourShading −> False, ContourStyle → {Thickness[.008], GrayLevel[1]},
> Contours → 12, ContourLines → True, Frame → False]

A program to try:

> ContourPlot[Mod[Floor[8 Abs[Log[Log[Tan[Sin[(x + I y)⁴]]]]]], 2], {x, −1.4, 1.4},
> {y, −1.4, 1.4}, Background → GrayLevel[0], AspectRatio → Automatic,
> ContourLines → True, Contours → 2, PlotPoints → 200, Frame → False,
> ContourShading → False, Background → GrayLevel[0], ContourStyle → GrayLevel[1]]

Exercise:

For each example given, try one of the other techniques described in 1.10.2. Try the techniques on other functions.

1.11 Solving Equations in one Variable

1.11.1 The Symbols = and ==

We used the symbol '=' in section 1.5, Naming Expressions and in section 1.9.4 in which an expression or constant was given a literal name such as 'a', and 'a' retained that value throughout the *Mathematica* session. In section 1.9.5, Stereograms, the command for a plot was assigned a literal name. In equation solving, the symbol '==' is used.

From Help - 1.5.5:
It is very important that you do not confuse $x = y$ with $x == y$. While $x = y$ is an imperative statement that actually causes an assignment to be done, $x == y$ merely tests whether x and y are equal, and causes no explicit action. If you have used the C programming language, you will recognize that the notation for assignment and testing in Mathematica is the same as in C.

An equation in one variable is written in the form $f[x] == 0$ or $f[x] == g[x]$. There are various methods of solving (finding the values of x for which the statement is true), depending on the type of the equation.

1.11.2 Exact Solutions of Algebraic Equations of Degree at most Four

From Help 1.5.7
Solve always tries to give you explicit formulas for the solutions to equations. However, it is a basic mathematical result that, for sufficiently complicated equations, explicit algebraic formulas cannot be given. If you have an algebraic equation in one variable, and the highest power of the variable is at most four, then Mathematica can always give you formulas for the solutions. However, if the highest power is five or more, it may be mathematically impossible to give explicit algebraic formulas for all the solutions.

Here is an example, which makes the syntax clear:

Solve[$x^4 + 3 x^3 + x^2 - 2 == 0$, x] $\{\{x \to -(-1)^{1/3}\}, \{x \to (-1)^{2/3}\}, \{x \to -1 - \sqrt{3}\}, \{x \to -1 + \sqrt{3}\}\}$

■ Making a List of Solutions

Solutions to **Solve** are not given as a list of numbers, and it is sometimes useful to have such a list. This can be achieved with the command **Replace** as the following example shows.

Replace[x, Solve[$x^4 + 3x^3 + x^2 - 2 == 0$, x]] $\{-(-1)^{1/3}, (-1)^{2/3}, -1 - \sqrt{3}, -1 + \sqrt{3}\}$

■ Expressing Solutions in Form a+I b

If a polynomial equation has non-real complex roots, *Mathematica* does not always give them in the form a+I b, a, b real. This can be achieved with the command **ExpToTrig.** We use this command to express the solutions to the previous equation in this form:

ExpToTrig[Replace[x, Solve[$x^4 + 3x^3 + x^2 - 2 == 0$, x]]]

$$\left\{-\frac{1}{2} - \frac{i\sqrt{3}}{2}, -\frac{1}{2} + \frac{i\sqrt{3}}{2}, -1 - \sqrt{3}, -1 + \sqrt{3}\right\}$$

Exact solutions can be found for some algebraic equations of higher degree.

Solve[$x^8 == 1$, x] $\{\{x \to -1\}, \{x \to -i\}, \{x \to i\}, \{x \to 1\}, \{x \to -(-1)^{1/4}\}, \{x \to (-1)^{1/4}\}, \{x \to -(-1)^{3/4}\}$

We use functional notation to define a procedure **unityRoots** for listing the nth roots of unity, each in the form a+I b:

unityRoots[n_] := ExpToTrig[Replace[z, Solve[$z^n = 1$, z]]];

We list the 5th roots of 1:

unityRoots[5] $\left\{1, -\frac{1}{4} - \frac{\sqrt{5}}{4} - \frac{1}{2}i\sqrt{\frac{1}{2}(5 - \sqrt{5})}, -\frac{1}{4} + \frac{\sqrt{5}}{4} + \frac{1}{2}i\sqrt{\frac{1}{2}(5 + \sqrt{5})}, \right.$

$$\left. -\frac{1}{4} + \frac{\sqrt{5}}{4} - \frac{1}{2}i\sqrt{\frac{1}{2}(5 + \sqrt{5})}, -\frac{1}{4} - \frac{\sqrt{5}}{4} + \frac{1}{2}i\sqrt{\frac{1}{2}(5 - \sqrt{5})}\right\}$$

Exact solutions are given. If you would like approximate solutions, you can use the command of the next section: **NSolve,** or proceed as follows:

unityRoots[5] // N $\{1., -0.809017 - 0.587785\,i, 0.309017 + 0.951057\,i, 0.309017 - 0.951057\,i, -0.809017$

Exercise:
1) Find the exact and approximate values of all solutions of the equation $x^4 - 2x^3 + 3x = 1$, and check one of your solutions using /..
2) Use **ListPlot** to write a procedure for plotting the nth roots of unity in the complex plane. Include a **PointSize** option, s, with a default value. Try out your procedure for various values of n and s.

1.11.3 Approximate Solutions of Algebraic Equations

Approximate solutions of an algebraic equation of higher degree than 4 can be found using **NSolve**. The example illustrates the syntax.

$$\text{NSolve}[x^5 - 2\,x^3 + x - 2 == 2\,x^4,\ x]$$

$$\{\{x \to -0.880625 - 0.622999\,i\},\ \{x \to -0.880625 + 0.622999\,i\},$$
$$\{x \to 0.519782 - 0.60111\,i\},\ \{x \to 0.519782 + 0.60111\,i\},\ \{x \to 2.72169\}\}$$

Another example:

$$\text{NSolve}\left[\tfrac{x^2 - 1}{5\,x^3 + 2} == x,\ x\right] \qquad \{\{x \to 0.540366 + 0.614048\,i\},\ \{x \to 0.540366 - 0.614048\,i\},$$
$$\{x \to -0.540366 + 0.0832806\,i\},\ \{x \to -0.540366 - 0.0832806\,i\}\}$$

Exercise:
Find approximate solutions of the equation $x^{10} + 9\,x^6 - 2\,x + 1 = 0$.

Lists of approximate solutions of algebraic equations can be found in the same way as was discussed in 1.11.2. Here is an example:

$$\text{Replace}[x,\ \text{NSolve}[24 - 19\,x - 14\,x^2 + 21\,x^3 - 8\,x^4 + x^5 == x,\ x]] \qquad \{-1.,\ 2.,\ 2.,\ 2.,\ 3.\}$$

In Chapter 7 we shall need to be able to find the absolute value of the derivative of an algebraic function f at points where f [x] = x.

$$\text{Replace}[x,\ \text{NSolve}[x^4 - 2\,x^3 + 1 == x,\ x]] \qquad \{-0.379567 - 0.76948\,i,\ -0.379567 + 0.76948\,i,\ 0.641445$$

$$\text{Abs}[D[x^4 - 2\,x^3 + 1,\ x]]\ /.\ x \rightarrow \% \qquad \{5.98058,\ 5.98058,\ 1.41302,\ 11.0804\}$$

Exercise:
1) Let $f\,[x] = -3 - 13\,x - 25\,x^2 - 20\,x^3 - 5\,x^4 + 2\,x^5 + x^6$, $x \in \mathbb{R}$. Find the absolute value of the derivative of f at the solutions of the equation f [x] = x.
2) Using **Replace**, make a list of the roots of the equation $1 - 2\,z + 3\,z^2 + 4\,z^3 - 5\,z^4 + 6\,z^5 = 0$, and then plot them in the complex plane.
3) Can you work out a routine **plotRoots**, say, for plotting the roots of the equation f [z] = 0, where f is an algebraic function?

1.11.4 Transcendental Equations

Suppose one wishes to find a solution to an equation of the form f [x] = 0. It can be proved that for some choices of x_1 and f, the sequence (x_n), defined by $x_{n+1} = x_n - \frac{f\,[x_n]}{f\,'[x_n]}$, may converge to a solution of the equation f [x] = 0. This method of solution is called Newton's method and can be applied to real and complex functions. The command **FindRoot** can be used to attempt to find a single approximate solution to the equation f [z] = 0.

? FindRoot

FindRoot[lhs==rhs, {x, x0}] searches for a numerical solution to the equation lhs==rhs, starting with x=x0.

Here is an example:

FindRoot[Sin[x] − x = I, {x, −0.5}]   {x → −1.5593 − 1.01376 *i*}

We check the result:

Sin[x] /. x −> −1.5592957325267205` − 1.0137606945691473` *i*   −1.5593 − 0.0137607 *i*

Sometimes, the sequence generated by the chosen starting point does not converge to a solution of the equation:

FindRoot[x³ − 3 x − 1, {x, 1}]

FindRoot::jsing : Encountered a singular Jacobian at the point x = 1.`. Try perturbing the initial point(s).

FindRoot[x³ − 3 x − 1, {x, 1}]

We try again:

FindRoot[x³ − 3 x − 1, {x, 1.1}]   {x → 1.87939}

Another example:

FindRoot[Cos[x] == x, {x, 9 I}]

FindRoot::cvnwt : Newton's method failed to converge to the prescribed accuracy after 15 iterations.

{x → 1.71758 − 0.41429 *i*}

The commands **AccuracyGoal** and **WorkingPrecision** can be used to obtain more accurate estimates for solutions of equations:

? AccuracyGoal

AccuracyGoal is an option for various numerical operations
    which specifies how many digits of accuracy should be sought in the final result.

? WorkingPrecision

WorkingPrecision is an option for various numerical operations which
    specifies how many digits of precision should be maintained in internal computations.

Here is an example:

$$\textbf{FindRoot}\left[\textbf{Sin[z]} + 1 + \frac{\textbf{z}}{\textbf{2}} = 0, \{\textbf{z, I}\}, \textbf{AccuracyGoal} \to 20, \textbf{WorkingPrecision} \to 30\right]$$

{z → −0.70457691292174592792030192896 + −0. × 10⁻³¹ *i*}

If f [z] has real coefficients and the starting point, z_0, is real, then **FindRoot** attempts to find real solutions to the equation f [z] = 0. However the equation may not have a real solution with this starting point. If you wish to find complex solutions in such a case, replace f [z] by f [z] + 0I.

Exercise:

1) Find approximate solutions of the equation Cos[z] == z^2 near 6 + 9I.
2) Find approximate solutions of the equation Cos[z] == z^2 near 1.2
3) Find approximate solutions of the equation Cos[z] == z^2 near 0.

1.11.5 Finding Co-ordinates of a Point on a 2D Plot

A suitable starting-point for the use of Newton's method for finding real solutions of the equation f [x] = 0 may sometimes be found by constructing the graph of f and using the mouse to find an approximate x-co-ordinate of a point (if there is one) where the graph crosses the x-axis. If there is no such point on your plot, try a different part of the domain of f.

Construct the plot, select it, move the mouse to the chosen point, hold down Control, the co-ordinates of the point will appear at the bottom of the screen.

Example: suppose you wish to find a positive solution to the equation: $3 x^3 - 3 - x \, Sin[x] = 0$. Construct part of the graph of f:

Plot[3 x^2 − 3 − x Sin[x], {x, −π, π}]

Using the method described above, we find, approximately, the x-co-ordinate of the point near 1 where the graph crosses the x-axis to be 1.19. We now apply Newton's method to obtain a better approximation:

FindRoot[3 x^3 − 3 − x Sin[x] == 0, {x, 1.19}]   {x → 1.09866}

Exercise:

Use the above method to find 2 negative roots of the equation: $x^6 - 4 x^3 + 1 - x^2 \, Sin[x] = 0$, and check you results.

In Chapter 7, a method is given for choosing starting points for solving complex equations.

Chapter 2

Using Color in Graphics

Introduction

This Chapter is based on the material of Chapter 1, in which graphics were constructed either in grayscale or in *Mathematica*'s default coloring. In this Chapter we demonstrate many techniques of assigning multiple colors to sequences of 2D graphics primitives, 2D plots, 2D parametric plots and 3D parametric curve plots. Practice is thus provided with lists, tables, plot options as well as coloring techniques. We also give methods of applying multiple colors to contour and density plots, 3D plots and 3D parametric surface plots.

Graphics can be colored in one of 2 ways: either using a graphics directive or using a graphics option. A graphics directive is paired with the definition of the graphic in the form {directive, graphic} or {graphic, directive}, for example: {**c, Disk[{0, 0}, 1]**}, where c is a color directive. The following types of graphics can be colored using graphics directives: 2D graphics primitives, 3D plots (using the command **Plot3D**) and 3D parametric plots. A graphics option is added after the definition of the graphic, using the command **PlotStyle** or **ColorFunction**, for example: **PlotStyle→c**, where c is a color directive.

Coloring plots constructed using the commands **Plot** and **ParametricPlot** require the use of the option **PlotStyle**. Coloring plots constructed using the commands **Plot3D**, **ContourPlot** and **DensityPlot** require the use of the option **ColorFunction**. This will become clearer once we go into details with each type of plot.

2.1 Selecting Colors

2.1.1 Using Color Selector

Click on the screen. Go to Input - Color Selector. Click on ? and then move the cursor onto the large multi-coloured square on the right of the color selector. Click anywhere there and you will be given instructions on how to choose a color. Having chosen a color, click on **OK** , and the name of the color appears at the place on the screen where you clicked. For example:

RGBColor[0.398444, 0.855482, 0.47657]

2.1.2 Using Color Charts

Following the path:

Help - Help-Browser - Getting Started/Demos - Graphics Gallery - Color Charts will lead to color charts for **GrayLevel, Hue** etc.

Suppose you are interested in one of the colors in the chart marked **RGBColor[r, g, 0]**. Select the chart, move the cursor over your chosen color, press **Control** and the **RGB** numbers for r and g for that particular color will appear at the bottom left of the screen.

From Mathematica's Help: Built-in Functions:

■*Hue[h] is a graphics directive which specifies that graphical objects which follow are to be displayed, if possible, in a color corresponding to hue h.*

■*Hue[h, s, b] specifies colors in terms of hue, saturation and brightness.*

The parameters h, s and b must all lie between 0 and 1. Values of s and b outside this range are clipped. Values of h outside this range are treated cyclically.

■ *As h varies from 0 to 1, the color corresponding to* **Hue[h]** *runs through red, yellow, green, cyan, blue, magenta and back to red again.*

■**Hue[h]** *is equivalent to* **Hue[h, 1, 1].**

2.2 Coloring 2D Graphics Primitives

2.2.1 Syntax for Coloring Graphics Primitives

In 1.8.3, we showed that a set of graphics primitives such as **Line**, **Circle** etc. can be shown on the same diagram, each grouped with one or more graphics directives such as **Thickness**, **GrayLevel** etc. In the same way each graphics primitive in such a set may be grouped with a color directive such as **Hue**, **RGBColor** etc.

We use the color selector to choose 4 colors for the following program. Notice that the color is paired with the corresponding graphics primitive Notice also that other graphics directives, such as **Thickness** can be included in the definition of the graphic. Also the graphics are executed in the order given, the second being placed on top of the first etc.

$$
\begin{aligned}
&\textbf{Show}\Big[\textbf{Graphics}\Big[\\
&\quad \big\{\{\textbf{Thickness[0.1]},\ \textbf{RGBColor[0.371099, 0.718761, 0.703136]},\ \textbf{Circle}\big[\{0, 0\},\ \sqrt{2}\,\big]\},\\
&\qquad \{\textbf{RGBColor[0.808606, 0.621103, 0.277348]},\ \textbf{Rectangle}[\{-1, -1\}, \{1, 1\}]\},\\
&\qquad \{\textbf{Thickness[0.05]},\ \textbf{RGBColor[0.722667, 0.73048, 0.855482]},\ \textbf{Circle}[\{0, 0\}, 1]\},\\
&\qquad \big\{\textbf{RGBColor[0.890639, 0.601572, 0.843763]},\\
&\qquad\quad \textbf{Rectangle}\Big[\big\{\tfrac{-1}{\sqrt{2}},\ \tfrac{-1}{\sqrt{2}}\big\},\ \big\{\tfrac{1}{\sqrt{2}},\ \tfrac{1}{\sqrt{2}}\big\}\Big]\big\}\big\}\Big],\\
&\quad \textbf{AspectRatio} \rightarrow \textbf{Automatic},\ \textbf{PlotRange} \rightarrow \textbf{All}\Big]
\end{aligned}
$$

2.2.2 Making Color Palettes by Coloring a Sequence of Rectangles

In Chapter 1 we showed how to construct a set of graphics primitives using the command **Table**. We now show how to assign colors to each element in such a table. We construct a table of pairs consisting of a graphics primitive and its specified color. We start with sequences of rectangles as these display clearly the sequence of colors that we choose.

■ Hue[h]

Our first example displays the full range of colors for the command **Hue**. Notice that $0 \leq \frac{n}{30} \leq 1$ if $1 \leq n \leq 30$. (Color Fig 2.1)

$$\text{Show}\Big[\text{Graphics}\Big[\text{Table}\Big[\Big\{\text{Hue}\Big[\frac{n}{30}\Big], \text{Rectangle}[\{n, 0\}, \{n + 0.8, 1\}]\Big\}, \{n, 1, 30\}\Big],$$

$$\text{AspectRatio} \rightarrow \text{Automatic}, \text{Axes} \rightarrow \text{False}\Big]\Big]$$

Many variations of the above **Hue** palette may be tried. Here are some suggestions:

1) Replace **Hue[$\frac{n}{30}$]** by **Hue[$1 - \frac{n}{30}$]**. This reverses the order of the colors.
2) Limit the **Hue** spectrum. For example, replace **Hue[$\frac{n}{30}$]** by **Hue[0.2 + 0.02 n]**.
3) Replace $\frac{n}{30}$ by f [n], where $0 \le f[n] \le 1$ for n beween 1 and 30. For example: **0.5 Sin[$\frac{n\pi}{30}$]**.

■ Hue[h, s, b]

The command **Hue[h, s, b]** can be used to obtain more variations:

1) A monochrome, or near monochrome sequence may be obtained by letting b = 1 (for maximum brightness), letting h be a constant, or letting h vary slightly from a fixed value, and letting s vary between 0 and 1 (or 1 and 0). For example, **Hue[0.9, $\frac{n}{30}$, 1]** or **Hue[0.9+0.004n]**.

2) One or more of the entries h, s or b may be selected in a random fashion. For example **Hue[1, 0.8-0.7(Random[]), 1]** or **Hue[Random[], 1, 1]**.

3) A sequence of pale shades may be obtained by giving s a low value, and a sequence of duller colors by giving b a low value.

■ RGBColor[r, g, b]

The command **RGBColor[r, g, b]** can be used in a similar fashion to the command **Hue[h, s, b]**.

1) A sequence of colors which do not differ much from a chosen color may be obtained as follows: choose a color using the Color Selector, say **RGBColor[0, 0.688, 0.538]**. Now vary slightly one or more of r, g and b.

For example: **RGBColor[0+0.0049n, 0.688-0.01n, 0.528+0.01n]**

2) Other interesting palettes consisting of m colors may be obtained in the following way: let **RGBColor[r, g, b]** be the nth color in the sequence. One can choose r, g and b to be not-necessarily affine functions of n satisfying $0 \le r[n], g[n], b[n] \le 1$ for $0 \le n \le m$. Quadratic or trigonometric functions are the simplest.

For example: **RGBColor[Cos[$\pi \frac{n}{60}$], $4\left(\frac{n}{30}\right)^2 - 4\frac{n}{30} + 1$, Sin[$\pi \frac{n}{30}$]]**.

3) Suppose one wishes to obtain a color palette with m colors which varies from **RGBColor[r1 , g1 , b1]** to **RGBColor[r2 , g2 , b2]** .

Here is a color directive for the nth color in the sequence:

$$\text{nthColor}[\text{RGBColor}[r1\_, g1\_, b1\_], \text{RGBColor}[r2\_, g2\_, b2\_], m\_] :=$$
$$\text{RGBColor}\left[r1 + \frac{r2 - r1}{m}\, n,\ g1 + \frac{g2 - g1}{m}\, n,\ b1 + \frac{b2 - b1}{m}\, n\right];$$

The above definition of the function colors must be implemented before it can be used in an example. Here is a program illustrating its use:

> Show[
>   Graphics[Table[{nthColor[RGBColor[1, 0, 0], RGBColor[0, 0, 1], 30], Rectangle[{n, 0},
>   {n + 0.8, 1}]}, {n, 1, 30}], AspectRatio → Automatic, Axes → False]]

▪ CMYKColor[c, m, y, k]

CMYKColor palettes can be chosen in a smilar way to RGBColor Palettes. For example:
CMYKColor[1-0.05n, 0.348-0.015n, 0.418+0.027n, 0.006n]

The number of rectangles in the palette can of course be varied in all the above cases.

Exercise:
Make some color palettes using the suggested examples and techniques.

2.2.3 Patterns made with Sequences of Graphics Primitives

Here is a plot of a family of disk-sectors, with center $\{0, 0\}$ and decreasing radii. The Hue spectrum has been restricted. Notice how gaps between the sectors were obtained. (Color Fig 2.2)

$$\text{Show}\left[\text{Graphics}\left[\text{Table}\left[\left\{\text{Hue}\left[0.4 + \frac{n}{72}, 0.5, 1\right],\ \text{Disk}\left[\{0, 0\},\ \frac{1}{n^{\frac{1}{2}}},\ \{n\,10°,\ (n + 0.8)\,10°\}\right]\right\},\right.\right.\right.$$
$$\left.\left.\left.\{n, 3, 360\}\right],\ \text{AspectRatio} \to \text{Automatic},\ \text{Axes} \to \text{False},\ \text{PlotRange} \to \text{All}\right]\right]$$

Exercise:
The above pattern can be varied by using a different color scheme, a different sector angle, a different number of iterations or different spacing between the sectors. A random factor can be included in one or more of the color function arguments.

Below is a pattern formed from filled ellipses. Recall that the command **Disk[{x, y}, {r_x, r_y}]** yields an elliptical disk with semi-axes r_x and r_y. We wish to construct a set of 20 elliptical disks of diminishing width and increasing length and with major axis alternating between the horizontal and vertical. Here is a formula for the nth disk:

$\text{Disk}[\{0, 0\}, \{1.1 - (-1)^n\, \frac{n}{20}, 1.1 + (-1)^n\, \frac{n}{20}\}] = \text{Disk}[\{0, 0\}, \{1.1 - \frac{n}{20}, 1.1 + \frac{n}{20}\}]$ if n is even;
$\qquad\qquad\qquad = \text{Disk}\,[\{0, 0\}, \{1.1 + \frac{n}{20}, 1.1 - \frac{n}{20}\}]$ if n is odd.

Notice that the semi-axis values are all positive for n between 1 and 20.

The pattern is formed from 20 filled ellipses with 10 different gray-scale shades. Note that the expression $\frac{1}{10}$ **Mod[n, 11]** takes the 10 different values $\frac{1}{10}$, $\frac{2}{10}$, ..., 1.

$$\mathbf{Show\Big[Graphics\Big[Table\Big[\Big\{GrayLevel\Big[\frac{1}{10}\,Mod[n,\,11]\Big],}$$

$$\mathbf{Disk\Big[\{0,\,0\},\,\Big\{1.2-(-1)^n\,\frac{n}{20},\,1.2+(-1)^n\,\frac{n}{20}\Big\}\Big]\Big\},\,\{n,\,20\}\Big],}$$

$$\mathbf{Background \to GrayLevel[0],\,AspectRatio \to Automatic\Big]}$$

Exercise:

1) Obtain variations on the above pattern by changing the number of disks or by changing 1 or more of the expressions $\frac{1}{10}$ Mod[n, 11] to $1 - \frac{1}{10}$ Mod[n, 11] or $\frac{1}{m}$ Mod[n, 11], where m > 10.

2) Adapt the program for a different number of colors.

3) Color the disks with **Hue[f[n]]** , where $0 \le f[n] \le 1$, for all n concerned.

4) Replace **'Disk'** by **'Circle'** in any of the above variations.

Here is the plot of a sequence of disks with centers and colors randomly chosen and with decreasing radii.

In this case, **CMYKColor** has been used. Replacing 'k' in **CMYKColor[c, m, y, k]** by 0 eliminates black and so we are more likely to obtain bright colors. (Color Fig 2.3)

$$\mathbf{Show\Big[Graphics\Big[Table\Big[\Big\{CMYKColor[Random[]\,,\,Random[],\,Random[]\,,\,0],}$$

$$\mathbf{Disk\Big[\{Random[]\,,\,Random[]\},\,1-\frac{n}{20}\Big]\Big\},\,\{n,\,0,\,20\}\Big],}$$

$$\mathbf{AspectRatio \to Automatic,\,Axes \to False\Big]\Big]}$$

Exercise:

1) The command **Rectangle** enables one to construct a rectangle with sides parallel to the co-ordinate axes and with given pair of diagonally opposite vertices. Construct, for p of your choice, a sequence of rectangles of different colors with vertices of form:

{{0, 0}, {Random[Real, {-p, p}], Random[Real, {-p, p}]}}.

2) Make some designs with other graphics primitives and various color schemes.

3) Make some 'palettes' with small rectangles as shown above.

4) Make some multi-colored curves, by constructing a table of differently colored points, each on a parametrically defined 2D curve.

5) Make a rainbow with colored concentric semicircles. [Use the built-in command **Circle** to construct a circular arc.]

6) Construct a set of differently colored points of increasing size on a spiral.

7) Color some of the sequences of graphics constructed in 1.8.2.

2.3 Coloring Sequences of 2D Curves Using the Command Plot

The method of assigning colors to a set of 2D plots differs in *Mathematica* from the method of assigning colors to a set of graphics primitives.

To color 2D plots one uses as option, the command **PlotStyle**.

In the following, a list consisting of 2 plots is followed, in the command **PlotStyle** by a list consisting of 2 colors. (Color Fig 2.4)

> **Plot[{1, 1.5}, {x, 0, 1}, PlotStyle → {Hue[0.3], Hue[0.8]}]**

If there are more plots than colours, colors are assigned cyclically. (Color Fig 2.5)

> **Plot[{0.5, 1, 1.5, 2, 2.5}, {x, 0, 1}, PlotStyle → {Hue[0.3], Hue[0.8]}]**

In Chapter 1 we showed how to make a list of functions using the command **Table**, and to plot them in one diagram. We now show how to color such a set of plots in varying colors. One can go to section 3.2.10 of the *Mathematica* Book in Help to find examples of interesting families of functions.

In the following example a table of functions, depending on the parameter n, is followed by a table of colors of the form Hue[0.3+0.02n]. Notice that the argument of the function **Hue** lies between 0 and 1 for n between 1 and 20. Notice also that we have limited the Hue spectrum to lie between **Hue[0.32]** and **Hue[0.7]**. We also have more plots than colors, so that the colors are repeated. When plotting a table of functions, it is advisable, and sometimes necessary, to use the command **Evaluate**, so that *Mathematica* evaluates the functions in the table before plotting the individual functions. (Color Fig 2.6)

> **Plot[Evaluate[Table[LegendreP[n, x], {n, 1, 40}], {x, −1, 1}],**
> **PlotStyle → Table[Hue[0.3 + 0.02 n], {n, 1, 20}], AspectRatio → 0.4, Axes → False]**

The following defines a 3-parameter family of curves:

$$\text{sumCosine}[n\_, \ x\_, \ m\_, \ p\_] := n \sum_{k=1}^{m} \frac{(\text{Cos}[p^k \, \pi \, x])}{2^k};$$

We choose m = 2 and p = 13, and give a command which can be used to generate a plot of 60 members of the family by letting n vary from 1 to 60:

$$\text{Plot}\Big[\text{Evaluate}\Big[\text{Table}[\text{sumCosine}[n, \ x, \ 2, \ 13], \ \{n, \ 1, \ 60\}], \ \Big\{x, \ \frac{-\pi}{16}, \ \frac{\pi}{16}\Big\}\Big],$$

$$\text{PlotStyle} \rightarrow \text{Table}\Big[\text{Hue}\Big[\frac{n}{30}\Big], \ \{n, \ 1, \ 30\}\Big], \ \text{AspectRatio} \rightarrow 0.3, \ \text{Axes} \rightarrow \text{False}\Big]$$

When the following program is implemented, parallel copies of a chosen single member of the above family are plotted. A random factor has been included to add interest to the plot:

$$\text{Plot}\Big[\text{Evaluate}\Big[\text{Table}\Big[\text{sumCosine}[1, \ x, \ 4, \ 3] \, \text{Random}[] + \frac{n}{10}, \ \{n, \ 1, \ 20\}\Big], \ \Big\{x, \ \frac{-\pi}{2}, \ \frac{\pi}{2}\Big\}\Big],$$

$$\text{PlotStyle} \rightarrow \text{Table}\Big[\text{Hue}\Big[\frac{n}{20}\Big], \ \{n, \ 1, \ 30\}\Big], \ \text{AspectRatio} \rightarrow 0.5, \ \text{Axes} \rightarrow \text{False}\Big]$$

This example is like the previous one, except that the plot is in white to give a lace-like effect, and a background color has been added.

$$\text{Plot}\Big[\text{Evaluate}\Big[\text{Table}\Big[\text{sumCosine}[1, \ x, \ 4, \ 3] \, \text{Random}[] + \frac{n}{10}, \ \{n, \ 1, \ 20\}\Big], \ \Big\{x, \ \frac{-\pi}{2}, \ \frac{\pi}{2}\Big\}\Big],$$

$$\text{PlotStyle} \rightarrow \text{Table}[\text{RGBColor}[1, \ 1, \ 1], \ \{n, \ 1, \ 30\}], \ \text{AspectRatio} \rightarrow 0.3,$$

$$\text{Axes} \rightarrow \text{False}, \ \text{Background} \rightarrow \text{RGBColor}[0, \ 0, \ 1]\Big]$$

Other interesting families to plot are:
1) Cos[nArcCos[x]]; (It is interesting that Cos[nArcCos[x]] is a polynomial in x for all $n \in \mathbb{N}$.)
2) ChebyshevU[n, x];
3) GegenbauerC[n, 1, x];
4) $100 \, \frac{x^2}{n} + \frac{n}{100}$;
5) LegendreP[n, 2, x];
6) Cos[$\frac{n}{2}$ + Sin[x^2]];
7) Cos[2.5 + 5 Sin[x^2]] + $\frac{n}{5}$.
In examples 1 − 5 use domain [−1, 1].

2.4 Coloring Sequences of 2D Parametric Curves

In this section, we use families of curves defined in 1.8.6. Remember to enter the definition of a family before using the definition in a plot.
2D parametric plots can be colored in a similar way to the plots in 2.3.
The following defines a 3-parameter family of curves:

family2[a_, b_, n_] := {n Sin[a t] Cos[b t], n Cos[a t] Cos[b t]};

If we choose fixed values for a and b, we obtain a sub-family of similar curves. We plot 1 member of the sub-family defined by $n = 1$, $a = \frac{3}{2}$, $b = \frac{7}{4}$. In order to obtain the complete curve we must let t range from 0 to 8π.

$$\textbf{ParametricPlot}\left[\textbf{Evaluate}\left[\textbf{family2}\left[\frac{3}{2}, \frac{7}{4}, 1\right]\right],\right.$$

$$\left.\textbf{\{t, 0, 8}\,\pi\textbf{\}, AspectRatio} \rightarrow \textbf{Automatic, Axes} \rightarrow \textbf{False}\right]$$

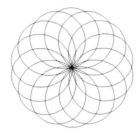

By varying n we now plot members of the above sub-family in multiple colors. We need a table of functions, and a table of colors. We could have the same number of colors as functions, but interesting effects can sometimes be obtained by having fewer colors than functions, as the list of colors is repeated cyclically. The sequence of 30 colors starts with a darker color ends with a pale color, so 91 curves are plotted to give a finished look to the 'flower'. A better image is sometimes obtained by increasing the range of values of the variable t. (Color Fig 2.7)

$$\textbf{ParametricPlot}\left[\textbf{Evaluate}\left[\textbf{Table}\left[\textbf{family2}\left[\frac{3}{2}, \frac{7}{4}, n\right], \{n, 1, 91\}\right]\right], \{t, 0, 24\,\pi\}\right],$$

$$\textbf{AspectRatio} \rightarrow \textbf{Automatic, Axes} \rightarrow \textbf{False,}$$

$$\textbf{PlotStyle} \rightarrow \textbf{Table[Hue[0.75} - \textbf{0.02 n], \{n, 1, 30\}]}\right]$$

Here is the plot of a different sub-family, differently colored. A background color has been introduced and we obtain a more monochrome effect by reducing the range of **Hue** values. (Color Fig 2.8)

$$\textbf{ParametricPlot}\left[\textbf{Evaluate}\left[\textbf{Table}\left[\textbf{family2}\left[\frac{3}{2}, 4, n\right], \{n, 1, 121\}\right]\right], \{t, 0, 4\,\pi\}\right],$$

$$\textbf{AspectRatio} \rightarrow \textbf{Automatic, Axes} \rightarrow \textbf{False,}$$

$$\textbf{PlotStyle} \rightarrow \textbf{Table[Hue[0.26} - \textbf{0.0025 n], \{n, 1, 60\}], Background} \rightarrow \textbf{Hue[0.35, 0.4, 0.6]}\right]$$

Other examples to try:
1) Change the family of curves:

family8[a_, n_] := {n t Sin[a t] Cos[t], n t Sin[a t] Sin[t]};

$$\text{ParametricPlot}\Big[\text{Evaluate}\Big[\text{Table}\Big[\text{family8}\Big[\frac{5}{2},\, n\Big],\, \{n,\, 1,\, 31\}\Big],\, \{t,\, 0,\, 6\pi\}\Big],$$

$$\text{AspectRatio} \to \text{Automatic, Axes} \to \text{False, PlotRange} \to \text{All},$$

$$\text{PlotStyle} \to \text{Table}\Big[\text{Hue}\Big[0.9 - \frac{n}{60}\Big],\, \{n,\, 1,\, 30\}\Big]\Big]$$

2) Using **family2**, change the parameters and the **PlotStyle**:

$$\text{ParametricPlot}\Big[\text{Evaluate}\Big[\text{Table}\Big[\text{family2}\Big[\frac{7}{2},\, 3,\, n\Big],\, \{n,\, 1,\, 31\}\Big],\, \{t,\, 0,\, 4\pi\}\Big],$$

$$\text{AspectRatio} \to \text{Automatic, Axes} \to \text{False},$$

$$\text{PlotStyle} \to \text{Table}\Big[\text{RGBColor}\Big[1,\, \frac{n}{30},\, \frac{n}{30}\Big],\, \{n,\, 1,\, 30\}\Big],$$

$$\text{Background} \to \text{Hue}[0.55,\, 0.2,\, 0.8]\Big]$$

Interesting effects can sometimes be obtained by plotting the curves in the reverse order, with the largest curve first. This can be done by allowing n to vary from -a to -b, where $0 \le b < a$. If b is chosen to be greater than 0 one sometimes obtains more detail at the centre of the 'flower'. In the following example, the r and g values in **RGBColor[r, g, b]** have been chosen as non-linear functions of n. (See section 2.2.2.) (Color Fig 2.9)

$$\text{ParametricPlot}\Big[\text{Evaluate}\Big[\text{Table}\Big[\text{family2}\Big[\frac{3}{2},\, 4,\, n\Big],\, \{n,\, -80,\, -20\}\Big],\, \{t,\, 0,\, 4\pi\}\Big],$$

$$\text{AspectRatio} \to \text{Automatic, Axes} \to \text{False},$$
$$\text{Background} \,-\!> \text{RGBColor}[0.203128,\, 0.578134,\, 0.136721],$$

$$\text{PlotStyle} \to \text{Table}\Big[\text{RGBColor}\Big[\text{Cos}\Big[\pi\,\frac{n}{120}\Big],\, 4\Big(\frac{n}{60}\Big)^2 - 4\,\frac{n}{60} + 1,\, \text{Sin}\Big[\pi\,\frac{n}{60}\Big]\Big],\, \{n,\, 1,\, 60\}\Big]\Big]$$

Suppose one wishes to obtain a table of m colors, starting with **RGBColor[r1, g1, b1]** and ending with **RGBColor[r2, g2, b2]**. The command **colors** defined below provides such a table:

$$\text{colors}[\text{RGBColor}[r1\_,\, g1\_,\, b1\_],\, \text{RGBColor}[r2\_,\, g2\_,\, b2\_],\, m\_] :=$$
$$\text{Table}\Big[\text{RGBColor}\Big[r1 + \frac{r2 - r1}{m}\, n,\, g1 + \frac{g2 - g1}{m}\, n,\, b1 + \frac{b2 - b1}{m}\, n\Big],\, \{n,\, 0,\, m\}\Big]$$

In the following example, we use the function **colors** defined above, applied to a set of members of the following family of curves: (Color Fig 2.10)

$$\text{family1}[a\_,\, b\_,\, c\_,\, n\_] :=$$
$$\Big\{n\Big(2 + \frac{\text{Sin}[a\,t]}{2}\Big)\text{Cos}\Big[t + \frac{\text{Sin}[b\,t]}{c}\Big],\, n\Big(2 + \frac{\text{Sin}[a\,t]}{2}\Big)\text{Sin}\Big[t + \frac{\text{Sin}[b\,t]}{c}\Big]\Big\};$$

ParametricPlot[Evaluate[Table[family1[7, 14, 2, n], {n, 1, 122}], {t, 0, 2π}],
AspectRatio → Automatic, Axes → False,
Background → RGBColor[0.121096, 0.652354, 0.187503],
PlotStyle → colors[RGBColor[0, 0, 1], RGBColor[1, 1, 1], 120]]

Interesting effects can some-times be obtained by multiplying one of the arguments in a color function by **Random[ ]** or by **Random[Real, {a, b}]**, where $0 \leq a \leq b \leq 1$. Here is an example chosen from **family1**: (Color Fig 2.11)

ParametricPlot$\Big[$Evaluate[Table[family1[15, 5, 1, n], {n, 1, 61}], {t, 0, 2π}],
AspectRatio → Automatic, Axes → False,
PlotStyle → Table$\Big[$RGBColor$\Big[$Random[Real, {0.6, 1}], $\dfrac{n}{30}$, $\dfrac{n}{30}\Big]$, {n, 1, 30}$\Big]\Big]$

Another example:

family3[a_, b_, c_, d_, f_, g_, n_] := { n (a Cos[b t] + Cos[c t]), n (d Sin[f t] + Sin[g t])};

ParametricPlot$\Big[$Evaluate[Table[family3[3, 1, 12, 3, 1, 12, n], {n, −100, −30}], {t, 0, 2π}],
AspectRatio → Automatic, Axes → False, PlotStyle →
Table$\Big[$RGBColor$\Big[$Cos$\Big[\pi \dfrac{n}{120}\Big]$ (Random[ ]), $4\left(\dfrac{n}{60}\right)^2 - 4\dfrac{n}{60} + 1$, Sin$\Big[\pi \dfrac{n}{60}\Big]\Big]$, {n, 1, 60}$\Big]\Big]$

Applying the above techniques to some sequences of members of the families 1, 3, 6, 9, 11 leads to plots which are not 'flower-like'.
In the following example, we let t vary from 0 to 20π which results in a 'smoother' image, where spaces between lines are eliminated. (Color Fig 2.12)

family11[a_, b_, n_] := {n Sin[a t] Cos[t], n Sin[b t] Sin[t]};

ParametricPlot$\Big[$Evaluate[Table[family11[5, 10, n], {n, 70}], {t, 0, 20π}],
AspectRatio → Automatic, Axes → False,
PlotStyle → Table$\Big[\Big\{$CMYKColor$\Big[$Sin$\Big[\dfrac{n\pi}{60}\Big]$, Cos$\Big[\dfrac{n\pi}{180}\Big]$, $1 - \dfrac{n}{15} + \dfrac{n^2}{900}$, 0$\Big\}\Big]$, {n, 1, 60}$\Big]\Big]$

Interesting effects can sometimes be obtained by using the fact that if $-1 \leq f[n] \leq 1$, then $0 \leq \text{Abs}[f[n]] \leq 1$. For example, if $1 \leq n \leq 80$, then $0 \leq \text{Abs}[\text{Cos}[\frac{mn\pi}{40}]] \leq 1$ and takes values between 0 and 1 more than once if $m > 0.5$. Here is an example: (Color Fig 2.13)

family6[a_, n_] := {n Cos[Cos[a t]] Cos[t], n Cos[Sin[a t]] Sin[t]};

ParametricPlot[Evaluate[Table[family6[11, n], {n, 1, 85}], {t, 0, 20π}],

AspectRatio → Automatic, Axes → False, PlotStyle →

Table[{CMYKColor[$\frac{n}{20} - \frac{n^2}{1600}$, Abs[Sin[$\frac{n\pi}{40}$]], Abs[$-1 + \frac{n}{40}$], 0]}, {n, 1, 80}]]

Another example: (Color Fig 2.14)

ParametricPlot[Evaluate[Table[family3[2, 1, 9, 1, 1, 7, n], {n, 1, 100}], {t, 0, 8π}],

AspectRatio → Automatic, Axes → False, PlotStyle →

Table[{RGBColor[Abs[Cos[$\frac{n\pi}{40}$]], $1 - \frac{n}{20} + \frac{n^2}{1600}$, Abs[$-1 + \frac{n}{40}$]]}, {n, 1, 80}]]

Programs for other examples:

ParametricPlot[Evaluate[Table[family3[3, 3, 10, 3, 3, 10, n], {n, 1, 85}], {t, 0, 4π}],

AspectRatio → Automatic, Axes → False, PlotStyle →

Table[{CMYKColor[Abs[$-1 + \frac{n}{25}$], $\frac{2n}{25} - \frac{n^2}{625}$, Abs[Sin[$\frac{n\pi}{25}$]], 0]}, {n, 1, 50}]]

ParametricPlot[Evaluate[Table[family3[3, 1, 7, 1, 1, 7, n], {n, 1, 100}], {t, 0, 8π}],

AspectRatio → Automatic, Axes → False, PlotStyle →

Table[{RGBColor[Abs[Cos[$\frac{n\pi}{40}$]], $1 - \frac{n}{20} + \frac{n^2}{1600}$, Abs[$-1 + \frac{n}{40}$]]}, {n, 1, 80}]]

Here is an example of the coloring of a sequence of non-similar curves constructed in 1.8.6. (Color Fig 2.15)

ParametricPlot[Evaluate[Table[family1[6, n, 5, 1], {n, 1, 30}], {t, 0, 2π}],

AspectRatio → Automatic, Axes → False,

PlotStyle → Table[RGBColor[Cos[$\pi \frac{n}{60}$], $4 \left(\frac{n}{30}\right)^2 - 4 \frac{n}{30} + 1$, Sin[$\pi \frac{n}{30}$]], {n, 1, 30}]]

Another example to try is:

ParametricPlot[Evaluate[Table[family3[6, 1, 4, 2, 4, n, 1], {n, 1, 25}], {t, 0, 2π}],
AspectRatio → Automatic, Axes → False, PlotRange → All,
PlotStyle → colors[RGBColor[1, 1, 0], RGBColor[0, 0, 1], 25]]

Many variations may be obtained by choosing other sub-families, other colorings and varying the number of family members to be plotted.

Exercise:

1) Try the above families for different values of the parameters, different numbers of curves

and different colorings.

2) Experiment with other families listed in 1.8.6.

Some programs to try:

family9[a_, b_, n_] := {n Sin[a t] Cos[t], n Sin[b t] Sin[t]};

ParametricPlot[Evaluate[Table[family1[3, 3, 1, n], {n, 1, 60}], {t, 0, 20 π}],

AspectRatio → Automatic, Axes → False, PlotStyle →

$$\textbf{Table}\Big[\Big\{\textbf{CMYKColor}\Big[\textbf{Abs}\Big[-1+\frac{n}{30}\Big], \frac{n}{15}-\frac{n^2}{900}, \textbf{Abs}\Big[\textbf{Sin}\Big[\frac{n\pi}{30}\Big]\Big], 0\Big]\Big\}, \{n, 1, 60\}\Big]\Big]$$

ParametricPlot[Evaluate[Table[family1[18, 4, 1, n], {n, 1, 31}], {t, 0, 2 π}],

AspectRatio → Automatic, Axes → False,

$$\textbf{PlotStyle → Table}\Big[\textbf{RGBColor}\Big[1, \frac{n}{30}, \frac{n}{30}\Big], \{n, 1, 30\}\Big],$$

Background → Hue[0.55, 0.2, 0.8]]

2.5 Coloring Sequences of 3D Parametric Curves

2.5.1 Coloring Sequences of Similar 3D Parametric Curves

Families of 3D parametric curves may be plotted in an analogous way to plotting families of 2D parametric curves. For example, a 3D curve can be formed from this parametric plot of a 2D curve:

ParametricPlot[{ Sin[8 π t] Cos[π t], Sin[8 π t] Sin[π t]},
{t, 0, 2 π}, AspectRatio → Automatic, Axes → False]

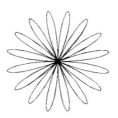

In order to form a 3D curve from the above plot, we insert a function of t as a z-co-ordinate. The distance from the point with parameter t on the above curve to the center of the curve is **Abs[Sin[8 π t]]**. If we choose z to be a positive (negative) function of this distance, then points equidistant from the center of the above 'daisy' are raised (lowered) to the same level, thus preseving the flower-like form. We choose z to be the cube of the above distance. This ensures that the 'petals' are cuved.

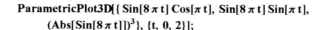

ParametricPlot3D[{ Sin[8 π t] Cos[π t], Sin[8 π t] Sin[π t],
(Abs[Sin[8 π t]])³}, {t, 0, 2}];

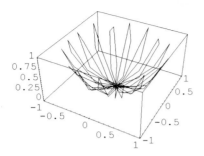

We now make a set of differently colored, similar copies of the above curve. The command **Plotstyle** is not an option for **ParametricPlot3D**. Instead, a graphics directive is paired with the co-ordinates of a point on the plot.

To color a 3D parametric plot we use the command **ParametricPlot3D[{f(t), g(t), h(t), c}...]** which colors the whole curve according to the color specification c. Thus, in the case of 3D parametric curves the color of the curve is included with the definition of the x, y, and z co-ordinates of the point with parameter t on the curve. (Color Fig 2.16)

ParametricPlot3D[
  Evaluate[Table[{n Sin[10 π t] Cos[π t], n Sin[10 π t] Sin[π t], n (Abs[Sin[10 π t]])³,
    Hue[0.9 − 0.025 n]}, {n, 1, 40}], {t, 0, 2}],
  Axes → False, Boxed → False, PlotPoints → 70]

Exercise:

The command:

ParametricPlot3D[{{ Sin[8 π t] Cos[π t], Sin[8 π t] Sin[π t],
  (Abs[Sin[8 π t]])³}, {1.5 Sin[8 π t] Cos[π t], 1.5 Sin[8 π t] Sin[π t],
  0.5 (Abs[Sin[8 π t]])³}}, {t, 0, 2}];

creates a flower-like form similar to the one in the second diagram above, but with longer 'petals' with lower curvature. Use the command **Join** whose use in graphics was illustrated in 1.8.4, to construct a version of the flower-like form above with a double row of 'petals'.

The above techniques can be applied to members of the families of 2D curves defined in the previous section. In the following example, the z-co-ordinate of the point P, with parameter t, on the nth curve is equal to the distance from the origin of the point times $\frac{1}{\pi}$ Sin[$\frac{n\pi}{80}$]. So as n increases from 0 to 60, the height of the point P first increases and then decreases. As the color of a curve is paired with the definition of the curve, we have exactly the same number of curves as colors. In order to repeat colors, the color of the nth curve is specified by:

RGBColor[$\frac{1}{30}$ **Mod[n, 30]**, $\frac{1}{30}$ **Mod[n, 30], 1**], so that as n increases from 31 to 60 the sequence of colors from 1 to 30 is repeated. (Color Fig 2.17)

ParametricPlot3D[Evaluate[

$$\text{Table}\left[\left\{n\,\text{Sin}\left[\frac{3}{2}\,\pi\,t\right]\,\text{Cos}[5\,\pi\,t],\ n\,\text{Cos}\left[\frac{3}{2}\,\pi\,t\right]\,\text{Cos}[5\,\pi\,t],\ \text{Sin}\left[n\,\frac{\pi}{80}\right]\,\text{Abs}[\text{Cos}[5\,\pi\,t]],\right.\right.$$

$$\left.\left.\text{RGBColor}\left[\frac{1}{30}\,\text{Mod}[n,30],\ \frac{1}{30}\,\text{Mod}[n,30],\ 1\right]\right\},\ \{n,0,60\}\right],\ \{t,0,4\}\right],$$

Axes → False, Boxed → False, PlotPoints → 90, ViewPoint −> {2.548, −0.090, 2.225},

BoxRatios → {2, 2, .6}, Background −> RGBColor[1, 0.835294, 0.905882]] (Fig2 .17)

Another example: (Color Fig 2.18)

ParametricPlot3D[Evaluate[Table[$\left\{n\left(\text{Sin}\left[\frac{3}{2}\,\pi\,t\right]+\text{Sin}[8\,\pi\,t]\right),\ n\left(\text{Cos}\left[\frac{3}{2}\,\pi\,t\right]+\text{Cos}[8\,\pi\,t]\right),\right.$

$$\left.\frac{n}{6}\,\text{Cos}\left[4\,\pi\left(\sqrt{2+\text{Cos}\left[\frac{13}{2}\,\pi\,t\right]}\right)\right]\right\},\ \text{RGBColor}\left[1,\ \frac{1}{30}\,\text{Mod}[n,30],\ \frac{1}{30}\,\text{Mod}[n,30]\right]\right\},$$

{n, 1, 60}], {t, 0, 2}], Axes → False, Boxed → False,

PlotPoints → 100, ViewPoint −> {3.104, −0.006, 1.555},

BoxRatios → {2, 1.5, 0.4}, ViewPoint −> {0.045, 0.001, 3.384}]

The following function **color12** can be used to obtain a color palette with m colors which varies from **RGBColor[r1, g1, b1]** to **RGBColor[r2, g2, b2]**:

color12[RGBColor[r1_, g1_, b1_], RGBColor[r2_, g2_, b2_], m_] := RGBColor[

$$r1+\left(\frac{r2-r1}{m}\right)\text{Mod}[n,m],\ g1+\left(\frac{g2-g1}{m}\right)\text{Mod}[n,m],\ b1+\left(\frac{b2-b1}{m}\right)\text{Mod}[n,m]];$$

In the following example, leaving out the first 9 values for n gives an interesting centre to the flower-like form. (Color Fig 2.19)

ParametricPlot3D[

$$\text{Evaluate}\left[\text{Table}\left[\left\{n\,(\text{Sin}[\pi\,t]+\text{Sin}[7\,\pi\,t]),\ n\,(\text{Cos}[\pi\,t]+\text{Cos}[7\,\pi\,t]),\ \frac{n}{9}\left(\sqrt{2+\text{Cos}[6\,\pi\,t]}\right)^{5},\right.\right.\right.$$

color12[RGBColor[1, 0, 0], RGBColor[1, 1, 0], 45]}, {n, 20, 90}], {t, 0, 2}],

Axes → False, Boxed → False, Background → RGBColor[0.492195, 0.859388, 0.386725],

PlotPoints → 90, ViewPoint −> {1.646, 0.045, 2.956}]

Other programs to try:

ParametricPlot3D$\Big[$

$\quad$**Evaluate**$\Big[$**Table**$\Big[\Big\{$**n** $(\mathbf{Sin}[\pi\, t] + \mathbf{Sin}[7\,\pi\, t])$, **n** $(\mathbf{Cos}[\pi\, t] + \mathbf{Cos}[7\,\pi\, t])$, $\dfrac{\mathbf{n}}{9}\,\big(\sqrt{2 + \mathbf{Cos}[6\,\pi\, t]}\,\big)^{5}$,

$\qquad$**color12[RGBColor[0.99, 0.36, 0.2], RGBColor[1, 1, 1], 30]**$\Big\}$, **{n, 10, 60}**$\Big]$, **{t, 0, 2}**$\Big]$,

$\quad$**Axes → False, Boxed → False, PlotPoints → 90, ViewPoint –> {1.126, 0.128, 3.224}**$\Big]$

ParametricPlot3D$\Big[$**Evaluate**$\Big[$

$\quad$**Table**$\Big[\Big\{$**n Sin**$\Big[\dfrac{7}{2}\,\pi\, t\Big]$ **Cos**$\Big[\dfrac{5}{2}\,\pi\, t\Big]$, **n Cos**$\Big[\dfrac{7}{2}\,\pi\, t\Big]$ **Cos**$\Big[\dfrac{5}{2}\,\pi\, t\Big]$, $\dfrac{\mathbf{n}}{2}\,\Big(\mathbf{Abs}\Big[\mathbf{Cos}\Big[\dfrac{5}{2}\,\pi\, t\Big]\Big]\Big)^{5}$,

$\qquad$**RGBColor**$\Big[1, \dfrac{1}{32}\,\mathbf{Mod[n, 32]}, \dfrac{1}{32}\,\mathbf{Mod[n, 32]}\Big]\Big\}$, **{n, 15, 64}**$\Big]$, **{t, 0, 4}**$\Big]$,

$\quad$**Axes → False, Boxed → False, PlotPoints → 170**$\Big]$

Exercise:

1) Vary the examples given by changing the coloring, the number of curves plotted or the view-point.

2) Apply the techniques of this section to some of the 2D parametric plots of the previous section.

One can create a 3D parametric curve by starting with a 2D parametric curve with equations: x = f [t], y = g [t] and choosing a function of t for the z-co-ordinate. For example, starting with a 2D curve:

$\quad$**ParametricPlot[{t Cos[t], t Sin[t]}, {t, 0, 8 π}, AspectRatio → Automatic]**

Include a z-co-ordinate:

$\quad$**ParametricPlot3D[{t Cos[t], t Sin[t], −t},**
$\quad$**{t, 0, 8 π}, AspectRatio → Automatic, PlotPoints → 100]**

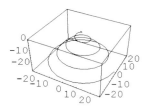

Plot a set of similar copies of the resulting curve and color them using the parameter t. (Color Fig 2.20)

$$\text{ParametricPlot3D}\Big[\text{Evaluate}\Big[\text{Table}\Big[\Big\{n\,t\,\text{Cos}[t],\ n\,t\,\text{Sin}[t],\ -60\,t,$$

$$\text{Hue}\Big[1-\frac{t}{8\,\pi}\Big]\Big\},\ \{n,\ 1,\ 60\}\Big],\ \{t,\ 0,\ 8\,\pi\}\Big],\ \text{Axes}\to\text{False},\ \text{Boxed}\to\text{False},\ \text{PlotPoints}\to$$

Programs for other examples:

$$\text{ParametricPlot3D}\Big[$$

$$\text{Evaluate}\Big[\text{Table}\Big[\Big\{4\,\text{Sin}[t],\ \frac{n}{15}\,(\text{Cos}[t]+\text{Cos}[3\,t]),\ \frac{n}{15}\,(\text{Cos}[t]-\text{Sin}[3\,t]),$$

$$\text{Hue}\Big[1-\frac{t}{2\,\pi}\Big]\Big\},\ \{n,\ 1,\ 20\}\Big],\ \{t,\ 0,\ 2\,\pi\}\Big],$$

$$\text{Axes}\to\text{False},\ \text{Boxed}\to\text{False},\ \text{PlotPoints}\to 200\Big]$$

$$\text{ParametricPlot3D}\Big[\text{Evaluate}\Big[$$

$$\text{Table}[\{2\,n+t^2,\ 5\,n\,\text{Cos}[t],\ 5\,n\,\text{Sin}[t],\ \text{Hue}[0.9-0.005\,n]\},\ \{n,\ 1,\ 20\}],\ \Big\{t,\ \frac{\pi}{2},\ 6\,\pi\Big\}\Big],$$

$$\text{Axes}\to\text{False},\ \text{Boxed}\to\text{False},\ \text{ViewPoint}\mathbin{-\!>}\{1.688,\ -0.855,\ 2.806\}\Big]$$

2.5.2 Sequences of Similar 2D Curves in Parallel Planes

Another method of constructing a graphic from a set of colored 2D plots is to construct a set of differently colored plane curves similar to each other and in parallel planes. Start with a 2D parametrically defined curve with equations x = f [t], y = g [t] and then construct a set of curves in 3D space with equations of the form x = n f [t], y = n g [t], z = h [n], n ∈ {1, ..., m}, for some m ∈ ℕ, and some function h. The function h determines the shape of the projection of the surface on the y-z plane. Here is an example: (Color Fig 2.21)

$$\text{ParametricPlot3D}\Big[\text{Evaluate}\Big[\text{Table}\Big[\Big\{n\,(6\,\text{Cos}[t]+\text{Cos}[6\,t]),\ n\,(6\,\text{Sin}[t]-\text{Sin}[6\,t]),\ 60\,(n)^{\frac{1}{2}},$$

$$\text{color12}[\text{RGBColor}[1,\ 1,\ 1],\ \text{RGBColor}[1,\ 0,\ .6],\ 100]\Big\},\ \{n,\ 1,\ 100\}\Big],\ \{t,\ 0,\ 2\,\pi\}\Big],$$

$$\text{ViewPoint}\mathbin{-\!>}\{-2.532,\ -5.750,\ 2.741\},\ \text{Axes}\to\text{False},\ \text{Boxed}\to\text{False},$$

$$\text{PlotPoints}\to 70,\ \text{Background}\to\text{RGBColor}[0.55079,\ 0.742199,\ 0.882826]\Big]$$

Another example: (Color Fig 2.22)

> **ParametricPlot3D[Evaluate[Table[{n (18 Cos[t] + Cos[18 t]),**
>   **n (18 Sin[t] − Sin[18 t]), −2 n, color12[RGBColor[0.700793, 0.285161, 0.285161],**
>   **RGBColor[0.957046, 0.859388, 0.859388], 60]}, {n, 1, 60}], {t, 0, 2 π}],**
>   **Axes → False, Boxed → False, PlotPoints → 70, ViewPoint → {−0.180, −2.724, 2.000},**
>   **BoxRatios → {1, 2, .5}]**

Exercise:

1) Experiment with this technique, using different starting curves, colorings.

2) Construct a representation of the flower of the Australian bottlebrush tree, starting with the curve with parametric equations:

   x = Sin[18π t]Cos[π t], y = Sin[18π t] Sin[π t].

3) Construct a leaf or a fern, starting with the curve below.

> **ParametricPlot[{t, Abs[t (t − 1)² (t + 1)²]},**
>   **{t, −0.7, 0.7}, AspectRatio → 0.2, Ticks → {Automatic, None}]**

2.5.3 3D Graphics Constructed by Rotating Plane Curves

Another method of constructing a multi-colored surface from a 2D parametric curve is to start with a 2D curve, C, in the y-z plane, and rotate it about the z-axis successively through the angles θ, 2θ,..... .

We load the package **Geometry `Rotations`** to calculate the required rotation. Before proceeding, we illustrate these rotations for a particular curve.

> **<< Geometry`Rotations`**

Color Fig 2.23 shows the positions taken up by the purple curve, C. It is shown in red after it has been rotated through $\frac{\pi}{4}$ about the z-axis, in blue after having been rotated through $-\frac{\pi}{4}$ about the x-axis and in green after having been rotated through $\frac{\pi}{4}$ about the y-axis. The x-axis is shown in black.

Here is an example starting with the curve in the y-z plane with parametric equations given by:

$\{y, z\} = \{Sin[2 (t + \pi)] Cos[t + \pi], (Sin[t + \pi])^2\}$.

In Color Fig 2.24, the y- and z-axes are shown in blue and red, respectively.

> **ParametricPlot3D[{{0, 0, t, {Hue[1], Thickness[0.001]}},**
>   **{0, −t, 0, {Hue[0.7], Thickness[0.001]}}, {0, Sin[2 (t + π)] Cos[t + π], (Sin[t + π])²}},**
>   **{t, 0, π − 2}, AspectRatio → Automatic, Boxed → True]**

The command:

Rotate3D[{0, y, z}, 0, 0, θ]   {y Sin[θ], y Cos[θ], z}

rotates the vector $(0, y, z)$ through θ radians about the z-axis, to give the vector $(y\mathrm{Sin}[\theta]$, $y\mathrm{Cos}[\theta]$, z). Using the command **/.**, we now replace y and z in the above vector with $\mathrm{Sin}[2\,(t+\pi)]\,\mathrm{Cos}[t+\pi]$ and $(\mathrm{Sin}[t+\pi])^2$ respectively. In order to rotate the curve through n 1 °, for n between 1 and 360, we replace θ (which is measured in radians) by n 1 °.

$\{\mathrm{y}\,\mathrm{Sin}[\theta], \mathrm{y}\,\mathrm{Cos}[\theta], \mathrm{z}\}$ /. $\{\mathrm{y} \,{-}{>}\, \mathrm{Sin}[2\,(t+\pi)]\,\mathrm{Cos}[t+\pi], \mathrm{z} \,{-}{>}\, (\mathrm{Sin}[t+\pi])^2, \theta \to (\mathrm{n}1°)\}$

$\{-\mathrm{Cos}[t]\,\mathrm{Sin}[\mathrm{n}1°]\,\mathrm{Sin}[2\,(\pi+t)],\ -\mathrm{Cos}[\mathrm{n}1°]\,\mathrm{Cos}[t]\,\mathrm{Sin}[2\,(\pi+t)],\ \mathrm{Sin}[t]^2\}$

We now use the command **ParametricPlot3D** applied to the above. A color directive which depends on t has been included. (Color Fig 2.25)

ParametricPlot3D[
  Evaluate[Table[{Sin[2 (t + π)] Cos[(t + π)] Sin[n 1 °], Sin[2 (t + π)] Cos[(t + π)] Cos[n 1 °],
  Sin[(t + π)] Sin[t + π], Hue[$\dfrac{0.7\,t}{\pi - 2}$]}, {n, 1, 360}],
  {t, 0, π – 2}], Axes $\to$ False, Boxed $\to$ False]

When implemented the following 2 commands display the above image as a 3D parametric surface plot, however, the transparent effect is lost:

b = ParametricPlot3D[{Sin[2 (t + π)] Cos[(t + π)] Sin[s], Sin[2 (t + π)] Cos[(t + π)] Cos[s],
  Sin[(t + π)] Sin[t + π], Hue[$\dfrac{0.7\,t}{\pi - 2}$]}, {s, 0, 2 π}, {t, 0, π – 2}, Lighting $\to$ False,
  PlotPoints $\to$ 40, Axes $\to$ False, Boxed $\to$ False, DisplayFunction $-{>}$ Identity]

  Show[b /. Polygon[p_] $-{>}$ {EdgeForm[], Polygon[p]},
  DisplayFunction $\to$ \$DisplayFunction]

Here is a program which is a variation on the program for the image above in that it has a different set of t values.

ParametricPlot3D[
  Evaluate[Table[{Sin[3 t] Cos[(t)] Sin[n 1 °], Sin[(3 t)] Cos[(t)] Cos[n 1 °], Sin[(t)] Sin[t],
  Hue[$\dfrac{2\,t}{\pi}$]}, {n, 1, 360}], {t, 0, $\dfrac{\pi}{2}$ – .3}], Axes $\to$ False, Boxed $\to$ False]

Exercise:

Adapt the program of the image above, by using different color schemes, starting angle, t value sets, box-ratios or line thickness directives.

In the following example, the starting angle has been increased to 36° and so n must vary from 0 to 10. A thickness directive which depends on t has been included. When two or more graphics directives are included, they must be enclosed in braces. (Color Fig 2.26)

$$\textbf{ParametricPlot3D}\Big[\textbf{Evaluate}\Big[$$
$$\textbf{Table}\Big[\Big\{\textbf{Sin}[2(t+\pi)]\ \textbf{Cos}[(t+\pi)]\ \textbf{Sin}[n\ 36\ °],\ \textbf{Sin}[2(t+\pi)]\ \textbf{Cos}[(t+\pi)]\ \textbf{Cos}[n\ 36\ °],$$
$$\textbf{Sin}[(t+\pi)]\ \textbf{Sin}[t+\pi],\ \Big\{\textbf{Thickness}\Big[0.1\ \textbf{Sin}\Big[\pi\ \frac{t}{\pi-2}\Big]\Big],\ \textbf{RGBColor}\Big[\Big(0.34+\frac{t}{3\,(\pi-2)}\Big),$$
$$\Big(0.55+\frac{t}{3.5\,(\pi-2)}\Big),\ \Big(0.12+\frac{t}{3.5\,(\pi-2)}\Big)\Big]\Big\}\Big\},\ \{n,\ 0,\ 10\}\Big],\ \{t,\ 0,\ \pi-2\}\Big],$$
$$\textbf{Axes} \rightarrow \textbf{False},\ \textbf{Boxed} \rightarrow \textbf{False},\ \textbf{PlotPoints} \rightarrow \textbf{200},\ \textbf{BoxRatios} \rightarrow \{1,\ 1,\ 1\},$$
$$\textbf{ViewPoint} -> \{0.862,\ -2.214,\ 2.409\}\Big]$$

Exercise:

Try the above techniques with other starting curves, number of rotations, thicknesses etc.

In order to construct the next image we start with a curve C_1 in the y-z plane and we rotate it through the angle ϕ about the x-axis to produce the curve C_2 in the y-z plane. Now, by rotating C_1 and C_2 alternately about the z- axis through the angles n θ, we obtain leaves of the plant lying on 2 different levels. Consider the command for rotating a curve in the y-z plane about the x-axis and then about the z-axis:

Rotate3D[{0, y, z}, 0, ϕ, θ]

$\{y\,Cos[\phi]\,Sin[\theta]+z\,Sin[\theta]\,Sin[\phi],\ y\,Cos[\theta]\,Cos[\phi]+z\,Cos[\theta]\,Sin[\phi],\ z\,Cos[\phi]-y\,Sin[\phi]\}$

Compare this with the command for rotating about the z-axis:

Rotate3D[{0, y, z}, 0, 0, θ] $\qquad$ $\{y\,Sin[\theta],\ y\,Cos[\theta],\ z\}$

So, it can be seen that if we multiply each occurence of ϕ by $\frac{(1+(-1)^n)}{2}$, then, for n odd, C_1 will be rotated about the z-axis and if n is even C_1 will first be rotated about the x-axis to produce C_2 and then C_2 will be rotated about the z-axis. Applying this technique to the curve of the previous example, we get:

$\{y \, \text{Cos}[\phi] \, \text{Sin}[\theta] + z \, \text{Sin}[\theta] \, \text{Sin}[\phi], \ y \, \text{Cos}[\theta] \, \text{Cos}[\phi] + z \, \text{Cos}[\theta] \, \text{Sin}[\phi], \ z \, \text{Cos}[\phi] - y \, \text{Sin}[\phi]\} \, /.$

$$\left\{y \rightarrow \text{Sin}[2 (t + \pi)] \, \text{Cos}[t + \pi], \ z \rightarrow (\text{Sin}[t + \pi])^2, \ \theta \rightarrow n \, 20°, \ \phi \rightarrow \frac{(1 + (-1)^n)}{2} \, 30°\right\}$$

$\{\text{Sin}[15 \, (1 + (-1)^n)°] \, \text{Sin}[20°n] \, \text{Sin}[t]^2 - \text{Cos}[15 \, (1 + (-1)^n)°] \, \text{Cos}[t] \, \text{Sin}[20°n] \, \text{Sin}[2 \, (\pi + t)],$
$\quad \text{Cos}[20°n] \, \text{Sin}[15 \, (1 + (-1)^n)°] \, \text{Sin}[t]^2 - \text{Cos}[15 \, (1 + (-1)^n)°] \, \text{Cos}[20°n] \, \text{Cos}[t] \, \text{Sin}[2 \, (\pi + t)],$
$\quad \text{Cos}[15 \, (1 + (-1)^n)°] \, \text{Sin}[t]^2 + \text{Cos}[t] \, \text{Sin}[15 \, (1 + (-1)^n)°] \, \text{Sin}[2 \, (\pi + t)]\}$

ParametricPlot3D$\big[$

$\quad$ **Evaluate**$\Big[$**Table**$\Big[\Big\{$**Sin**$[15 \, (1 + (-1)^n)°]$ **Sin**$[20°n]$ **Sin**$[t]^2 -$ **Cos**$[15 \, (1 + (-1)^n)°]$ **Cos**$[t]$

$\qquad$ **Sin**$[20°n]$ **Sin**$[2 \, (\pi + t)]$, **Cos**$[20°n]$ **Sin**$[15 \, (1 + (-1)^n)°]$ **Sin**$[t]^2 -$

$\qquad$ **Cos**$[15 \, (1 + (-1)^n)°]$ **Cos**$[20°n]$ **Cos**$[t]$ **Sin**$[2 \, (\pi + t)]$,

$\qquad$ **Cos**$[15 \, (1 + (-1)^n)°]$ **Sin**$[t]^2 +$ **Cos**$[t]$ **Sin**$[15 \, (1 + (-1)^n)°]$ **Sin**$[2 \, (\pi + t)]$,

$\qquad \Big\{$ **Thickness**$\Big[0.07 - \dfrac{0.12 \, t}{\pi - 2}\Big]$, **RGBColor**$\Big[\Big(0.535164 + \dfrac{t}{3 \, (\pi - 2.5)}\Big)$,

$\qquad\quad \Big(0.742199 - \dfrac{n}{144}\Big)$ **Random[Real, {0.7, 1}]**,

$\qquad\quad \Big(0.0568757 + \dfrac{t}{4 \, (\pi - 2.5)}\Big)$ **Random[Real, {0.7, 1}]**$\Big]\Big\}\Big\}$, $\{$**n, 1, 18**$\}\Big]$,

$\quad$ $\{$**t, 0,** π **− 2.5**$\}\Big]$, **Axes → False, Boxed → False, BoxRatios →**

$\quad$ $\{$**1, 1, .5**$\}\Big]$ (Color Fig 2.27)

In the following example, the curve with parametric equations x = 0, y = t, z = 1 − Cos[t] is first rotated through a randomly chosen angle about the x-axis and then through n 3° about the z-axis, with n ranging from 1 to 360. (Color Fig 2.28)

Rotate3D[{0, y, z}, 0, ϕ**,** θ**]** /. {**y → t, z →** (**1 − Cos[t]**)**,** θ **→ n 3°**}

$\{$t Cos$[\phi]$ Sin$[3°n]$ + (1 − Cos$[t]$) Sin$[3°n]$ Sin$[\phi]$,
$\quad$ t Cos$[3°n]$ Cos$[\phi]$ + Cos$[3°n]$ (1 − Cos$[t]$) Sin$[\phi]$, (1 − Cos$[t]$) Cos$[\phi]$ − t Sin$[\phi]\}$

ParametricPlot3D$\big[$**Evaluate**$\big[$**Table**$\big[\big\{$**t Cos[(Random[Integer, {−60, 50}])°] Sin[3°n] +**

$\quad$ (**1 − Cos[t]**) **Sin[3°n] Sin[(Random[Integer, {−60, 50}])°]**,

$\quad$ **t Cos[3°n] Cos[(Random[Integer, {−60, 50}])°] +**

$\qquad$ **Cos[3°n] (1 − Cos[t]) Sin[(Random[Integer, {−60, 50}])°]**,

$\quad$ (**1 − Cos[t]**) **Cos[(Random[Integer, {−60, 50}])°] −**

$\qquad$ **t Sin[(Random[Integer, {−60, 50}])°]**,

$\quad \Big\{$**Thickness[0.005], Hue**$\Big[0.05 + \dfrac{0.7 \, t}{2 \, \pi}\Big]\Big\}\Big\}$, $\{$**n, 1, 360**$\}\big]$, $\{$**t, 0,** $\pi$$\}\big]$,

$\quad$ **Axes → False, Boxed → False, BoxRatios → {1, 1, 1},**

$\quad$ **ViewPoint −> {1.573, −2.904, 0.738}**$\big]$

Exercise:

Adapt the above programs to construct images of bulbous plants, by choosing a different initial curve, a different color scheme, a thickness directive, different number of curves plotted and/or different angles of rotation.

In the following example, we start with a curve given by the command:

ParametricPlot[{3 Sin[t] − Sin[3 t], −(3 Cos[t] − Cos[3 t])}, {t, 0, π}, AspectRatio → 1.7]

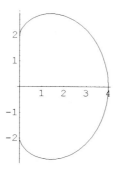

We then add terms to the x- and y-co-ordinates of the point with parameter t to get a curve which is narrower at the lower end. We also construct a curve which we will use to form a stem. We then apply the above rotatory construction. Random factors have been introduced into the coloring, to produce a striped effect. We use the command **Join** to display the 'apple' and the 'stem'.

$$\textbf{ParametricPlot}\left[\left\{\left\{\frac{t}{2.2}, 2.1\,\textbf{Sin}\left[\frac{t}{2.2}\right] + 2.2\right\}, \left\{3\,\textbf{Sin[t]} - \textbf{Sin[3 t]} + \frac{1}{4}\,\textbf{Sin[t]},\right.\right.\right.$$

$$\left.\left.\left. -\left(3\,\textbf{Cos[t]} - \textbf{Cos[3 t]} - \frac{1}{1.2}\,\textbf{Sin[t]}\right)\right\}\right\}, \{t, 0, \pi\}, \textbf{AspectRatio} \to 1.7\right]$$

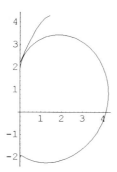

ParametricPlot3D[

$$\text{Evaluate}\left[\text{Join}\left[\left\{\left\{0,\ \frac{t}{2.2},\ 2.1\ \text{Sin}\left[\frac{t}{2.2}\right]\right\}+2.2,\ \left\{\text{Thickness}[0.02],\ \text{RGBColor}\left[\right.\right.\right.\right.$$

$$\left.\left. 0.62,\ \frac{t}{\pi}\ \text{Random}[],\ 0.33\ \text{Random}[\text{Real},\ \{0.8,\ 1\}]\right]\right\}\right\}\right\},$$

$$\text{Table}\left[\left\{\left(3\ \text{Sin}[t]-\text{Sin}[3\ t]+\frac{1}{4}\ \text{Sin}[t]\right)\text{Sin}[n\ 2\ ^{\circ}],\ \left(3\ \text{Sin}[t]-\text{Sin}[3\ t]+\frac{1}{4}\ \text{Sin}[t]\right)\right.\right.$$

$$\text{Cos}[n\ 2\ ^{\circ}],\ -\left(3\ \text{Cos}[t]-\text{Cos}[3\ t]-\frac{1}{1.5}\ \text{Sin}[t]\right\},\ \left\{\text{Thickness}[0.01],\right.$$

$$\left.\left. \text{RGBColor}\left[0.94,\ \frac{t}{\pi}\ \text{Random}[],\ 0.54\ \text{Random}[\text{Real},\ \{0.8,\ 1\}]\right]\right\}\right\},\ \{n,\ 1,\ 180\}\right],$$

$$\{t,\ 0,\ \pi\}\bigg],\ \text{Axes}\to\text{False},\ \text{Boxed}\to\text{False},\ \text{BoxRatios}\to\{1,\ 1,\ 1.4\},$$

$$\text{ViewPoint}->\{1.613,\ -1.744,\ 2.409\}\bigg]\ (\text{Color Fig 2.29})$$

In the program below we choose a color scheme which ends with the color with which it started:

$$\text{ParametricPlot3D}\left[\text{Evaluate}\left[\text{Table}\left[\left\{\left(3\ \text{Sin}[t]-\text{Sin}[3\ t]+\frac{1}{4}\ \text{Sin}[t]\right)\text{Sin}[n\ 2\ ^{\circ}],\right.\right.\right.\right.$$

$$\left(3\ \text{Sin}[t]-\text{Sin}[3\ t]+\frac{1}{4}\ \text{Sin}[t]\right)\text{Cos}[n\ 2\ ^{\circ}],\ -\left(3\ \text{Cos}[t]-\text{Cos}[3\ t]-\frac{1}{1.5}\ \text{Sin}[t]\right),$$

$$\left\{\text{Thickness}[.015],\ \text{RGBColor}\left[0.94,\ 1-\frac{\text{Sin}\left[n\ \frac{\pi}{180}\right]}{2},\ 0.54\right]\right\}\right\},\ \{n,\ 1,\ 180\}\bigg],\ \{t,\ 0,\ \pi\}\bigg],$$

$$\text{Axes}\to\text{False},\ \text{Boxed}\to\text{False},\ \text{BoxRatios}\to\{1,\ 1,\ 1\},$$

$$\text{ViewPoint}->\{1.927,\ 1.753,\ 2.160\}\bigg]$$

Exercise:
Construct images of vegetables and other fruit e.g. carrot, cucumber, pear, peach.

In the following example, we use a table with 2 parameters to construct a plant-like form. The following command is used to construct an image of a circlet of leaves:

```
ParametricPlot3D[Evaluate[
    Table[{24 Sin[(t + π)] Cos[(t + π)] Sin[n 40°], 24 Sin[(t + π)] Cos[(t + π)] Cos[n 40°],
        15 Sin[(t + π)] Sin[t + π], {Thickness[0.035 Sin[5 t]]}}, {n, 0, 12}],
    {t, 0, π − 2.6}], Axes → False, Boxed → False, PlotPoints → 200]
```

We now wish to construct a sequence of 6 circlets, similar to the above, of increasing height and decreasing diameter, with centers on the curve, C, where C has parametric equation $\{x, y, z\} = \{30\,\text{Sin}[2\,t], 30\,\text{Cos}[2\,t] - 1, 107\,t\}$. We need to use a 2-parameter table.

Consider the following table with 2 parameters:

   Table[{n, 2 n, m}, {n, 1, 3}, {m, 1, 2}]

   {{{1, 2, 1}, {1, 2, 2}}, {{2, 4, 1}, {2, 4, 2}}, {{3, 6, 1}, {3, 6, 2}}}

Notice that we do not obtain a list of co-ordinates of points. The command **Flatten** can be used to obtain a list of point co-ordinates:

   Flatten[Table[{n, 2 n, m}, {n, 1, 3}, {m, 1, 2}], 1]

   {{1, 2, 1}, {1, 2, 2}, {2, 4, 1}, {2, 4, 2}, {3, 6, 1}, {3, 6, 2}}

We choose co-ordinates $\left\{30\,\text{Sin}\left[0.56\,\sqrt{m+1}\right],\ 30\,\text{Cos}\left[0.56\,\sqrt{m+1}\right] - 1,\ 29.9\,\sqrt{m+1}\right\}$ for the center of the mth circlet. This ensures that it lies on the curve C, and that the distances between successive centers decreases as m increases. The first 5 circlets will be in shades of green and the sixth in shades of red. The command **Flatten** is used in order to obtain a single list of points with parameter t. The following routine is used to construct the 6 circlets. We use the command **DisplayFunction->Identity** as we wish to construct a 'stem' and display the circlets and the stem together. (Color Fig 2.30)

```
p2 = ParametricPlot3D[ Evaluate[
      Flatten[ Table[ {4 (6 − m) Sin[(t + π)] Cos[(t + π)] Sin[n 40 °] + 30 Sin[0.56 √(m + 1)],
         4 (6 − m) Sin[(t + π)] Cos[(t + π)] Cos[n 40 °] + 30 Cos[0.56 √(m + 1)] − 1,
         15 Sin[(t + π)] Sin[t + π] + 29.9 √(m + 1), {Thickness[0.035 Sin[5 t] ∗ (1 − 0.1 m)],
            Hue[If[m ≥ 5, 1, 0.3], If[m ≥ 5, 0.2 + 0.6 t, 0.4 + 0.7 t], 0.6 + 0.05 m]}},
         {n, 0, 9}, {m, 0, 5}], 1], {t, 0, π/5}, Boxed → False,
      Axes → False, DisplayFunction → Identity]]

r2 = ParametricPlot3D[
      {30 Sin[2 t], 30 Cos[2 t] − 1, 107 t, {Thickness[0.018 − 0.01 t], Hue[0.2, 0.4, 0.6]}},
      {t, 0.28, 0.68}, Boxed → False, Axes → False, DisplayFunction → Identity]

Show[p2, r2, DisplayFunction −> $DisplayFunction]
```

Exercise:

Use the techniques of the above example to construct other plant-like forms.

In the following example, the command **Join** is used to construct 2 sequences of 3D parametric curves, one representing petals and the other stamens of a flower. Random factors have been used to introduce irregularities into the flower-like form: (Color Fig 2.31)

$$\text{ParametricPlot3D}\Big[\text{Evaluate}\Big[$$

$$\text{Join}\Big[\text{Table}\Big[\Big\{\frac{1}{4}\,(\text{Sin}[10\,°]\,\text{Sin}[16\,°\,n]\,\text{Sin}[t]^2 - \text{Cos}[10\,°]\,\text{Cos}[t]\,\text{Sin}[16\,°\,n]\,\text{Sin}[2\,(\pi + t)]),$$

$$\frac{1}{4}\,(\text{Cos}[16\,°\,n]\,\text{Sin}[10\,°]\,\text{Sin}[t]^2 - \text{Cos}[10\,°]\,\text{Cos}[16\,°\,n]\,\text{Cos}[t]\,\text{Sin}[2\,(\pi + t)]),$$

$$\frac{\text{Random[Real, }\{1, 1.7\}]}{4}\,(\text{Cos}[2\,°]\,\text{Sin}[t]^2 + \text{Cos}[t]\,\text{Sin}[10\,°]\,\text{Sin}[2\,(\pi + t)]),$$

$$\{\text{Thickness}[0.005\,\text{Sin}[9.1\,t]], \text{Hue}[0.18,\ 1 - 2.\,t,\ 1]\}\Big\}, \{n, 1, 22\}\Big],$$

$$\text{Table}[\{-\text{Cos}[t]\,\text{Sin}[8\,°\,(1 + 2\,n)]\,\text{Sin}[2\,(\pi + t)],$$

$$- \text{Cos}[8\,°\,(1 + 2\,n)]\,\text{Cos}[t]\,\text{Sin}[2\,(\pi + t)],\ \text{Random[Real, }\{1, 1.8\}]\,(\text{Sin}[t]^2),$$

$$\{\text{Thickness}[0.025\,\text{Sin}[9.1\,t]], \text{Hue}[0.99,\ 1 - 2\,t,\ 1]\}\}, \{n, 1, 22\}]\Big],$$

$$\{t, 0, \pi - 2.8\}\Big], \text{Axes} \rightarrow \text{False}, \text{Boxed} \rightarrow \text{False}, \text{BoxRatios} \rightarrow \{1, 1, .2\}\Big]$$

Exercise:

1) Use the above technique to construct a flower-like form with more than one row of petals.

2) Experiment with the construction methods described in 2.5.3. Try starting with parts of 2D parametric curves used earlier in this chapter. When plotted in the x-y plane your curve should satisfy the following condition: any line parallel to the x-axis should not intersect the curve on both sides of the y-axis.

3) Construct some 2-image stereograms using images from 2.5.1, 2.5.2 and 2.5.3.

2.5.4 Plane Patterns Constructed from Curves with Parametric Equations of the Form: { 0, f[t], g[t] }

In *Mathematica*, 3D parametric curves can be colored in such a way that the color varies along the curve. This enables us to construct plane patterns with such curves. We start with a set of straight lines in the y-z plane, with equations: x = 0, y = t, z = n. If the coloring system we choose depends only on n or only on t, we will obtain only stripes or checks. In order to obtain a more interesting pattern the coloring must depend on n and t. Here is an example: (Color Fig 2.32)

```
ParametricPlot3D[Evaluate[Table[{0,  t,  n,
        {Hue[Abs[0.6 + 0.25 Sin[0.2 π t n] − 0.000626 n ∗ t]], Thickness[0.025]}}, {n, 1, 40}],
    {t, 2, 40}], Axes → False, Boxed → False, PlotPoints → 80, ViewPoint −> {−2,  0,  0}]
```

In the following example, we color a set of sine curves. A 2-color pattern is achieved in the following way:

two colors: **RGBColor[0.74, 0.57, 0.35]** and **RGBColor[0.97, 0.95, 0.93]** are chosen. In **RGBColor[r, g, b]**, the r is replaced by 0.74 **Mod[Floor[n t],2]+ 0.97 Mod[Floor[n t+1], 2]**. If **Floor[n t]** is even, then **Floor[n t +1]** is odd and so the value of r is 0.97 etc. The expression 'nt' can be replaced by a different expression in n and t. (Color Fig 2.33)

$$\textbf{ParametricPlot3D}\Big[\textbf{Evaluate}\Big[\textbf{Table}\Big[\Big\{0,\ \ \textbf{t, n}\ +\textbf{Sin}\Big[\pi\ \frac{t}{4}\Big],$$

$$\{\textbf{RGBColor}[0.74\,\textbf{Mod}[\textbf{Floor}[\textbf{n t}],\ 2]+0.97\,\textbf{Mod}[\textbf{Floor}[\textbf{n t}+1],\ 2],$$
$$0.57\,\textbf{Mod}[\ \textbf{Floor}[\textbf{n t}],\ 2]+0.95\,\textbf{Mod}[\ \textbf{Floor}[\textbf{n t}+1],\ 2],$$
$$0.35\,\textbf{Mod}[\textbf{Floor}[\textbf{n t}],\ 2]+0.93\,\textbf{Mod}[\textbf{Floor}[\textbf{n t}+1],\ 2]],\ \textbf{Thickness}[0.03]\}\Big\},$$

$$\{\textbf{n, 1, 30}\}\Big],\ \{\textbf{t, 0, 30}\}\Big],\ \textbf{Axes}\rightarrow\textbf{False, Boxed}\rightarrow\textbf{False,}$$

$$\textbf{PlotPoints}\rightarrow\textbf{180, ViewPoint}->\{\textbf{3.384,}\ -\textbf{0.040,}\ -\textbf{0.026}\}\Big]$$

In the following program, extra small terms with a random factor are added to r, g or b in **RGBColor[r, g, b]** to vary the coloring.

$$\textbf{ParametricPlot3D}\Big[\textbf{Evaluate}\Big[\textbf{Table}\Big[\Big\{0,\ \ \textbf{t, n}\ +\textbf{Sin}\Big[\pi\ \frac{t}{4}\Big]-\textbf{Cos}\Big[\pi\ \frac{t}{4}\Big],$$

$$\Big\{\textbf{RGBColor}\Big[0.74\,\textbf{Mod}\Big[\textbf{Floor}\Big[\textbf{n}\ \frac{t}{2}\Big],\ 2\Big]+0.97\,\textbf{Mod}\Big[\textbf{Floor}\Big[\textbf{n}\ \frac{t}{2}+1\Big],\ 2\Big]-\frac{n}{80}\,\textbf{Random}[$$

$$0.57\,\textbf{Mod}\Big[\ \textbf{Floor}\Big[\textbf{n}\ \frac{t}{2}\Big],\ 2\Big]+0.95\,\textbf{Mod}\Big[\ \textbf{Floor}\Big[\textbf{n}\ \frac{t}{2}+1\Big],\ 2\Big]-\frac{n}{80},$$

$$0.35\,\textbf{Mod}\Big[\textbf{Floor}\Big[\textbf{n}\ \frac{t}{2}\Big],\ 2\Big]+0.93\,\textbf{Mod}\Big[\textbf{Floor}\Big[\textbf{n}\ \frac{t}{2}+1\Big],\ 2\Big]-\frac{n}{160}\Big],$$

$$\textbf{Thickness}[0.025]\Big\}\Big\},\ \{\textbf{n, 1, 40}\}\Big],\ \{\textbf{t, 0, 40}\}\Big],$$

$$\textbf{Axes}\rightarrow\textbf{False, Boxed}\rightarrow\textbf{False, PlotPoints}\rightarrow\textbf{200,}$$

$$\textbf{ViewPoint}->\{\textbf{3.384,}\ -\textbf{0.040,}\ -\textbf{0.026}\}\Big]$$

A program for another example:

$$\textbf{ParametricPlot3D}\Big[\textbf{Evaluate}\Big[\textbf{Table}\Big[\Big\{0,\ \ \textbf{t, n,}$$

$$\Big\{\textbf{Hue}\Big[\textbf{Abs}\Big[\frac{t+t^2*n+n^3}{280}\Big]\Big],\ \textbf{Thickness}[0.025]\Big\}\Big\},\ \{\textbf{n, 1, 40}\}\Big],\ \{\textbf{t, 2, 40}\}\Big],\ \textbf{Axes}\rightarrow\textbf{False}$$

$$\textbf{Boxed}\rightarrow\textbf{False, PlotPoints}\rightarrow\textbf{90, ViewPoint}->\{\textbf{3.384,}\ -\textbf{0.040,}\ -\textbf{0.026}\}\Big]$$

Another example: (Color Fig 2.34)

$$\textbf{ParametricPlot3D} \Big[\textbf{Evaluate} \Big[\textbf{Table} \Big[\Big\{ 0, \textbf{ n Cos}\Big[\frac{t\pi}{20} \Big], \textbf{ n Sin}\Big[\frac{t\pi}{20} \Big],$$

$$\Big\{ \textbf{RGBColor}\Big[0.84 \textbf{ Mod}\Big[\textbf{Floor}\Big[\textbf{n}\frac{t}{2} \Big], 2 \Big] + 0.57 \textbf{ Mod}\Big[\textbf{Floor}\Big[\textbf{n}\frac{t}{2} + 1 \Big], 2 \Big] - \frac{n}{80} \textbf{ Random}[$$

$$0.57 \textbf{ Mod}\Big[\textbf{Floor}\Big[\textbf{n}\frac{t}{2} \Big], 2 \Big] + 0.95 \textbf{ Mod}\Big[\textbf{Floor}\Big[\textbf{n}\frac{t}{2} + 1 \Big], 2 \Big] - \frac{n}{80},$$

$$0.35 \textbf{ Mod}\Big[\textbf{Floor}\Big[\textbf{n}\frac{t}{2} \Big], 2 \Big] + 0.93 \textbf{ Mod}\Big[\textbf{Floor}\Big[\textbf{n}\frac{t}{2} + 1 \Big], 2 \Big] - \frac{n}{160} \Big],$$

$$\textbf{Thickness[0.025]} \Big\} \Big\}, \{n, 1, 40\} \Big], \{t, 0, 40\} \Big],$$

$$\textbf{Axes} \rightarrow \textbf{False, Boxed} \rightarrow \textbf{False, PlotPoints} \rightarrow \textbf{160,}$$

$$\textbf{ViewPoint} -> \{3.384, \ -0.040, \ -0.026\} \Big]$$

We construct a basket-like form from a circular form such as the above, constructing the circular form in the x-y plane, and including a z-co-ordinate which depends on n. In the example below, the basket is hemi-spherical: (Color Fig 2.35)

$$\textbf{ParametricPlot3D} \Big[\textbf{Evaluate} \Big[\textbf{Table} \Big[\Big\{ \textbf{n Cos}\Big[\frac{t\pi}{20} \Big], \textbf{ n Sin}\Big[\frac{t\pi}{20} \Big], -\sqrt{1600 - (n)^2},$$

$$\Big\{ \textbf{RGBColor}\Big[$$

$$0.74 \textbf{ Mod}\Big[\textbf{Floor}\Big[\textbf{n}\frac{t}{2} \Big], 2 \Big] + 0.97 \textbf{ Mod}\Big[\textbf{Floor}\Big[\textbf{n}\frac{t}{2} + 1 \Big], 2 \Big] - \frac{n}{80} \textbf{ Random[]},$$

$$0.57 \textbf{ Mod}\Big[\textbf{Floor}\Big[\textbf{n}\frac{t}{2} \Big], 2 \Big] + 0.95 \textbf{ Mod}\Big[\textbf{Floor}\Big[\textbf{n}\frac{t}{2} + 1 \Big], 2 \Big] - \frac{n}{80},$$

$$0.35 \textbf{ Mod}\Big[\textbf{Floor}\Big[\textbf{n}\frac{t}{2} \Big], 2 \Big] + 0.93 \textbf{ Mod}\Big[\textbf{Floor}\Big[\textbf{n}\frac{t}{2} + 1 \Big], 2 \Big] - \frac{n}{160} \Big],$$

$$\textbf{Thickness[0.028]} \Big\} \Big\}, \{n, 1, 39\} \Big], \{t, 0, 40\} \Big],$$

$$\textbf{Axes} \rightarrow \textbf{False, Boxed} \rightarrow \textbf{False, PlotPoints} \rightarrow \textbf{260} \Big]$$

Here is another example: (Color Fig 2.36)

$$\textbf{ParametricPlot3D} \Big[\textbf{Evaluate} \Big[\textbf{Table} \Big[\Big\{ 0, \textbf{ n t Sin}\Big[\frac{4t}{3} \Big] \textbf{Sin[t], n t Sin}\Big[\frac{4t}{3} \Big] \textbf{Cos[t],}$$

$$\Big\{ \textbf{Hue}\Big[\frac{t}{10\pi} \Big], \textbf{Thickness[0.01]} \Big\} \Big\}, \{n, 50, 60\} \Big], \{t, 0, 3\pi\} \Big], \textbf{Axes} \rightarrow \textbf{False,}$$

$$\textbf{Boxed} \rightarrow \textbf{False, PlotPoints} \rightarrow \textbf{180, ViewPoint} -> \{-2, \ 0, \ 0\}, \textbf{PlotRange} \rightarrow \textbf{All} \Big]$$

Exercise:

1) Try other color schemes and thickness directives in the above examples.

2) Apply the above techniques to other sets of lines, for example lines not parallel to an axis.

3) Apply the above techniques to other sets of curves, such as more complicated trigonometric curves, semi-circles, closed curves other than circles.

4) Construct a set of circles with centers on the positive z-axis, touching the y-axis and with increasing radii, and experiment with coloring techniques.

2.6 Coloring 3D Parametric Surface Plots

A 3D parametric surface plot can be colored with a graphics directive as in 2.5. In the following examples a surface considered in section 1.9.6 is colored firstly according to the t parameter, secondly according to the θ parameter. Thirdly, a different example is colored according to both parameters. The option: **Lighting→False** must be included if a color directive is used in **ParametricPlot3D**. (Color Fig 2.37)

Coloring according to parameter t:

$$\textbf{ParametricPlot3D}\Big[\{(9\,\textbf{Cos}[\theta]-\textbf{Cos}[9\,\theta])\,\textbf{Cos}[t]-5\,t,\ (9\,\textbf{Sin}[\theta]-\textbf{Sin}[9\,\theta])\,\textbf{Cos}[t]-7\,t,$$

$$7\,t,\ \textbf{RGBColor}\Big[0.52+\textbf{Abs}\Big[\frac{t}{2\,\pi}\Big],\ 0.8-\textbf{Abs}\Big[\frac{t}{2\,\pi}\Big],\ 0.54\Big]\},$$

$$\Big\{t,\ \frac{-\pi}{2},\ \frac{\pi}{2}\Big\},\ \{\theta,\ 0,\ 2\,\pi\},\ \textbf{AspectRatio}\to\textbf{Automatic},\ \textbf{PlotPoints}\to 50,$$

$$\textbf{ViewPoint}->\{2.870,\ 1.385,\ 1.137\},\ \textbf{Axes}\to\textbf{False},\ \textbf{Boxed}\to\textbf{False},\ \textbf{Lighting}\to\textbf{False}\Big]$$

Noticing that $0.74\textbf{Mod}[\textbf{Floor}[5\theta],\ 2]+0.97\textbf{Mod}[\textbf{Floor}[5\theta+1],\ 2] = 0.97$ if **Floor[5 θ]** is even
$$= 0.74 \text{ if } \textbf{Floor[5 } \theta\textbf{]} \text{ is odd,}$$
we give a programme for coloring according to the parameter θ:

$$\textbf{ParametricPlot3D}\Big[\{(9\,\textbf{Cos}[\theta]-\textbf{Cos}[9\,\theta])\,\textbf{Cos}[t]-5\,t,\ (9\,\textbf{Sin}[\theta]-\textbf{Sin}[9\,\theta])\,\textbf{Cos}[t]-7\,t,$$

$$7\,t,\ \textbf{RGBColor}[0.74\,\textbf{Mod}[\textbf{Floor}[5\,\theta],\ 2]+0.97\,\textbf{Mod}[\textbf{Floor}[5\,\theta+1],\ 2],$$

$$0.57\,\textbf{Mod}[\textbf{Floor}[5\,\theta],\ 2]+0.95\,\textbf{Mod}[\textbf{Floor}[5\,\theta+1],\ 2],$$

$$0.35\,\textbf{Mod}[\textbf{Floor}[5\,\theta],\ 2]+0.93\,\textbf{Mod}[\textbf{Floor}[5\,\theta+1],\ 2]]\},$$

$$\Big\{t,\ \frac{-\pi}{2},\ \frac{\pi}{2}\Big\},\ \{\theta,\ 0,\ 2\,\pi\},\ \textbf{AspectRatio}\to\textbf{Automatic},\ \textbf{PlotPoints}\to 80,$$

$$\textbf{ViewPoint}->\{2.870,\ 1.385,\ 1.137\},$$

$$\textbf{Axes}\to\textbf{False},\ \textbf{Boxed}\to\textbf{False},\ \textbf{Lighting}\to\textbf{False}\Big]$$

In the following example, we color part of an ellipsoid: (Color Fig 2.38)

$$\textbf{basket}=\textbf{ParametricPlot3D}\Big[\Big\{-9\,\textbf{Sin}\Big[\frac{\pi}{2}\,s\Big]\,\textbf{Cos}[2\,\pi\,t],\ -9\,\textbf{Sin}\Big[\frac{\pi}{2}\,s\Big]\,\textbf{Sin}[2\,\pi\,t],$$

$$-4.5\,\textbf{Cos}\Big[\frac{\pi}{2}\,s\Big],\ \textbf{Hue}[0.25\,\textbf{Mod}[\textbf{Floor}[10\,(s+t)],\ 2]]\Big\},\ \{s,\ 0,\ 1\},\ \{t,\ 0,\ 1\},$$

$$\textbf{PlotPoints}\to 30,\ \textbf{Lighting}\to\textbf{False},\ \textbf{Axes}->\textbf{False},\ \textbf{Boxed}->\textbf{False},$$

$$\textbf{PlotRegion}\to\{\{-0.15,\ 1.1\},\ \{-0.4,\ 1.1\}\},\ \textbf{ViewPoint}->\{1.405,\ -2.593,\ 1.659\}\Big]$$

The edges of the defining polygons of a 3D Parametric surface plot can be removed, as was shown in 1.9.5. The following commands will achieve this:

Show[basket /. Polygon[p_] –> {EdgeForm[ ], Polygon[p]}]

In the following program, we adapt **basket** to obtain a circular pattern by choosing a view point directly above the surface:

$$\textbf{ParametricPlot3D}\!\left[\left\{-9\,\textbf{Sin}\!\left[\frac{\pi}{2}\,\textbf{s}\right]\textbf{Cos}[2\,\pi\,\textbf{t}],\ -9\,\textbf{Sin}\!\left[\frac{\pi}{2}\,\textbf{s}\right]\textbf{Sin}[2\,\pi\,\textbf{t}],\right.\right.$$

$$-2.5\,\textbf{Cos}\!\left[\frac{\pi}{2}\,\textbf{s}\right],\ \textbf{RGBColor}\!\left[1-\frac{1}{2}\,\textbf{Mod}[\textbf{Floor}[2\,\textbf{Sin}[2\,\pi\,(\textbf{s}+\textbf{t})]],\,2],$$

$$\left.\frac{1}{2}\,\textbf{Mod}[\textbf{Floor}[2\,\textbf{Cos}[2\,\pi\,(\textbf{s}+\textbf{t})]],\,2],\ 1-\frac{1}{2}\,\textbf{Mod}[\textbf{Floor}[2\,\textbf{Sin}[2\,\pi\,(\textbf{s}+\textbf{t})]],\,2]\right]\right\},$$

{s, 0, 1}, {t, 0, 1}, PlotPoints → 30, Lighting → False, Axes –> False,

$$\textbf{Boxed –> False, ViewPoint –> \{0, 0, 2\}}\bigg]$$

In the following example, color is applied to a torus. A background color is included, and *Mathematica* uses a complementary color for the mesh lines. (Color Fig 2.39)

$$\textbf{ParametricPlot3D}\!\left[\left\{(2+1\,\textbf{Cos}[2\,\pi\,\textbf{s}])\,\textbf{Sin}[2\,\pi\,\textbf{t}],\ (2+1\,\textbf{Cos}[2\,\pi\,\textbf{s}]\,)\,\textbf{Cos}[2\,\pi\,\textbf{t}],\right.\right.$$

$$\textbf{Sin}[2\,\pi\,\textbf{s}],\ \textbf{RGBColor}\!\left[\frac{1}{2}\,\textbf{Mod}[\textbf{Floor}[8\,\textbf{Sin}[2\,\pi\,(\textbf{s}+\textbf{t})]],\,3],\right.$$

$$\left.1-\frac{1}{2}\,\textbf{Mod}[\textbf{Floor}[6\,\textbf{Sin}[2\,\pi\,(\textbf{s}+\textbf{t})]],\,3],\,1]\right\},\ \textbf{\{s, 0, .5\}, \{t, 0, 1\},}$$

PlotPoints → {20, 60}, Background → RGBColor[0.580392, 0.996078, 0.8],

$$\textbf{Lighting → False, Axes –> False, Boxed –> False}\bigg]$$

The following program gives another example of the coloring of a torus. A 2D effect is obtained by viewing the torus from a point on the z-axis.

ParametricPlot3D[{(1.2 + 1 Cos[2 π s]) Sin[2 π t], (1.2 + 1 Cos[2 π s]) Cos[2 π t],
Sin[2 π s], Hue[(0.7 – 0.5 Mod[Floor[6 (s + t)], 2])]}, {s, 0, 0.5}, {t, 0, 1},
PlotPoints → {20, 60}, Background → RGBColor[0.0117189, 0.121096, 0.882826],
Lighting → False, Axes –> False, Boxed –> False, ViewPoint –> {0, 0, 3.383}]

Exercise:
Use the above techniques to color parametric surfaces such as cylinders, paraboloids, hyperboloids and the surfaces encountered in Chapter 1, including section 1.9.6.

2.7 Coloring Density and Contour Plots

In Chapter 1 the command **DensityPlot[f[x, y], {x, a, b}, {y, c, d}, PlotPoints→m]** was discussed.

Recall that, the rectangle defined by the range of x- and y- values is divided into m×m sub-rectangles, each of which is assigned a number between 0 and 1, according to its height. Similarly when using the command **ContourPlot[f[x, y], {x, a, b}, {y, c, d}, Contours→m}** the region between a pair of succesive contour lines is assigned a height number between 0 and 1.

Suppose the nth sub-rectangle/sub-region is assigned the number t_n which we shall call its height number. To color the sub-rectangles/sub-regions, we use the command **ColorFunction**. If we wish to work with **RGBColor**, we choose 3 functions r, g and b such that r [t], g [t] and b [t] lie between 0 and 1 for t between 0 and 1. The functions r, g and b can be constant, linear or non-linear. Examples of non-linear functions are Sin, Cos and quadratic functions. The pure function **RGBColor[r[#], g[#], b[#]] &** assigns the color **RGBColor[r[t_n], g[t_n], b[t_n]]** to the nth sub-rectangle/sub-region. We can then add the command **ColorFunction→(RGB-Color[r[#], g[#], b[#]] &)** as an option to the plot. We use analogous techniques for **CMYK-Color** and **Hue**. The option **Background** can be used to place a frame round the image.

2.7.1 Making Palettes for the Use of ColorFunction in Density, Contour and 3D Plots

In 2.3.2, we constructed color palettes for use with coloring 2D graphics primitives and 3D parametric plots. We can also construct color palettes for use with the commands: **DensityPlot**, **ContourPlot** and **Plot3D**. Here is an example using the option **ColorFunction→ RGBColor[r[#], g[#], b[#]]&** in which the functions r, g and b are non linear: (Color Fig 2.40)

$$\text{ContourPlot}\left[x, \{x,\ 0,\ 20\},\ \{y,\ 0,\ 2\},\right.$$
$$\left.\text{ColorFunction} \to \left(\text{RGBColor}\left[\text{Cos}\left[\pi\ \frac{\#}{2}\right],\ 4\,\#^2 - 4\,\# + 1,\ \text{Sin}[\pi\,\#]\right]\ \&\right),\right.$$
$$\left.\text{AspectRatio} \to \text{Automatic},\ \text{Contours} \to 20\right]$$

Interesting effects can sometimes be obtained by choosing **Abs[2# -1]**, or **Abs[Sin[m π #]]**, m > 1, or **Abs[Cos[m π #]]**, m > 0.5, as these functions take values beween 0 and 1 more than once. Here is the program for an example:

> **ContourPlot[x, {x, 0, 20}, {y, 0, 2},**
> **ColorFunction → (RGBColor[1 − Abs[Cos[π#]], 1 − Abs[Sin[2π#]], 1 − 4# + 4#^{2}] &),**
> **AspectRatio → Automatic, Contours → 20, Frame → False]**

2.7.2 Contour Plots

In our first example, we use the above color scheme: (Color Fig 2.41)

ContourPlot[Abs[Log[Cos[Log[Cos[2 / (x + I (y))^2]]]]], {x, −1.4, 1.4},
{y, −1.4, 1.4}, AspectRatio → Automatic, PlotPoints → 150, Contours → 45,
Frame → False, ContourShading → True, Axes → False, ContourLines → False,
ColorFunction –> (RGBColor[1 − Abs[Cos[π #]], 1 − Abs[Sin[2 π #]], 1 − 4 # + 4 #^{2}] &)]

The contour plot of **Mod[IntegerPart[k f[x,y],n]** or **Mod[Floor[k f[x,y],n]** is sometimes interesting. We call this the 'Mod-Floor Process'. We apply this process to the contour plot of the previous example, with k = 10 and n = 20. (Color Fig 2.42)

ContourPlot[Mod[Floor[10 Abs[Log[Cos[Log[Cos[2 / (x + I (y))^2]]]]]], 20], {x, −1.4, 1.4},
{y, −1.4, 1.4}, AspectRatio → Automatic, PlotPoints → 150, Contours → 45,
Frame → False, ContourShading → True, Axes → False, ContourLines → False,
ColorFunction –> (RGBColor[1 − Abs[Cos[π #]], 1 − Abs[Sin[2 π #]], 1 − 4 # + 4 #^{2}] &)]

Other examples:

ContourPlot[− Im[Log[Cos[Log[Tan[(x + I y)^4]]]]], {x, −.9, .9}, {y, −.9, .9},
AspectRatio → Automatic, ContourLines → False, Contours → 10, PlotPoints → 200,
Frame → False, Background → RGBColor[0.851575, 0.570321, 0.699229],
ColorFunction –> (CMYKColor[(4 #^2 − 4 # + 1), Abs[Cos[π #]], Abs[2 # − 1], 0] &)]

ContourPlot[Mod[Floor[−25 Im[Log[Log[Sin[Cos[x^2 + I y] + Cos[y^2 + I x]]]]]], 37],
{x, −4, −1.4}, {y, 1.4, 4}, AspectRatio → Automatic, PlotPoints → 180,
Contours → 27, Frame → False, ContourLines → False, Axes → False,
Background –> RGBColor[0.0823529, 0.380392, 0.866667], ColorFunction –>
(CMYKColor[Abs[2 # − 1], (4 #^2 − 4 # + 1), 1 − Abs[Cos[2 π #]], 0] &)]

The following examples are illustrated in the color pages.

ContourPlot[Mod[Floor[2 Re[(Log[Log[Cos[Log[Cos[(x − I y^2)] + Sin[y − I x^2]]]]])^3]], 42],
{y, −1.8, 1.6}, {x, −1.3, 1.7}, ContourLines → False,
PlotPoints → 310, Contours → 42, Frame → False,
Background → RGBColor[0.105882, 0.741176, 0.878431], ColorFunction –>
(RGBColor[1 − Abs[Cos[π #]], 1 − Abs[Sin[2 π #]], 1 − 4 # + 4 #^{2}] &)] (Color Fig 2.43)

ContourPlot[Mod[Floor[−5 Im[Log[Log[Sin[Sin[x − I y] + Sin[y + I x]]]]]], 40],
{x, −2, 3.5}, {y, −2, 3.5}, PlotPoints → 220, ContourLines → False,
AspectRatio → Automatic, Frame → False, Contours → 40,
Background → RGBColor[0.47657, 0.83595, 0.531258],
ColorFunction → (RGBColor[1 − #^{2}, Abs[Sin[2 π #]], 4 #^{2} − 4 # + 1] &)] (Color Fig 2.44)

ContourPlot$\Big[$Mod[IntegerPart[3 (Cos[4 x] + Cos[4 y])], 4],

{x, -9π, 9π}, {y, -9π, 9π}, Contours $\to$ 10, Frame $\to$ False,
Background $->$ RGBColor[0.671885, 0.300786, 0.300786],

ColorFunction $\to \Big($If$\Big[$# $\le$ 0, RGBColor[0, 0, 0],

CMYKColor$\Big[\dfrac{\#}{3}$, 4 #^2 $-$ 4 # + 1, 1 $-$ Sin[π#], 0$\Big]\Big]$ &$\Big)\Big]$ (Color Fig 2.45)

Some programs to try:

ContourPlot[Mod[Floor[10 Re[Log[Sin[Cos[x^2 + I y] + Cos[y^2 + I x]]]]], 35],
{y, $-3.$, 3.}, {x, $-3.$, 3.}, PlotPoints $\to$ 200, ContourLines $\to$ False,
Frame $\to$ False, AspectRatio $\to$ Automatic, ColorFunction $->$
(RGBColor[1 $-$ Abs[Sin[2 π#]], Abs[Cos[2 π#]], #] &), Contours $\to$ 30]

ContourPlot[$-$Mod[Floor[25 Abs[Log[Tan[x^2 + I y] + Sin[y^2 $-$ I x^3]]]], 15],
{y, -1.8, 2.1}, {x, -1.5, 1.5}, AspectRatio $\to$ Automatic,
ContourLines $\to$ False, Contours $\to$ 15, PlotPoints $\to$ 150, ColorFunction $\to$
(RGBColor[1 $-$ (4$\#^2$ $-$ 4 # + 1), Sin[π#], 1 $-$ Cos[π # / 4]] &), Frame $\to$ False]

ContourPlot[$-$Abs[Mod[Floor[15 Tan[(Sin[x + I y])^2]], 20]],
{x, -2.15, 3.15}, {y, -2.15, 2.15}, PlotPoints $\to$ 200,
ContourLines $\to$ False, Frame $\to$ False, ColorFunction $\to$
(CMYKColor[Abs[2 # $-$ 1], Abs[Cos[π#]], 4 #^2 $-$ 4 # + 1, 0] &), Contours $\to$ 30]

ContourPlot[$-$Mod[Floor[3 Im[Log[Log[Sin[Cos[x^2 + I y] + Cos[y^2 + I x]]]]]], 12],
{x, -3, 3}, {y, -3, 3}, AspectRatio $\to$ Automatic, ContourLines $\to$ False,
Contours $\to$ 25, PlotPoints $\to$ 200, Frame $\to$ False,
ColorFunction $\to$ (RGBColor[1 $-$ #^4, 1 $-$ #^2, 4 #^2 $-$ 4 # + 1] &),
Background $\to$ RGBColor[0, 0, 1]]

ContourPlot[$-$Mod[Floor[15 Abs[Log[Log[Cos[Sin[(x + I (y))^4]]]]]], 28],
{x, -1.5, 1.5}, {y, -1.5, 1.5}, ContourLines $\to$ False, Contours $\to$ 23,
PlotPoints $\to$ 200, Frame $\to$ False, ColorFunction $\to$
(CMYKColor[Abs[2 # $-$ 1], (4 #^2 $-$ 4 # + 1), 1 $-$ Abs[Cos[2 π#]], 0] &)]

The option **ColorFunction$\to$Hue** can also be used, as illustrated in the following program, in which the contour lines are not included.

ContourPlot[Floor[15 Sin[x^2 + y^2] Sin[x^4 + y^4]], {x, -5, 5}, {y, -5, 5},
ColorFunction $\to$ Hue, Background $\to$ RGBColor[0.125002, 0.281254, 0.820325],
Frame $\to$ False, Axes $\to$ False, Contours $\to$ 31]

■ Choosing Specific Colors for a Plot

One can assign a particular color to each of a set of height numbers, using the command **Which**. This technique is particularly useful when constructing the contour plot of a function of the form **Mod[IntegerPart[f[x]],n]**, for small integral n, as the plot will have exactly n height numbers.

We first choose the colors, and then construct a palette: (Color Fig 2.46)

```
ContourPlot[x, {x, 0, 20}, {y, 0, 2},
   ColorFunction → (Which[# < 0.1, RGBColor[1, 0, 0], 0.1 ≤ # && # ≤ 0.5,
         RGBColor[0.917983, 0.871107, 0.00390631], 0.5 < # && # ≤ 0.75, RGBColor[
            0.792981, 0.753918, 0.00390631], True, RGBColor[0, 0.484382, 0.972671]] &),
   AspectRatio → Automatic, Contours → 20]

ContourPlot[Mod[IntegerPart[3 (Sin[3 x] + Sin[3 y])], 4],
   {x, −9 π, 9 π}, {y, −9 π, 9 π}, Contours → 10, Frame → False,
   Background −> RGBColor[0.792981, 0.753918, 0.00390631],
   ColorFunction → (Which[# < 0.1, RGBColor[1, 0, 0],
         0.1 ≤ # && # ≤ 0.5, RGBColor[0.917983, 0.871107, 0.00390631],
         0.5 < # && # ≤ 0.75, RGBColor[0.792981, 0.753918, 0.00390631],
         True, RGBColor[0, 0.484382, 0.972671]] &)] (Color Fig 2.47)
```

In the next example, 4 colors have been chosen to obtain an impression of polished granite: (Color Fig 2.48)

```
ContourPlot[−Mod[IntegerPart[3 Re[ChebyshevU[6, x + y I]]], 4], {x, 2, 4},
   {y, −0.25, 1.75}, PlotPoints −> 150, ContourLines → False, Frame → False,
   ColorFunction → (Which[# ≤ 0.25, RGBColor[0.972671, 0.523445, 0.500008],
         0.25 < # && # ≤ 0.5, RGBColor[0.855482, 0.855482, 0.871107],
         0.5 < # && # ≤ 0.75, RGBColor[0.609384, 0.613291, 0.660166],
         True, RGBColor[0.453132, 0.457038, 0.519539]] &)]
```

■ Random Choice of Colors

One can let *Mathematica* choose the coloring for a contour plot by choosing, in **CMYKColor[c, m, y, k]**, random values for c, y, and m and zero for k, in order to obtain a range of bright colors. (Color Fig 2.49)

```
ContourPlot[Im[Log[Log[Sin[x + I y²]]]], {x, −3, 6}, {y, −3, 3}, AspectRatio → Automatic,
   ContourLines → False, PlotPoints → 200, Contours → 25, Frame → False,
   ColorFunction −> (CMYKColor[Random[], Random[], Random[], 0] &)]
```

An example to try:

```
ContourPlot[Abs[Log[Log[Sin[Cos[x + I y]]]]], {x, −3, 3},
    {y, −3, 3}, AspectRatio → Automatic, ContourLines → True,
    PlotPoints → 200, Contours → 25, Frame → False,
    ColorFunction −> (CMYKColor[Random[], Random[], Random[], 0] &)]
```

In the following program, the randomness color spectrum has been limited by choosing the entry m in **CMYKColor[c, m, y, k]** as **0.6Random[]**. The entry k is chosen as 0.1, in order to obtain a slightly darker set of colors.

```
ContourPlot[Abs[Log[Log[Cos[Log[Cos[(x + I y²)] + Sin[y + I x²]]]]]],
    {y, −5.5, 3.5}, {x, −2.3, 2.2}, AspectRatio → Automatic,
    ContourLines → False, PlotPoints → 200, Contours → 10, Frame → False,
    ColorFunction −> (CMYKColor[Random[], 0.6 Random[], Random[], 0.1] &)]
```

If you do not like the coloring, you can press shift-enter to obtain another one.

■ Repeating a Sequence of Colors

When using a table of colors to color a sequence of 2D curves in section 2.3, if there were more curves than colors, the colors were repeated, which in some cases resulted in attractive effects. Colors are not repeated with **ColorFunction**. We can achieve repetition of colors as follows: using the function **UnitStep**, we construct a function which takes the values between 0 and 1 twice in the interval $[0, 1]$.

Plot[2 x − UnitStep[x − 0.5], {x, 0, 1}, AspectRatio → Automatic]

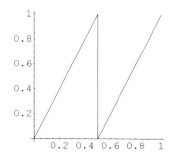

We choose the pure function **CMYKColor$\left[1 - \text{Cos}[\frac{\pi}{3}\right], 4\# - 4\#^2, \text{Sin}\left[\pi \frac{\#}{4}\right], 0\right]$ &**, and replace each occurence of # by **(2# - UnitStep[# - 0.5])**, using the replacement command **/.**.

CMYKColor[1 − #², 1 − #, Sin[π #], 0] /. # → 2 # − UnitStep[# − 0.5]

CMYKColor[1 − (2 #1 − UnitStep[−0.5 + #1])²,
    1 − 2 #1 + UnitStep[−0.5 + #1], Sin[π (2 #1 − UnitStep[−0.5 + #1])], 0]

Here is the program for the resulting palette:

```
ContourPlot[x, {x, 0, 20}, {y, 0, 2},
   ColorFunction → (CMYKColor[1 − (2 #1 − UnitStep[−0.5` + #1])²,
        1 − 2 #1 + UnitStep[−0.5` + #1], Sin[π (2 #1 − UnitStep[−0.5` + #1])], 0] &),
   AspectRatio → Automatic, Contours → 20]
```

Here is an example of a contour plot using the above palette: (Color Fig 2.50)

```
ContourPlot[Im[Log[Log[Cos[Sin[x + I y] + Cos[y + I x]]]]], {x, −3, 3}, {y, −0, 6},
   AspectRatio → Automatic, ContourLines → False, PlotPoints → 300, Contours → 25,
   Frame → False, ColorFunction → (CMYKColor[1 − (2 #1 − UnitStep[−0.5` + #1])²,
        1 − (2 #1 − UnitStep[−0.5` + #1]), Sin[π (2 #1 − UnitStep[−0.5` + #1])], 0] &)]
```

Other programs to try:

```
ContourPlot[−Re[Log[Log[Sin[Sin[x + I y] + Sin[y − I x]]]]],
   {x, −3.5, 3.5}, {y, −3.5, 3.5}, AspectRatio → Automatic, PlotPoints → 200,
   ContourShading → True, Frame → False, ContourLines → False, ColorFunction →
   (CMYKColor[1 − (2 #1 − UnitStep[−0.5` + #1])², 2 #1 − UnitStep[−0.5` + #1],
        Sin[π (2 #1 − UnitStep[−0.5` + #1])], 0] &), Contours → 15]
```

$$\text{ContourPlot}\left[\text{Cos}\left[\frac{1}{x\,y}\right]\right], \{x, -3, 3\}, \{y, -3, 3\},$$

```
   AspectRatio → Automatic, ContourLines → True, PlotPoints → 100,
   Contours → 20, Frame → False, ColorFunction →
   (CMYKColor[1 − (2 #1 − UnitStep[−0.5` + #1])², 2 #1 − UnitStep[−0.5` + #1],
        Sin[π (2 #1 − UnitStep[−0.5` + #1])], 0] &), ContourStyle →
```

$$\text{Table}\left[\left\{\text{CMYKColor}\left[1 - \text{Cos}\left[\frac{n\,\pi}{40}\right], 1 - \frac{n}{5} + \frac{n^2}{100}, 1 - \text{Sin}\left[\frac{n\,\pi}{20}\right], 0\right]\right\}, \{n, 1, 20\}\right]\right]$$

■ Coloring the Contour Lines

In a contour plot, one can omit the contour shading, and color the contour lines, using the command **ContourStyle**, with a table of colors. In the following example, there are more contours than colors, so the sequence of colors is repeated: (Color Fig 2.51)

$$\text{ContourPlot}\left[\text{Re}[\text{Log}[\text{Cos}[\text{Log}[\text{Cos}[1/((x + y + I (x - y))^4 + y + I x)]]]]],\right.$$

```
   {x, −0.75, 0.75}, {y, −0.61, 0.85}, AspectRatio → Automatic, ContourLines → True,
   PlotPoints → 250, Contours → 35, Frame → False, ContourShading → False,
```

$$\text{ContourStyle} \rightarrow \text{Table}\left[\left\{\text{RGBColor}\left[1 - \left(\frac{n}{20}\right)^2, 1 - \frac{n}{5} + \frac{n^2}{100}, 1 - \frac{n}{20}\right]\right\}, \{n, 1, 20\}\right]\right]$$

A program to try:

$$\text{ContourPlot}\Big[-\text{Re}\Big[\text{Sin}\Big[\text{Log}\Big[\frac{1}{4}\,(x+y+I\,(x-y))^4\Big]\Big]\Big],\ \{x,\,-1.05,\,1.05\},$$

$$\{y,\,-1.05,\,1.05\},\ \text{PlotPoints}\to 250,\ \text{ContourShading}\to\text{False},\ \text{Contours}\to 55,$$

$$\text{ContourLines}\to\text{True},\ \text{AspectRatio}\to\text{Automatic},\ \text{Frame}\to\text{False},\ \text{ContourStyle}\to$$

$$\text{Table}\Big[\Big\{\text{RGBColor}\Big[\Big(\frac{n}{20}\Big)^2,\ 1-\text{Abs}\Big[\text{Cos}\Big[\pi\,\frac{n}{10}\Big]\Big],\ 1-\Big(\frac{n}{5}-\frac{n^2}{100}\Big)\Big]\Big\},\ \{n,\,1,\,20\}\Big]\Big]$$

In the following example, we have included a background color which contrasts with the colors of the lines: (Color Fig 2.52)

$$\text{ContourPlot}\Big[-\text{Abs}[\text{Log}[\text{Log}[\text{Sin}[\text{Cos}[x^2+I\,y^2]+\text{Sin}[y+I\,x]]]]],$$

$$\{x,\,-4,\,1\},\ \{y,\,-1,\,3\},\ \text{AspectRatio}\to\text{Automatic},\ \text{PlotPoints}\to 200,$$

$$\text{ContourShading}\to\text{False},\ \text{Frame}\to\text{False},\ \text{ContourStyle}\to$$

$$\text{Table}\Big[\Big\{\text{Thickness}[0.004],\ \text{RGBColor}\Big[1-\Big(\frac{n}{20}\Big),\ 1-\Big(\frac{n}{20}\Big)^4,\ 4\Big(\frac{n}{20}\Big)^2-4\Big(\frac{n}{20}\Big)+1\Big]\Big\},$$

$$\{n,\,1,\,20\}\Big],\ \text{Contours}\to 50,\ \text{Background}\to\text{RGBColor}[1,\,0,\,0]\Big]$$

Another program to try:

$$\text{ContourPlot}\Big[-\text{Im}[\text{Log}[\text{Sin}[\text{Log}[\text{Sin}[x+I\,y]+\text{Cos}[y+I\,x]]]]],$$

$$\{x,\,-5,\,5\},\ \{y,\,-4,\,4\},\ \text{AspectRatio}\to\text{Automatic},\ \text{ContourLines}\to\text{True},$$

$$\text{Contours}\to 170,\ \text{PlotPoints}\to 140,\ \text{Frame}\to\text{False},\ \text{ContourShading}\to\text{False},$$

$$\text{Background}\to\text{RGBColor}[0,\,0.5,\,0.5],\ \text{ContourStyle}\to$$

$$\text{Table}\Big[\Big\{\text{CMYKColor}\Big[\Big(1-\frac{n}{20}+\frac{n^2}{1600}\Big),\ \frac{n}{80},\ 1-\text{Sin}\Big[\frac{n\pi}{80}\Big],\ 0\Big]\Big\},\ \{n,\,1,\,80\}\Big]\Big]$$

■ Coloring the Contours and the Contour Lines

We can color the contours as well as the contour lines.

The command:

$$\text{ColorFunction}\to\big(\text{CMYKColor}[1-\text{Sin}[\pi\,\#],\ (4\,\#^\wedge 2-4\,\#+1),\ 1-\text{Cos}\big[\pi\,\tfrac{\#}{3}\big],\ 0\big]\ \&\big)$$

is used to color the contours. We choose to plot 20 contours. For the contour lines, we choose a sequence of colors which contrast with the colors in the above **ColorFunction** command. We choose the color function: $(\text{CMYKColor}[\#^2,\ 4\,\#-4\,\#^2+1,\ \text{Sin}[\pi\,\#],\ 0]\ \&)$, and replace # by $\frac{n}{20}$, using the command /. .

$$\text{CMYKColor}[\#^2,\ 4\,\#-4\,\#^2+1,\ \text{Sin}[\pi\,\#],\ 0]\ /.\ \#\to\tfrac{n}{20}\qquad \text{CMYKColor}\Big[\tfrac{n^2}{400},\ 1+\tfrac{n}{5}-\tfrac{n^2}{100},\ \text{Sin}[\tfrac{n\pi}{20}],\ 0\Big]$$

Palettes showing the above color schemes can be seen in Color Fig 2.53 and Color Fig 2.54.

$$\mathbf{Show}\Big[\mathbf{Graphics}\Big[\mathbf{Table}\Big[\Big\{\mathbf{CMYKColor}\Big[\frac{n^2}{400},\; 1-\Big(1-\frac{n}{5}+\frac{n^2}{100}\Big),\; 1-\mathbf{Sin}\Big[\frac{n\pi}{20}\Big],\; 0\Big],$$

$$\mathbf{Rectangle}[\{n,\,0\},\,\{n+0.8,\,1\}]\Big\},\;\{n,\,1,\,20\}\Big],$$

$$\mathbf{AspectRatio}\to\mathbf{Automatic},\;\mathbf{Axes}\to\mathbf{False}\Big]\Big]$$

$$\mathbf{ContourPlot}\Big[-x,\;\{x,\,0,\,20\},\;\{y,\,0,\,2\},$$

$$\mathbf{ColorFunction}\to\Big(\mathbf{CMYKColor}\Big[1-\mathbf{Sin}[\pi\#],\,(4\#\wedge2-4\#+1),\,1-\mathbf{Cos}\Big[\pi\frac{\#}{3}\Big],\,0\Big]\,\&\Big),$$

$$\mathbf{AspectRatio}\to\mathbf{Automatic},\;\mathbf{Contours}\to20\Big]$$

The image generated by the following program is colored using the color scheme defined above:

$$\mathbf{ContourPlot}\Big[\mathbf{Im}[\mathbf{Log}[\mathbf{Log}[\mathbf{Sin}[\mathbf{Sin}[(x+\mathbf{I}\,y)^4]]]]],\;\{x,\,0.5,\,1.5\},\;\{y,\,0.5,\,1.5\},\;\mathbf{Contours}\to43,$$

$$\mathbf{PlotPoints}\to100,\;\mathbf{Frame}\to\mathbf{False},\;\mathbf{Background}\to\mathbf{RGBColor}[0.5,\,0.6,\,0.9],$$

$$\mathbf{ColorFunction}\to\Big(\mathbf{CMYKColor}\Big[1-\mathbf{Sin}[\pi\#],\,(4\#\wedge2-4\#+1),\,1-\mathbf{Cos}\Big[\pi\frac{\#}{3}\Big],\,0\Big]\,\&\Big),$$

$$\mathbf{ContourStyle}\to\mathbf{Table}\Big[\Big\{\mathbf{CMYKColor}\Big[\frac{n^2}{400},\,1-\Big(1-\frac{n}{5}+\frac{n^2}{100}\Big),\,1-\mathbf{Sin}\Big[\frac{n\pi}{20}\Big],\,0\Big]\Big\},$$

$$\{n,\,1,\,20\}\Big]\Big]\;(\text{Color Fig 2.55})$$

Notice the **ContourStyle** command in the following example. This technique enables one to change the number of elements in the table of colors and also to change a **ColorFunction** option to a **ContourStyle** option more easily. (Color Fig 2.56)

$$\mathbf{ContourPlot}\Big[\mathbf{Abs}[\mathbf{Log}[\mathbf{Sin}[\mathbf{Log}[(x+\mathbf{I}\,y)^4]]]],\;\{x,\,-0.7,\,0.7\},$$

$$\{y,\,-0.7,\,0.7\},\;\mathbf{Contours}\to25,\;\mathbf{PlotPoints}\to200,\;\mathbf{Frame}\to\mathbf{False},$$

$$\mathbf{ColorFunction}\to\Big(\mathbf{CMYKColor}\Big[1-\mathbf{Cos}\Big[\pi\frac{\#}{3}\Big],\,4\#\wedge2-4\#+1,\,\mathbf{Sin}[\pi\#],\,0\Big]\,\&\Big),$$

$$\mathbf{Background}\to\mathbf{RGBColor}[0.9,\,0.6,\,0.53],\;\mathbf{ContourStyle}\to$$

$$\mathbf{Table}\Big[\Big\{\mathbf{CMYKColor}[1-4\#+4\#^2,\,\#,\,1-\mathbf{Sin}[\pi\#],\,0]\;/.\;\#\to\frac{n}{20}\Big\},\;\{n,\,1,\,20\}\Big]\Big]$$

Another example: (Color Fig 2.57)

$$\mathbf{ContourPlot}[\mathbf{Abs}[\mathbf{Log}[\mathbf{Cos}[\mathbf{Sin}[x^2+\mathbf{I}\,y]-\mathbf{Sin}[y+\mathbf{I}\,x]]]],\;\{y,\,-3,\,3\},\;\{x,\,-3,\,3\},$$

$$\mathbf{ContourLines}\to\mathbf{True},\;\mathbf{Contours}\to15,\;\mathbf{PlotPoints}\to250,\;\mathbf{Frame}\to\mathbf{False},$$

$$\mathbf{ColorFunction}\to(\mathbf{CMYKColor}[\mathbf{Cos}[0.5\,\pi\#],\,4\#^2-4\#+1,\,\mathbf{Sin}[\pi\#],\,0]\,\&),$$

$$\mathbf{Background}\to\mathbf{RGBColor}[0.132815,\,0.699229,\,0.273442],$$

$$\mathbf{ContourStyle}\to\mathbf{Table}[\{\mathbf{Thickness}[0.007],$$

$$\mathbf{CMYKColor}[1-\mathbf{Cos}[0.025\,n\,\pi],\,0.2\,n-0.01\,n^2,\,1-\mathbf{Sin}[0.05\,n\,\pi],\,0]\},\;\{n,\,1,\,20\}]]$$

Some more programs to try:

ContourPlot[Im[Exp[Cos[Log[Sin[x + I y] + Cos[y + I x]]]]]], {x, −4, 4},
{y, −4, 4}, ContourLines → True, Contours → 25, PlotPoints → 200,
Frame → False, Background → RGBColor[0.63, 0.41, 0.9], ColorFunction −>
(CMYKColor[1 − Sin[π #], (4 #^2 − 4 # + 1), 1 − Cos[π #/3], 0] &), ContourStyle →
Table[{CMYKColor[n² / 400, 1 − (1 − n/5 + n²/100), 1 − Sin[n π/20], 0]}, {n, 1, 20}]]

$$ContourPlot\left[-Re[Log[Log[Cos[Log[Tan[(x + I y)^6]]]]]], \{y, -1.1, 1.1\},\right.$$

{x, −1.1, 1.1}, AspectRatio → Automatic, ContourLines → True,
PlotPoints → 100, Contours → 20, Frame → False, ColorFunction →
(CMYKColor[(2 #1 − UnitStep[−0.5` + #1])², (2 #1 − UnitStep[−0.5` + #1]),
1 − Sin[π (2 #1 − UnitStep[−0.5` + #1])], 0] &), ContourStyle →

$$Table\left[\left\{CMYKColor\left[Cos\left[\frac{n\pi}{40}\right], \left(1 - \frac{n}{5} + \frac{n^2}{100}\right), Sin\left[\frac{n\pi}{20}\right], 0\right]\right\}, \{n, 1, 20\}\right],$$

$$Background \to RGBColor[0.128908, 0.742199, 0.480476]\Big]$$

In the following examples, we combine the 'Mod-Floor process' with the above technique: (Color Fig 2.58)

$$ContourPlot\Big[Mod[Floor[3 Im[Log[Log[Sin[Sin[(x + I y)^4]]]]]], 10]$$

, {x, 0.15, 1.3}, {y, 0.15, 1.3}, PlotPoints → 280, ContourLines → True,
AspectRatio → Automatic, Frame → False, Contours → 15, ContourStyle →

$$Table\left[\left\{Thickness[0.005], CMYKColor\left[\frac{n^2}{400}, \left(1 - \frac{n}{5} + \frac{n^2}{100}\right), 1 - Cos\left[\frac{n\pi}{40}\right], 0\right]\right\},$$

{n, 1, 20}], Background → RGBColor[0.722667, 0.308598, 0.308598],

$$ColorFunction \to (RGBColor[\#, 1 - Sin[\pi \#], 1 - \#] \&)\Big]$$

The following command gives a different coloring for the above graphic:

$$ContourPlot\Big[Mod[Floor[-2 Im[Log[Log[Sin[Sin[(x + I y)^4]]]]]], 6]$$

, {x, 0.1, 1.3}, {y, 0.1, 1.3}, PlotPoints → 250, ContourLines → True,
AspectRatio → Automatic, Frame → False, Contours → 20, ContourStyle →

$$Table\left[\left\{CMYKColor\left[\frac{n^2}{400}, 1 - \left(1 - \frac{n}{5} + \frac{n^2}{100}\right), Cos\left[\frac{n\pi}{40}\right], 0\right]\right\}, \{n, 1, 20\}\right],$$

Background → RGBColor[0.160159, 0.644541, 0.0781262],

$$ColorFunction \to (RGBColor[\#, Sin[\pi \#], \#] \&)\Big]$$

■ Color Sequences Starting with Color 1 and Ending with Color 2

Suppose one wishes to obtain a color sequence which varies from **RGBColor[r1, g1, b1]** to **RGBColor[r2, g2, b2]**. Here is a pure function which enables one to do this:

> **cf[RGBColor[r1_, g1_, b1_], RGBColor[r2_, g2_, b2_]] :=**
> **(RGBColor[r1 + (r2 − r1) #, g1 + (g2 − g1) #, b1 + (b2 − b1) #]) &;**

Here is an example of its use: (Color Fig 2.59)

> **ContourPlot[Mod[−Im[2 Log[Log[Sin[Tan[(x + I y)]]]]], 3], {x, 1.36, 2.14},**
> **{y, −0.3, 0.3}, PlotPoints → 200, ContourLines → False, AspectRatio → Automatic,**
> **Frame → False, Contours → 15, Background → RGBColor[1, 1, 0],**
> **ColorFunction → cf[RGBColor[1, 1, 0], RGBColor[0, 0, 1]]]**

For the following image, we use a variation of the function cf defined above, in order to include an extra color: (Color Fig 2.60)

> **cf2[RGBColor[r1_, g1_, b1_], RGBColor[r2_, g2_, b2_]] :=**
> **(If[# ≥ 0.9, RGBColor[0.648447, 0.792981, 0.914076],**
> **RGBColor[r1 + (r2 − r1) #, g1 + (g2 − g1) #, b1 + (b2 − b1) #]]) &;**

> **ContourPlot[−Mod[Floor[9 Re[(Log[Cos[Log[Sin[(x + I y)⁴]]]])²]], 50],**
> **{x, −1.4, 1.4}, {y, −1.4, 1.4}, ContourLines → False, Contours → 40,**
> **PlotPoints → 200, Frame → False, Background → RGBColor[0.87, 0.81, 0.1],**
> **ColorFunction → cf2[RGBColor[1., 1., 0], RGBColor[0, 0, 0.]]]**

The program for another example:

> **ContourPlot[Mod[IntegerPart[3 Im[Log[Log[Cos[x² + I y] + Sin[x + I y²]]]]], 20],**
> **{x, −6.5, 2.5}, {y, −1, 3}, AspectRatio → Automatic, PlotPoints → 80, Contours → 20,**
> **Frame → False, ContourStyle → Table[{GrayLevel[Abs[Sin[n π /30]]]}, {n, 1, 10}],**
> **ColorFunction → cf2[RGBColor[0.53, 0.44, 0.26], RGBColor[0.83, 0.77, 0.66]]]**

Consider the pure function below:

> **If[# ≤ 0, RGBColor[1, 1, 1],**
> **CMYKColor[1 − Cos[π #/3], (4 #² − 4 # + 1), 1 − Sin[π #], 0]] &**

Let P be a point on the plot with height number h.

If h ≤ 0, then P is assigned the color **RGBColor[1, 1, 1]**, while, if h > 0, then P is assigned the color **CMYKColor[1 − Cos[π h / 3], (4 h² − 4 h + 1), 1 − Sin[π h], 0]**. The above command is used to color the image shown in Color Fig 2.61.

$$\text{ContourPlot}\Big[\text{Mod}[\text{IntegerPart}[-3\,\text{Im}[\text{Log}[\text{Log}[\text{Cos}[x^2 + I\,y] - \text{Sin}[x + I\,y^2]]]]], 10],$$

$$\{x, -6.5, 2.5\}, \{y, -1, 2\}, \text{AspectRatio} \to 0.4, \text{PlotPoints} \to 30, \text{Contours} \to 10,$$

$$\text{Frame} \to \text{False}, \text{Axes} \to \text{False}, \text{Background} \to \text{RGBColor}[1, 0, 0],$$

$$\text{ContourStyle} \to \text{Table}\Big[\Big\{\text{CMYKColor}\Big[\Big(1 - \frac{n}{5} + \frac{n^2}{100}\Big), 1 - \frac{n}{20}, \text{Sin}\Big[\frac{n\,\pi}{20}\Big], 0\Big]\Big\}, \{n, 1, 10\}\Big],$$

$$\text{ColorFunction} \to \Big(\text{If}\Big[\# \le 0, \text{RGBColor}[1, 1, 1],$$

$$\text{CMYKColor}\Big[1 - \text{Cos}\Big[\pi\,\frac{\#}{3}\Big], (4\,\#^\wedge 2 - 4\,\# + 1), 1 - \text{Sin}[\pi\,\#], 0\Big]\Big] \&\Big)\Big]$$

Another example: (Color Fig 2.62)

$$\text{ContourPlot}\Big[-\text{Im}\big[(\text{Log}[\text{Log}[\text{Sin}[\text{Log}[\text{Cos}[(x + I\,y^2)] + \text{Sin}[y + I\,x^2]]]]])^2\big],$$

$$\{y, -4.5, 3.5\}, \{x, -.3, 3.4\}, \text{PlotPoints} \to 250, \text{ContourLines} \to \text{True},$$

$$\text{AspectRatio} \to 0.4, \text{Frame} \to \text{False}, \text{Contours} \to 20, \text{ContourStyle} \to$$

$$\text{Table}[\{\text{CMYKColor}[0.0025\,n^2, 0.2\,n - 0.01\,n^2, \text{Cos}[0.025\,n\,\pi], 0]\}, \{n, 1, 20\}],$$

$$\text{Background} \to \text{RGBColor}[0.160159, 0.644541, 0.0781262],$$

$$\text{ColorFunction} \to (\text{RGBColor}[\text{Abs}[2\,\# - 1], \text{Abs}[\text{Sin}[2\,\pi\,\#]], \#] \&)\Big]$$

The program for another example:

$$\text{ContourPlot}\Big[\text{Mod}[\text{Floor}[2\,\text{Arg}[\text{Log}[\text{Log}[\text{Sin}[\text{Sin}[(x + I\,(\,y))^4]]]]]], 4],$$

$$\{x, -1.4, 1.4\}, \{y, -1.4, 1.4\}, \text{ContourLines} \to \text{True}, \text{Contours} \to 25,$$

$$\text{PlotPoints} \to 150, \text{Frame} \to \text{False}, \text{Background} \to \text{RGBColor}[0, 0, 1],$$

$$\text{ColorFunction} \to \Big(\text{CMYKColor}\Big[1 - \text{Sin}[\pi\,\#], (4\,\#^\wedge 2 - 4\,\# + 1), 1 - \text{Cos}\Big[\pi\,\frac{\#}{3}\Big], 0\Big] \&\Big),$$

$$\text{ContourStyle} \to \text{Table}\Big[\Big\{\text{CMYKColor}\Big[\frac{n^2}{400}, \frac{n}{5} - \frac{n^2}{100}, 1 - \text{Sin}\Big[n\,\frac{\pi}{20}\Big], 0\Big]\Big\}, \{n, 1, 20\}\Big]\Big]$$

The transformation $\{x, y\} \to \frac{1}{\sqrt{2}}\,\{x - y, x + y\}$ rotates the surface with equation $z = f\,[x, y]$ through $\frac{\pi}{4}$ about the z-axis. In the following program, we apply this transformation to the above contour plot, using a different color scheme:

$$\text{ContourPlot}\Big[-\text{Mod}[\text{Floor}[2\,\text{Arg}[\text{Log}[\text{Log}[\text{Sin}[\text{Sin}[1\,/\,4\,(x + y + I\,(x - y))^4]]]]]], 4],$$

$$\{x, -1.4, 1.4\}, \{y, -1.4, 1.4\}, \text{Contours} \to 25, \text{PlotPoints} \to 150,$$

$$\text{Frame} \to \text{False}, \text{Background} \to \text{RGBColor}[0.8, 0.72, 0.77],$$

$$\text{ColorFunction} \to (\text{CMYKColor}[(4\,\#^\wedge 2 - 4\,\# + 1), 1 - \text{Cos}[\pi\,\#\,/\,3], 1 - \text{Sin}[\pi\,\#], 0] \&),$$

$$\text{ContourStyle} \to \text{Table}\Big[\Big\{\text{CMYKColor}\Big[\frac{n^2}{400}, \frac{n}{5} - \frac{n^2}{100}, \text{Sin}\Big[n\,\frac{\pi}{20}\Big], 0\Big]\Big\}, \{n, 1, 20\}\Big]\Big]$$

A program to try:

```
ContourPlot[Mod[Floor[-3 Re[Log[Cos[Log[Sin[(x + I y)^4]]]]]], 8],
   {x, -1.2, 1.2}, {y, -1.2, 1.2}, AspectRatio → Automatic,
   ContourLines → True, Contours → 20, PlotPoints → 150, Frame → False,
   ColorFunction -> (CMYKColor[1 - Cos[π #/3], (4 #^2 - 4 # + 1), 1 - Sin[π #], 0] &),
   Background → RGBColor[1, 0, 0], ContourStyle →
     Table[{CMYKColor[(1 - n/5 + n^2/100), n/20, 1 - Sin[n π/20], 0]}, {n, 1, 20}]]
```

In the following program, the command **ColorFunction** is omitted and so *Mathematica* uses the default **GrayLevel** coloring.

```
ContourPlot[-Im[Log[Log[Sin[Log[Cos[(x - I y^3)] + Sin[y + I x^3]]]]]],
   {y, -2, 2}, {x, -2, 2}, ContourLines → True, PlotPoints → 100,
   Contours → 35, Frame → False, AspectRatio → .6, ContourStyle →
     Table[{CMYKColor[Abs[Cos[0.05 n π]], (1 - 0.2 n + 0.01 n^2), 1 - Sin[.05 n π], 0]},
       {n, 1, 20}], Background -> RGBColor[0.128908, 0.742199, 0.480476]]
```

In the following example, 'E.T 's Egg Box', we use the technique described above of repeating the sequence of contour colors and we also color the contour lines: (Color Fig 2.63)

```
ContourPlot[-Abs[Log[Sin[Log[Sin[x - I y] + Cos[y + I x]]]]],
   {y, -10.6, 11}, {x, -1.3, 11}, AspectRatio → Automatic,
   ContourLines → True, PlotPoints → 100, Contours → 25, Frame → False,
   ColorFunction → (CMYKColor[1 - Cos[1/3 π (2 #1 - UnitStep[-0.5` + #1])],
       4 (2 #1 - UnitStep[-0.5` + #1]) - 4 (2 #1 - UnitStep[-0.5` + #1])^2,
       Sin[0.25 π (2 #1 - UnitStep[-0.5` + #1])], 0] &),
   ContourStyle → Table[{Thickness[.005], CMYKColor[1 - Cos[0.025 n π],
       1 - 0.2 n + 0.01 n^2, 1 - Sin[0.05 n π], 0]}, {n, 1, 20}],
   Background -> RGBColor[0.576471, 0.843137, 0.4]]
```

Another program to try:

```
ContourPlot[-Re[Log[Log[Sin[Sin[x + I y] + Sin[y - I x]]]]],
   {y, -3, 3}, {x, -3, 3}, AspectRatio → Automatic, ContourLines → True,
   PlotPoints → 200, Contours → 30, Frame → False, ColorFunction →
     (CMYKColor[1 - (2 #1 - UnitStep[-0.5` + #1])^2, 2 #1 - UnitStep[-0.5` + #1],
       Sin[π (2 #1 - UnitStep[-0.5` + #1])], 0] &), ContourStyle → Table[
       {CMYKColor[1 - Cos[0.025 n π], 1 - 0.2 n + 0.01 n^2, 1 - Sin[0.05 n π], 0]}, {n, 1, 20}],
   Background -> RGBColor[0.128908, 0.742199, 0.480476]]
```

■ Constructing Circular Contour Plots.

We illustrate this technique by means of an example. Suppose we have constructed the contour plot of F[x, y] = Im[f[x+Iy]] for | x + Iy | ≤ 1.34, and we wish to restrict the plot to lie within the circle center the origin radius 1.34. We use **Plot3D** to plot the graph of F for $-1.34 \leq x \leq 1.34$ and $-1.34 \leq y \leq 1.34$. We need to find a lower bound not much smaller than the minimum for the plotted values of F[x, y], so we choose (approximately) the smallest tick mark on the z-axis.

> **Plot3D[Im[Log[Log[Cos[Log[Sin[(x + I y)^8]]]]]], {y, −1.34, 1.34},**
> **{x, −1.34, 1.34}, AspectRatio → Automatic, Boxed → True,**
> **BoxRatios → {1, 1, 2}, Ticks → {Automatic, Automatic, {−3.29, −1, 0, 1, 2}}]**

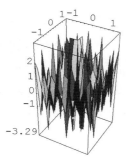

We estimate from the graph that a lower bound for F is −3.3. We define a function h as follows:

> **h[z_] := If[Abs[z] ≥ 1.34, −3.3, Im[Log[Log[Cos[Log[Sin[(z)^8]]]]]]];**

We plot the graph of h:

> **Plot3D[h[x + I y], {y, −1.34, 1.34}, {x, −1.34, 1.34},**
> **AspectRatio → Automatic, Boxed → True, BoxRatios → {1, 1, 2},**
> **Ticks → {Automatic, Automatic, {−3.2, −1, 0, 1, 2}}]**

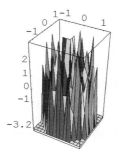

In the following contour plot of h the points in the plane where $|h[z]| = -3.3$ are colored white. (Color Fig 2.64)

> ContourPlot[−h[x + I y], {y, −1.34, 1.34}, {x, −1.34, 1.34}, AspectRatio → Automatic,
> PlotPoints → 250, Contours → 10, Frame → False, Axes → False, ColorFunction −>
> (If[# ≥ 1, RGBColor[1, 1, 1], Hue[0.65 − 0.3 Abs[Cos[π #]], 1 − 0.8 #, 1]] &)]

If too many contours are chosen, it may happen that the contour with height value 0 is not assigned the color white. In such a case one can replace the command **(If[#≤0, RGBColor[1,1,1]**,.....**)** with **(If[#≤a, RGBColor[1,1,1]**,.....**)**, where a is a small positive number.

> c[z_] := If[−Abs[z] ≤ −1.31, −3.85, −Abs[Log[Cos[Log[Sin[(z)^6]]]]]];

> ContourPlot$\Big[$c[x + I y], {x, −1.31, 1.31}, {y, −1.31, 1.31}, AspectRatio → Automatic,
> PlotPoints → 100, Contours → 70, Frame → False, Axes → False, ContourStyle →
> Table$\Big[\Big\{$CMYKColor$\Big[$1 − $\Big(1 - \frac{n}{5} + \frac{n^2}{100}\Big)$, $\frac{n}{20}$, 1 − Sin$\Big[\frac{n\pi}{20}\Big]$, 0$\Big]\Big\}$, {n, 1, 20}$\Big]$,
> ColorFunction −> $\Big($If$\Big[$# ≤ .05, RGBColor[1, 1, 1],
> CMYKColor$\Big[$Cos$\Big[\pi \frac{\#}{3}\Big]$, 4 #^2 − 4 # + 1, Sin[π #], 0$\Big]\Big]$ &$\Big)\Big]$ (Color Fig 2.65)

In the following program, the contours are colored, but the contour lines are not: (Color Fig 2.66)

> r[z_] := If[Abs[z − 1.85] ≥ .29, −3.2, Im[Log[Log[Sin[Tan[(x + I y)]]]]]];

> ContourPlot[r[x + I y], {x, 1.55, 2.15}, {y, −0.3, 0.3}, PlotPoints → 200,
> ContourLines → False, AspectRatio → Automatic, Frame → False,
> Contours → 25, ColorFunction → (If[# ≤ 0, RGBColor[1., 1., 1.],
> RGBColor[1 − #^2, Abs[Cos[4 π #]], 4 #^2 − 4 # + 1]] &)]

In the following example, the contour lines are colored, but not the contours, and the command **AspectRatio** is used to obtain an elliptical plot: (Color Fig 2.67)

> w[z_] := If[Abs[z] ≥ 1.3, −0.01, Abs[Log[Log[Sin[Log[Sin[(x + I y)^8]]]]]]];

> ContourPlot$\Big[$−w[x + I y], {x, −1.3, 1.3}, {y, −1.3, 1.3}, AspectRatio → 0.7,
> PlotPoints → 200, ContourShading → False, Frame → False, ContourStyle →
> Table$\Big[\Big\{$RGBColor$\Big[$1 − $\frac{n}{20}$, $\Big(1 - \frac{n}{20}\Big)^2$, $\frac{n}{5} - \frac{n^2}{100}\Big]\Big\}$, {n, 1, 20}$\Big]$, Contours → 10$\Big]$

We change the definition of w to obtain a ring-shaped contour plot. In this case, it is usually necessary to increase the number of contours to obtain an interesting plot. Here is the program:

$$\text{w2[z\_] := If[Abs[z]} \geq 1.3 \,||\, \text{Abs[z]} \leq 0.4, \, -10, \, \text{Abs[Log[Log[Sin[Log[Sin[(x + I\,y)}^8]]]]]]];$$

$$\text{ContourPlot}\Big[-\text{w2[x + I\,y], \{x, } -1.3, 1.3\}, \{y, -1.3, 1.3\}, \text{AspectRatio} \rightarrow \text{Automatic},$$

PlotPoints → 200, ContourShading → False, Frame → False, ContourStyle →

$$\text{Table}\Big[\Big\{\text{RGBColor}\Big[1 - \frac{n}{20}, \Big(1 - \frac{n}{20}\Big)^2, \frac{n}{5} - \frac{n^2}{100}\Big]\Big\}, \{n, 1, 20\}\Big], \text{Contours} \rightarrow 105\Big]$$

Exercise:

1) Construct variations on the examples of this section, by using different color schemes, number of contours, number of plot-points, varying the ranges of the x- and y- values and applying the different techniques described in this section.

2) Try coloring the contour plots of a combination of the functions Sin, Cos, Tan, Log, Exp and polynomials in x and y, using addition, multiplication and composition. If your chosen function F involves a polynomial with complex coefficients, prefix F with **Abs**, **Re**, **Arg** or **Im**.

3) Try coloring contour plots of Sin[f [x, y]], Cos[f [x, y]], where f [x, y] is a rational function of x and y, such as:

$$\frac{1}{x^2+y^2}, \; \frac{1}{x\,y}, \; \frac{x^2+y^2}{x\,y}, \; \frac{x+y}{x^2+y^2}, \; \frac{x}{y^2}.$$

2.7.3 Density Plots

In many cases, contour and density plots of the same function are similar if many plot-points are used and mesh and contour lines are omitted.

The following 2 examples show the distinctive character of density plots, as relatively few plot-points are used. In the first case, mesh lines are retained. (Color Fig 2.68)

$$\text{DensityPlot}\Big[\text{Mod[Floor[}-2\,\text{Im[Log[Log[Sin[Sin[x } - \text{I\,y] + Sin[}-y - \text{I\,x]]]]]]], 6],}$$

{x, −4., 4.}, {y, −4.2, 2.7}, AspectRatio → Automatic, PlotPoints → 45, Mesh → True, Frame → False, Background → RGBColor[0.52549, 0.572549, 0.203922],

$$\text{ColorFunction} \rightarrow \Big(\text{RGBColor}\Big[1 - \#^5, 1 - (1 - 4\,\# + 4\,\#^2), \text{Cos}\Big[\pi \, \frac{\#}{2}\Big]\Big] \&\Big)\Big]$$

DensityPlot[Abs[Log[Log[Cos[Log[Tan[(x + I y)⁴]]]]]],
{y, −1.1, 1.1}, {x, −1.1, 1.1}, AspectRatio → Automatic,
Mesh → False, PlotPoints → 30, Frame → False, ColorFunction →
(RGBColor[1 − (2#1 − UnitStep[−0.5` + #1])², 1 − Sin[π (2#1 − UnitStep[−0.5` + #1])],
1 − 4 (2#1 − UnitStep[−0.5` + #1]) + 4 (2#1 − UnitStep[−0.5` + #1])²] &),
Background −> RGBColor[0.91017, 0.378912, 0.378912]]

In the following plots, the mesh is omitted and 235 plot-points are used. A very smooth effect is achieved. (Color Fig 2.69)

```
DensityPlot[Mod[Floor[−12 Im[Log[Log[Sin[Cos[x − I y] + Sin[−y − I x]]]]]], 16],
    {x, −4., 4.}, {y, −4.2, 1.5}, AspectRatio → Automatic, PlotPoints → 235, Mesh → False,
    Frame → False, Background → RGBColor[0.776471, 0.639216, 0.215686],
    ColorFunction → (RGBColor[1 − Abs[Cos[π #]], 1 − Abs[Sin[2 π #]], (1 − 4 # + 4 #²)] &)]
```

Programs to try:

```
DensityPlot[−Im[Log[Log[Cos[Cos[(x² + I y)] + Sin[(y − I x)²]]]]], {x, −2, 1.9},
    {y, −1, 1.1}, AspectRatio → Automatic, PlotPoints → 200, Mesh → False,
    Frame → False, ColorFunction → (RGBColor[Sin[π #1], #1⁴, 1 − (1 − 4 #1 + 4 #1²)] &),
    Background → RGBColor[0, 0, 0]]
```

$$
\text{DensityPlot}\left[\text{Mod}[\text{IntegerPart}[10\,\text{Exp}[(\text{Sin}[7\,x^2] * \text{Sin}[7\,y^2])]], 10], \left\{x, \frac{-\pi}{2}, \frac{\pi}{2}\right\},\right.
$$
$$
\left\{y, \frac{-\pi}{2}, \frac{\pi}{2}\right\}, \text{PlotPoints} \to 250, \text{Mesh} \to \text{False}, \text{AspectRatio} \to \text{Automatic},
$$
$$
\text{Frame} \to \text{False}, \text{Axes} \to \text{False}, \text{Background} \to \text{RGBColor}[0, 0.996109, 0],
$$
$$
\left.\text{ColorFunction} \to (\text{CMYKColor}[1 − \text{Sin}[\pi\#], \#^2, 4\#^2 − 4\# + 1, 0] \&)\right]
$$

Density plots with relatively few plot points can be used to design embroidery patterns. For example: (Color Fig 2.70)

```
DensityPlot[Mod[Floor[9 Re[Log[Log[Cos[Log[Sin[(x + I y)⁴]]]]]]], 11], {y, −1.4, 1.4},
    {x, −1.4, 1.4}, PlotPoints → 100, AspectRatio → Automatic, Frame → False,
    Mesh → False, Background → RGBColor[0.47657, 0.83595, 0.531258],
    ColorFunction → (RGBColor[1 − #^2, Abs[Sin[2 π #]], 4 #^2 − 4 # + 1] &),
    PlotRegion → {{0, 1}, {0.05, 1}}, Epilog → Text["Fig 2.70", {0, −1.51}]]
```

In 2.7.2 we showed how to choose specific colors for a plot. This technique can be applied to density plots. Some plots which have bi-lateral symmetry can be adapted to produce a carpet-like image by changing the aspect ratio. Here is an example: (Color Fig 2.71)

```
DensityPlot[Mod[Floor[29 Abs[Log[Log[Sin[Log[Tan[(x + I y)⁴]]]]]]], 10],
    {y, −1.22, 1.22}, {x, −1.22, 1.22}, AspectRatio → 0.6, Mesh → False, PlotPoints → 180,
    Frame → False, Background → RGBColor[0.886732, 0.609384, 0.531258],
    ColorFunction → (Which[# ≤ 0.1, RGBColor[0.453132, 0.703136, 0.785168],
            0.1 ≤ # && # ≤ 0.3, RGBColor[0.816419, 0.460945, 0.246098],
            0.3 < # && # ≤ 0.5, RGBColor[0.664073, 0.750011, 0.503914],
            0.5 < # && # ≤ 0.75, RGBColor[0.0898451, 0.40235, 0.65626],
            True, RGBColor[0.933608, 0.839857, 0.792981]] &)]
```

The program for another example:

```
DensityPlot[Mod[Floor[20 Abs[Log[Log[Sin[Log[Sin[(x + I y)^8]]]]]]], 20],
    {y, −1, 1}, {x, −1, 1}, AspectRatio → 0.6, Mesh → False, PlotPoints → 80,
    Frame → False, Background → RGBColor[0.820325, 0, 0.472663],
    ColorFunction → (Which[# ≤ 0.3, RGBColor[0.820325, 0, 0.472663],
            0.3 < # && # ≤ 0.65, RGBColor[0.875013, 0.812512, 0.742199],
            0.65 < # && # ≤ 0.75, RGBColor[0.062501, 0.417975, 0.859388],
            True, RGBColor[0.324224, 0.324224, 0.367193]] &)]
```

In 2.7.2, we showed how to color a plot using a color sequence which varied from a first color to a second color, using the command:

```
cf[RGBColor[r1_, g1_, b1_], RGBColor[r2_, g2_, b2_]] :=
    (RGBColor[r1 + (r2 − r1) #, g1 + (g2 − g1) #, b1 + (b2 − b1) #]) &;
```

In the following example, we apply the above technique to a density plot: (Color Fig 2.72)

$$DensityPlot\left[Mod\left[Floor\left[8\ Abs\left[Log\left[Log\left[Sin\left[Sin\left[\frac{1}{4}\ (x + y + I\ (x − y))^4\right]\right]\right]\right]\right]\right], 8\right],$$

```
    {x, −1.5, 1.5}, {y, −1.5, 1.5}, AspectRatio → Automatic,
    PlotPoints → 250, Frame → False, Axes → False, Mesh → False,
    Background –> RGBColor[0.760784, 0.643137, 0.984314],
    ColorFunction → cf[RGBColor[0.403922, 0, 0.85098], RGBColor[1, 1, 1]]]
```

Other programs to try:

```
DensityPlot[−Im[Log[Sin[x^2 + I y^2] − Sin[x + I y]]], {x, −6, 5},
    {y, −2, 3}, AspectRatio → Automatic, PlotPoints → 50, Frame → False,
    Mesh → False, Background → RGBColor[0.839857, 0.812512, 0.738293],
    ColorFunction → cf[RGBColor[0.527352, 0.437507, 0.261723], RGBColor[1, 1, 1]]]
```

$$DensityPlot\left[Cos\left[\frac{1}{x\ y}\right], \{x, −1.5, 1.5\}, \{y, −1.5, 1.5\}, PlotPoints → 500,\right.$$

```
    Mesh → False, Axes → False, Frame → False, Background → RGBColor[0, 0, 1],
    ColorFunction → cf[RGBColor[1, 1, 0], RGBColor[0, 0, 1]]]
```

In the following program, we need the definition of the following function:

$$sumCosine[n\_, x\_, m\_, p\_] := n \sum_{k=1}^{m} \frac{(Cos[p^k \pi x])}{2^k};$$

```
DensityPlot[Mod[IntegerPart[5 (sumCosine[1, x, 2, 3] + sumCosine[1, y, 4, 3])], 5],
    {x, −π/16, π/16}, {y, −π/8, π/8}, PlotPoints → 100,
    Mesh → False, AspectRatio → Automatic, Frame → False,
    ColorFunction → cf[RGBColor[.85, 0, 0], RGBColor[0, 0, 1]]]
```

2.8 Coloring 3D Surface Plots

Points on 3D Surface plots can be colored according to their x- and y-co-ordinates. A color directive can be included in the specification for the surface. (Color Fig 2.73)

$$\textbf{Plot3D}[\{1 - \textbf{Sin}[x^2 + y^2]/(x^2 + y^2),\ \textbf{Hue}[(x^2 + y^2)/8]\},\ \{x, -2\pi, 2\pi\},\ \{y, -2\pi, 2\pi\},$$
$$\textbf{PlotPoints} \rightarrow 40,\ \textbf{Boxed} \rightarrow \textbf{False},\ \textbf{Axes} \rightarrow \textbf{False},\ \textbf{ViewPoint} -> \{-0.009,\ -2.561,\ 2.212\}]$$

We can also use the command **ColorFunction** to color 3D plots. In this case, the points are colored according to their z-co-ordinates. Each point on the plot is assigned a height number between 0 and 1 in a similar way to that in **ContourPlot**. In the example below, the pure function $\left(\textbf{Hue}\!\left[0.3 + \frac{\#}{2}\right] \&\right)$ assigns the color $\textbf{Hue}\!\left[0.3 + \frac{h}{2}\right]$ to the point with height number h. (Color Fig 2.74)

$$\textbf{Plot3D}[0.5\,(\textbf{Abs}[\textbf{Abs}[x] - \textbf{Abs}[y]]) - 0.5\,(\textbf{Abs}[\textbf{Abs}[x] + \textbf{Abs}[y]]),\ \{x, -2, 2\},\ \{y, -2, 2\},$$
$$\textbf{PlotPoints} \rightarrow 40,\ \textbf{ColorFunction} \rightarrow (\textbf{Hue}[0.9 - 0.5\#]\ \&),\ \textbf{Boxed} \rightarrow \textbf{False},\ \textbf{Axes} \rightarrow \textbf{False}]$$

Here is the command for another example, with the mesh lines excluded:

$$\textbf{Plot3D}[-\textbf{Re}[\textbf{Log}[\textbf{Log}[\textbf{Cos}[\textbf{Sin}[x + I\,y] + \textbf{Cos}[y + I\,x]]]]],$$
$$\{x, -3, 3\},\ \{y, -3, 6\},\ \textbf{AspectRatio} \rightarrow \textbf{Automatic},\ \textbf{Boxed} \rightarrow \textbf{False},$$
$$\textbf{PlotPoints} \rightarrow 70,\ \textbf{Mesh} \rightarrow \textbf{False},\ \textbf{Axes} \rightarrow \textbf{False},\ \textbf{Axes} \rightarrow \textbf{False},$$
$$\textbf{ColorFunction} \rightarrow (\textbf{RGBColor}[\textbf{Sin}[(\#1^2)],\ 1 - \textbf{Abs}[\textbf{Cos}[3\pi\#1]],\ (1 - 4\#1 + 4\#1^2)]\ \&)]$$

$$\textbf{Multipole}[\textbf{n\_}] := \sum_{i=1}^{n} (-1)^i \Big/ \sqrt{\left(x - \textbf{Cos}\!\left[\frac{2\pi i}{n}\right]\right)^2 + \left(y - \textbf{Sin}\!\left[\frac{2\pi i}{n}\right]\right)^2}$$

Below is the program for a plot of the function, **Multipole[9]**, defined above, with various options.

$$\textbf{Plot3D}[\textbf{Evaluate}[-\textbf{Multipole}[9]],\ \{x, -2, 2\},\ \{y, -2, 2\},\ \textbf{PlotPoints} \rightarrow 40,$$
$$\textbf{BoxRatios} \rightarrow \{2, 1, 0.37\},\ \textbf{ViewPoint} -> \{0.168,\ -2.840,\ 1.833\},$$
$$\textbf{Mesh} \rightarrow \textbf{False},\ \textbf{PlotRange} \rightarrow \{-2, 4\},\ \textbf{Boxed} \rightarrow \textbf{False},\ \textbf{Axes} \rightarrow \textbf{False},$$
$$\textbf{ColorFunction} -> (\textbf{If}[\# \leq 0.4,\ \textbf{RGBColor}[0.40235,\ 0.648447,\ 0.726574],$$
$$\textbf{RGBColor}[0.60 - 0.25\#,\ 0.77 - 0.35\#,\ 0.12 + 0.1\#]]\ \&)]$$

The following example shows a view of a surface from a point directly above it. The surface is colored using the command **ColorOutput->CMYKColor**. (Color Fig 2.75)

$$\textbf{Plot3D}[-\textbf{Mod}[\textbf{Floor}[2\,\textbf{Abs}[(\textbf{Log}[\textbf{Sin}[4\,\textbf{Log}[(y + I\,x)]]])^2]],\ 25],\ \{x, -0.7, 0.7\},$$
$$\{y, -0.7, 0.7\},\ \textbf{PlotPoints} \rightarrow 250,\ \textbf{Mesh} -> \textbf{False},\ \textbf{BoxRatios} \rightarrow \{1, 1, 0.4\},$$
$$\textbf{Axes} \rightarrow \textbf{False},\ \textbf{Boxed} \rightarrow \textbf{False},\ \textbf{ViewPoint} \rightarrow \{0, 0, 8\},\ \textbf{ColorOutput} \rightarrow \textbf{CMYKColor}]$$

Programs for views directly below surfaces: (In the third example we use the color function **cf2** defined in 2.7.2)

Plot3D$\left[-\text{Mod}\left[\text{Floor}\left[40\,\text{Abs}\left[\left(\text{Log}\left[\text{Log}\left[\text{Sin}\left[\text{Log}\left[(x^2+\text{I}\,(y))^4\right]\right]\right]\right]\right)\right]\right],40\right]\right.$,
{y, −4.6, 4.6}, {x, −3.5, 3.5}, PlotPoints → 300, Mesh → False,
Boxed → False, Axes → False, ViewPoint → {0, 0, −3},
ColorFunction → (CMYKColor[(Abs[−1 + 2 #1] − Abs[−1 + 2 #1]^2),
  Abs[−1 + 2 #1], 1 − Cos[0.5 π Abs[−1 + 2 #1]], 0] &)$\Big]$

Plot3D$\left[\text{Mod}\left[\text{Floor}\left[8\,\text{Re}\left[\left(\text{Log}\left[\text{Cos}\left[\text{Log}\left[\text{Sin}[1/(x+\text{I}\,y)^4]\right]\right]\right]\right)^2\right]\right],50\right]\right.$,
{x, −1.1, 1.1}, {y, −1.1, 1.1}, Mesh → False, PlotPoints → 300,
Boxed → False, ViewPoint −> {0, 0, −3}, Axes → False,
ColorFunction −> (CMYKColor[(4 #^2 − 4 # + 1), Abs[Cos[π #]], Abs[2 # − 1], 0] &)$\Big]$

Plot3D$\left[\text{Mod}\left[\text{Floor}\left[36\,\text{Re}\left[\text{Log}\left[\text{Sin}\left[\text{Tan}\left[(x+y)^2/2+\text{I}\,(x-y)/\sqrt{2}\,\right]+\text{Tan}[y^2+\text{I}\,x]\right]\right]\right]\right],45\right]\right.$,
{y, −3, 3}, {x, −3, 3.}, PlotPoints → 200, BoxRatios → {1, 1, 0.2},
ColorFunction → cf2[RGBColor[0, 1, 0], RGBColor[0, 0, 0]],
Mesh → False, Boxed → False, Axes → False, ViewPoint → {0, 0, −3}$\Big]$

Elevation views of surfaces from a *Mathematica* view point of the form {a, 0, 0} or {0, a, 0} are sometimes interesting. When using the view point {0, a, 0}, it is sometimes a good idea to choose a surface with a small range of y values, as large parts of the surface may otherwise be obscured by regions with large z-co-ordinates. This method can be used to obtain elevation views of strips of a surface parallel to the x-axis. Here is an example:

Plot3D[Abs[Log[Sin[Log[Log[Cos[(y + I x)^6]]]]]], {x, −1.6, 1.6}, {y, −1.6, −1.2},
  PlotPoints → 250, Mesh −> False, Axes → False, Boxed → False, ViewPoint → {0, 2, 0}]

The program for another example: (In this case a good vertical view was obtained without resorting to a narrow strip.)

Plot3D$\left[\text{Mod}\left[\text{Floor}\left[4\,\text{Abs}\left[\text{Log}\left[\text{Log}\left[\text{Sin}\left[\text{Tan}\left[(x^2+\text{I}\,y)^3\right]+\text{Cos}[y^3-\text{I}\,x]\right]\right]\right]\right]\right],20\right]\right.$,
{x, −1.8, 1.9}, {y, −1.2, 1.2}, Mesh → False, PlotPoints → 200,
Boxed → False, Axes → False, ViewPoint → {2, 0, 0}$\Big]$

■ 3D Surface Plots Derived from Examples in 2.7.2

It is interesting to compare the contour plot of a function with a view from above or below of a 3D plot of the same function. The following example shows a view from below of a 3D plot of the function whose contour plot is shown in Color Fig. 2.60. The mesh has been omitted. (Color Fig 2.76)

$$\text{Plot3D}\left[\text{Mod}\left[\text{Floor}\left[9\,\text{Re}\left[(\text{Log}[\text{Cos}[\text{Log}[\text{Sin}[(x+I\,y)^4]]]])^2\right]\right],\,50\right],\right.$$
$$\{x,\,-1.4,\,1.4\},\,\{y,\,-1.4,\,1.4\},\,\text{AspectRatio}\to\text{Automatic, Mesh}\to\text{False},$$
$$\text{PlotPoints}\to 200,\,\text{Boxed}\to\text{False, Axes}\to\text{False, ViewPoint}\to\{0,\,0,\,-4\},$$
$$\left.\text{ColorFunction}\to\text{cf2[RGBColor[1, 1, 0], RGBColor[0, 0, 0]]}\right]$$

■ Plotting a Sum of Functions with Randomly Chosen Parameters

We cannot plot on one diagram a list of more than one 3D surface without using the command **Show**. However, given a family of surfaces, we can plot the sum of some members of the family, which can lead to interesting results.

Here is the equation for a family of surfaces with parameters a, b, c, d and f.

$$\text{mount}[a\_,\,b\_,\,c\_,\,d\_,\,f\_] := \frac{a}{1 + (b\,x + c)^2 + (d\,y + f)^2};$$

Here is a plot of one member of the family:

$$\text{Plot3D[mount[1, 2, 0, 3, 0], \{x, -2, 2\}, \{y, -2, 2\}, PlotRange}\to\text{All, Axes}\to\text{False]}$$

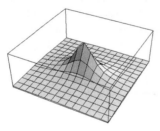

Now we plot the sum of 2 members of the family:

$$\text{Plot3D[mount[1, 2, 0, 2, 0] + mount[1.5, 1, -1, 2, -2],}$$
$$\{x,\,-1,\,3\},\,\{y,\,-2,\,2\},\,\text{PlotRange}\to\text{All, Axes}\to\text{False]}$$

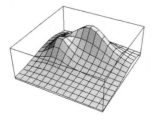

Now we plot the sum of 100 members of the family with the numbers a, b, c, d and f chosen at random (within certain limits) by *Mathematica*. We use the command **Which** in the color scheme in order to obtain a green base and white tips to the image. Again, we obtain a different picture every time we implement the command. (Color Fig 2.77)

```
Plot3D[Evaluate[Apply[Plus, Table[mount[Random[Real, {1, 5}], Random[Real, {1, 4}],
        Random[Real, {-15, 30}], Random[Real, {1, 4}], Random[Real, {-15, 15}]],
    {n, 1, 100}]]], {x, -9, 13}, {y, -9, 10}, PlotRange → All,
    Background -> RGBColor[0.812512, 0.87892, 0.906264], Boxed → False,
    PlotPoints → 100, ViewPoint -> {-0.012, -3.384, -0.026},
    Mesh → False, Axes → False, BoxRatios → {22, 12, 2},
    ColorFunction → (Which[# ≤ 0.1, RGBColor[0.527, 0.625, 0.437],
        # > 0.6, RGBColor[1, 1, 1], True, Hue[0.57, 0.8 # + 0.2, 0.8]] &)]
```

Here is a plot with *Mathematica*'s coloring: (Color Fig 2.78)

```
Plot3D[Evaluate[Apply[Plus, Table[mount[Random[Real, {1, 2}], Random[Real, {1, 4}],
        Random[Real, {-10, 10}], Random[Real, {1, 4}], Random[Real, {-10, 10}]],
    {n, 1, 100}]]], {x, -9, 13}, {y, -9, 10}, PlotRange → All, Boxed → False,
    PlotPoints → 100, ViewPoint -> {-0.012, -3.384, -0.026},
    Mesh → False, Axes → False, BoxRatios → {10, 10, 2}]
```

Here is the command for another example using different coloring, view-point and ranges of the x and y variables.

```
Plot3D[Evaluate[Apply[Plus, Table[mount[Random[Real, {1, 5}], Random[Real, {1, 4}],
        Random[Real, {-15, 30}], Random[Real, {1, 4}], Random[Real, {-15, 15}]],
    {n, 1, 100}]]], {x, -9, 9}, {y, -9, 9}, PlotRange → All,
    Background -> RGBColor[0.812512, 0.87892, 0.906264], Boxed → False,
    PlotPoints → 30, ViewPoint -> {3.366, 0.105, 0.329},
    Mesh → False, Axes → False, BoxRatios → {22, 12, 2},
    ColorFunction → (RGBColor[0.64 - 0.45 #, 0.72 - 0.45 #, 0.33 + 0.4 #] &)]
```

Exercise:

Vary the above program by changing the color option, the x- and y- ranges, the limitations of the random choices, the view-point or the number of terms in the sum.

■ Contour Plots Derived from the Above Section

Interesting images can be obtained by making contour plots of sums of randomly chosen functions. In the following examples we use the technique of repeating colors, discussed in 2.7.2 and apply it to the family **bivnorm** defined below:

```
bivnorm[c_, w_, a_, b_] := c Exp[-((x - a)² + (y - b)²) / w];
```

In the following program, contour lines are included:

```
ContourPlot[
    Evaluate[Apply[Plus, Table[−bivnorm[Random[Real, {1, 5}], Random[Real, {1, 4}],
            Random[Real, {−10, 10}], Random[Real, {−10, 10}]], {n, 1, 30}]]], {x, −12, 12},
    {y, −12, 12}, PlotRange → All, ContourLines → True, Contours → 20, Frame → False,
    ColorFunction → (CMYKColor[1 − Cos[(1 / 3) π (2 #1 − UnitStep[−0.5` + #1])],
            4 (2 #1 − UnitStep[−0.5` + #1])² − 4 (2 #1 − UnitStep[−0.5` + #1]) + 1,
            1 − Sin[π (2 #1 − UnitStep[−0.5` + #1])], 0] &),
    PlotPoints →
        200]
```

Some color sequences result in a 3D effect if the contour lines are omitted. Here is an example: (Color Fig 2.79)

```
ContourPlot[
    Evaluate[Apply[Plus, Table[−bivnorm[Random[Real, {1, 5}], Random[Real, {1, 4}],
            Random[Real, {−10, 10}], Random[Real, {−10, 10}]], {n, 1, 40}]]], {x, −12, 12},
    {y, −12, 12}, PlotRange → All, ContourLines → False, Contours → 30, Frame → False,
    ColorFunction → (CMYKColor[1 − Cos[(1 / 3) π (2 #1 − UnitStep[−0.5` + #1])],
            4 (2 #1 − UnitStep[−0.5` + #1]) − 4 (2 #1 − UnitStep[−0.5` + #1])²,
            Sin[0.25 π (2 #1 − UnitStep[−0.5` + #1])], 0] &),
    PlotPoints → 100, Background –> RGBColor[0.515633, 0.91017, 0.445319]]
```

Exercise:

1) Try the above techniques for generating 3D and contour plots with the families **bivnorm** and **mount** defined above and to the family **hill** defined by:

hill[a_, b_, c_, d_, f_, g_, h_] := a (Exp[−(b x + c y + d)²] + Exp[−(f x + g y + h)²]).

2) Apply the above methods to other families of surfaces.

One way to generate a family of surfaces is to start with a surface with equation $z = f [x, y]$.

Then $F[a, b, c, d, e] = af [bx+c, dy+e]$ and $G[a, b, c, d, e, g, h,] = af [bx + cy + d, ex + gy + h]$ form families of affine transformations of the original surface.

Some suggestions for equations of surfaces are: $z = Abs[Gamma[x + I y]]$;

$$z = Im[EllipticF[x + I y, 2]];$$

$$z = Cos[x] Cos[y] E^{\frac{-\sqrt{x^2+y^2}}{2}} .$$

Chapter 3

Patterns Constructed from Straight Lines

Introduction

In this Chapter, we discuss 2 methods of forming patterns by joining points on 2D parametric curves. In the first method, the line segments joining the points form a polygonal line. In the second method, the line segments joining the points generally do not have common end-points. We show how to color the plots with a single color and with multiple colors.

3.1 First Method of Construction

The idea underlying this construction is due to Maurer (1987).

Let $x = f[t]$, $y = g[t]$ be parametric equations for a curve C. We shall be mostly concerned with the case where f and g are trigonometric functions. Let d be a fixed angle. Attractive patterns can sometimes be constructed by joining points on C with parameters d, $2 d$, $3 d$, and so on. To construct such a pattern, we need to construct lines joining the points with co-ordinates $\{f[r d], g[r d]\}$ and $\{f[(r + 1) d], g[(r + 1) d]\}$ for r in a suitable range. We will need to make a table of such points.

We start with the family of curves having parametric equations $x = \mathrm{Sin}[nt]\, \mathrm{Sin}[mt]$, $y = \mathrm{Cos}[nt]\, \mathrm{Sin}[mt]$. The following represents the line joining the rth to the (r+1)th point on the curve with parameters m and n. As we will be working in degrees, not radians, we must specify this.

> **Line[{{Sin[n r d°] Sin[m r d°], Cos[n r d°] Sin[m r d°]},**
> **{Sin[n (r + 1) d°] Sin[m (r + 1) d°], Cos[n (r + 1) d°] Sin[m (r + 1) d°]}}]**

We now make a table of such lines:

Table[Line[{{Sin[n r d°] Sin[m r d°], Cos[n r d°] Sin[m r d°]},
 {Sin[n (r + 1) d°] Sin[m (r + 1) d°], Cos[n (r + 1) d°] Sin[m (r + 1) d°]}}], {r, 0, k}]

We use the commands **Show** and **Graphics** to ask *Mathematica* to display the table of lines.
We also use the option **AspectRatio→Automatic**.

Show[Graphics[Table[Line[{{Sin[n r d°] Sin[m r d°], Cos[n r d°] Sin[m r d°]},
 {Sin[n (r + 1) d°] Sin[m (r + 1) d°], Cos[n (r + 1) d°] Sin[m (r + 1) d°]}}],
 {r, 0, k}]], AspectRatio → Automatic]

The above is a procedure for constructing graphics involving the parameters n, m, d°, and k,
so, as discussed in 1.7.5, we define a function des with arguments n, m, d°, and k which will
instruct *Mathematica* to implement the procedure:

des[n_, m_, d°_, k_] :=
 Show[Graphics[Table[Line[{{Sin[n r d°] Sin[m r d°], Cos[n r d°] Sin[m r d°]},
 {Sin[n (r + 1) d°] Sin[m (r + 1) d°], Cos[n (r + 1) d°] Sin[m (r + 1) d°]}}],
 {r, 0, k}]], AspectRatio → Automatic];

Here is a plot of the curve with parametric equations: x = Sin[3 t] Sin[4 t], y = Cos[3 t] Sin[4 t] :

ParametricPlot[{Sin[3 t] Sin[4 t], Cos[3 t] Sin[4 t]},
 {t, 0, 2 π}, AspectRatio –> Automatic, Axes –> False]

In the following example, the points with parameters n(37) °, 0 ≤ n ≤ 360 on the above curve
have been plotted and joined in order.

des[3, 4, 37 °, 360]

The above plots can be uniformly colored by introducing one extra variable:

design1[n_, m_, d°_, k_, x_] :=
   Show[Graphics[Table[{x, Line[{{Sin[n r d°] Sin[m r d°], Cos[n r d°] Sin[m r d°]},
     {Sin[n (r + 1) d°] Sin[m (r + 1) d°], Cos[n (r + 1) d°] Sin[m (r + 1) d°]}}]},
   {r, 0, k}]], AspectRatio → Automatic];

Now the parameter x can be replaced by a color command: (Color Fig 3.1)

design1[2, 5, 119°, 360, RGBColor[0.867201, 0.0117189, 0.503914]]

Exercise:

Experiment with **design1** for various values of the parameters, for example: m = 3, 5, n = 16, 1, d° = 37°, 163° and k = 360.

Beautiful patterns can sometimes be obtained by giving either one or both the parameters m and n fractional values. The parameter k must be increased accordingly in order to obtain a complete pattern. The picture is often very detailed and the size may need to be enlarged in order to get the best effect.

Here is an example from a different family of curves, **design2**, specified as follows:

design2[n_, d°_, k_, x_] :=
   Show[Graphics[Table[{x, Line[{{Sin[n r d°] Cos[r d°], Sin[n r d°] Sin[r d°]},
     {Sin[n (r + 1) d°] Cos[(r + 1) d°], Sin[n (r + 1) d°] Sin[(r + 1) d°]}}]},
   {r, 0, k}]], AspectRatio → Automatic]

If the argument d° in the function **design1** or **design2** is chosen so that d is not prime to 360, then a simpler pattern is usually obtained.

$$design2\left[\frac{3}{2}, 222°, 180, RGBColor[0, 0, 0]\right]$$

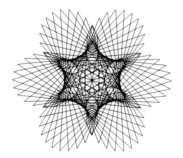

Exercise:

1) Experiment with **design2**. Some suggested values for the parameters are given below:

a) m = $\frac{4}{3}$, d° = 121° and k = 1080;

b) m = $\frac{3}{2}$, d° = 222° and k = 720;

c) m = $\frac{2}{3}$, d° = 129° and k = 1080;

d) m = $\frac{2}{3}$, d° = 89° and k = 1080.

2) Experiment with **design1**. Some suggested values for the parameters are given below:

a) m = $\frac{3}{2}$, n = 2, d° = 79° and k = 720;

b) m = $\frac{3}{2}$, n = 8, d° = 49° and k = 720;

c) m = $\frac{3}{2}$, n = 1, d° = 111° and k = 720;

d) m = $\frac{5}{2}$, n = 3, d° = 111° and k = 720;

e) m = $\frac{5}{2}$, n = 3, d° = 21° and k = 720;

f) m = $\frac{5}{2}$, n = 3, d° = 66° and k = 180.

Some patterns are created by the spaces between the lines, and sometimes the pattern changes with very small adjustments to the size:

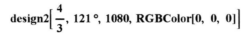

Exercise:

Experiment with **design1**. Some suggested values for the parameters are given below:

1) m = $\frac{1}{2}$, n = 1, d° = 169° and k = 720.

2) m = $\frac{3}{2}$, n = 2, d° = 79° and k = 720.

3) m = $\frac{3}{2}$, n = 1, d° = 149° and k = 720.

A background color can be introduced to **design1** by introducing another parameter as follows: (Color Fig 3.2)

```
design1B[n_, m_, d°_, k_, x_, y_] :=
    Show[Graphics[Table[{x, Line[{{Sin[n r d°] Sin[m r d°], Cos[n r d°] Sin[m r d°]},
            {Sin[n (r + 1) d°] Sin[m (r + 1) d°], Cos[n ( r + 1) d°] Sin[m (r + 1) d°]}}]},
        {r, 0, k}]], AspectRatio → Automatic, Background → y];
```

$$\text{design1B}\left[\frac{5}{2},\ 2,\ 26°,\ 1080,\right.$$

$$\left.\text{RGBColor}[0.337255,\ 0.00392157,\ 0.745098],\ \text{RGBColor}[1,\ 0.92549,\ 0.980392]\right]$$

The following routine results in patterns which are not 'flower-like':

```
design3[a_, b_, d°_, k_, x_] :=
    Show[Graphics[Table[{x, Line[{{ Sin[a r d°] Cos[r d°], Sin[b r d°] Sin[r d°]},
            { Sin[a (r + 1) d°] Cos[(r + 1) d°], Sin[b (r + 1) d°] Sin[(r + 1) d°]}}}]},
        {r, 0, k}]], AspectRatio → Automatic, PlotRange → All];

design3[5, 2, 29°, 360, RGBColor[0, 0, 0]]
```

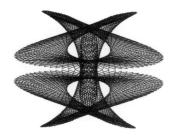

Exercise:

1) Experiment with **design3**. Some suggested values for the parameters are given below:

a) a = 3, b = 2, d° = 37° and k = 360;

b) a = 4, b = 8, d° = 17° and k = 360;

c) a = 5, b = 6, d° = 23° and k = 360;

d) a = 2, b = 8, d° = 17° and k = 360.

2) Experiment with:

a) The hypocycloid of 4 cusps with parametric equations $x = (Cos[\theta])^3$, $y = (Sin[\theta])^3$;

b) Other families of curves defined in 1.8.6.

3) Define a function which will enable you to construct regular pentagons and regular 'star-shaped' polygons by applying the above technique to the circle with parametric equations $x = Cos[t]$, $y = Sin[t]$ and choosing d to be of the form $\frac{r}{s}\pi$ where r, s are natural numbers.

3.2 Second Method of Construction

Here is another technique for producing patterns from parametric plots. Start with the circle with parametric equations: $x = Cos[t]$, $y = Sin[t]$. Choose a small angle $\theta° = 1°$, $2°$ or $3°$ and choose a natural number, n, greater than 1. Now construct the line segments joining the points on the circle with co-ordinates $\{Cos[r\theta°], Sin[r\theta°]\}$ and $\{Cos[n r \theta°], Sin[n r \theta°]\}$ for $0 \le r \le k$ (where k depends on n).

For example, choosing $\theta = 3$ and n = 2:

Show[
  Graphics[Table[Line[{{Cos[r 3 °], Sin[r 3 °]}, {Cos[2 r 3 °], Sin[2 r 3 °]}}]], {r, 0, 180}]],
  AspectRatio → Automatic]

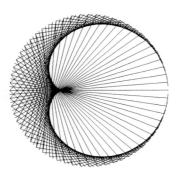

A Cardioid appears - although the above construction consists of straight lines alone. It can easily be shown that each of the straight lines is tangent to a cardioid. The cardioid is called the envelope of the family of straight lines.

Show[Graphics[Table[Line[{{Cos[r 3 °], Sin[r 3 °]}, {Cos[2 r 3 °], Sin[2 r 3 °]}}]], {r, 0, 30}]],
  AspectRatio → Automatic]

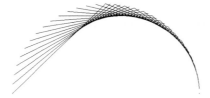

We now define a function which depends on the variables $\theta°$, n , k and color x.

pattern1[$\theta°$_, n_, k_, x_] :=
  Show[Graphics[Table[{x, Line[{{ Cos[r $\theta°$], Sin[r $\theta°$]}, {Cos[n r $\theta°$], Sin[n r $\theta°$]}}]},
      {r, 0, k}]], AspectRatio → Automatic];

To obtain a complete pattern, k should be equal to $\frac{360}{\theta}$. Once the program has been implemented, it may be necessary to scale the picture in order to obtain a better effect. Generally the size of the picture needs to be reduced for θ values of 2, 3 and 4 and increased for θ value of 0.5. (Color Fig 3.3)

pattern1[1 °, 37, 360, RGBColor[0, 0, 0.9]]

Exercise:
Experiment with **pattern1**. Some suggested values for the parameters are given below:
1) $\theta° = 1°$, n = 137 and k = 360.
2) $\theta° = 1°$, n = 141 and k = 360.

3) $\theta^o = 1°$, n = 76 and k = 360.
4) $\theta^o = 1°$, n = 53 and k = 360.
5) $\theta^o = 2°$, n = 19 and k = 720.
6) $\theta^o = 0.5°$, n = 139 and k = 720.
7) $\theta^o = 0.5°$, n = 113 and k = 720.

The following variation on **pattern1** has a thickness option for the lines of the pattern which allows for bolder, less delicate patterns.

pattern1T[θ^o_, n_, k_, x_, y_] := Show[Graphics[
Table[{x, Thickness[y], Line[{{ Cos[r θ^o], Sin[r θ^o]}, {Cos[n r θ^o], Sin[n r θ^o]}}]}},
{r, 0, k}]], AspectRatio → Automatic];

pattern1T[1 °, 19, 360, RGBColor[0, 0, 0], 0.01]

Sometimes an interesting effect can be obtained by generating an incomplete pattern:

pattern1[1 °, 33, 180, RGBColor[0, 0, 0]]

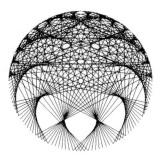

Exercise:
1) Experiment with **pattern1B**.
2) Adapt the program for curves other than the circle. The technique is best applied to curves

with a simple shape. Examples are:

  a) the Limacon with polar equation r = 2Cos[θ] - 1;

  b) the hypocycloid with parametric equations: x = a(Cos[t])^3, y = b (Sin[t])^3;

  c) the cardioid with parametric equations x = (1 − Cos[θ]) Cos[θ], y = (1 − Cos[θ]) Sin[θ].

3.3 Assigning Multiple Colors to the Designs

All of the above patterns may be given multiple colors by assigning colors to the lines separately as follows.

Consider the following program, which was used in 3.1:

```
design1[n_, m_, d°_, k_, x_] :=
    Show[Graphics[Table[{x, Line[{{Sin[n r d°] Sin[m r d°], Cos[n r d°] Sin[m r d°]},
                {Sin[n (r + 1) d°] Sin[m (r + 1) d°], Cos[n ( r + 1) d°] Sin[m (r + 1) d°]}}]},
            {r, 0, k}]], AspectRatio → Automatic];
```

In the program itself, replace x by **Hue[a + b r]** where a, b are parameters. In the name of the program, replace x by the two parameters a and b. Now a and b must be chosen so that 0 ≤ a + b r ≤ 1 if 0 ≤ r ≤ k. In this way, the color of the rth line depends on r. It is advisable, if the acute angle between successive lines is relatively large, to limit the range of colors considerably. So the program would look like this:

```
design1MC[n_, m_, d°_, k_, a_, b_] := Show[
    Graphics[Table[{Hue[a + b r], Line[{{Sin[n r d°] Sin[m r d°], Cos[n r d°] Sin[m r d°]},
                {Sin[n (r + 1) d°] Sin[m (r + 1) d°], Cos[n ( r + 1) d°] Sin[m (r + 1) d°]}}]},
            {r, 0, k}]], AspectRatio → Automatic];
```

Exercise:

Experiment with **design1MC**. Some suggested values for the parameters n, m, d°, k, a, and b are given in the following table:

1) 0.5, 5, 83 °, 720, 1, −1 / 3960.

2) 0.5, 4, 211 °, 720, 0.5, 1 / 3960.

3) 2.5, 4, 37 °, 720, 1, −1 / 3960.

4) 1.5, 2, 67 °, 720, 0.7, 1 / 3960.

5) 3, 16, 37 °, 360, 0.5, 1 / 720.

6) 2, 5, 29 °, 360, 0.9, −1 / 1440.

Here is an example of a daisy pattern: (Color Fig 3.4)

daisy3MC[n_, d°_, k_, a_, b_] :=
  Show[Graphics[Table[{Hue[a + b r], Line[{{Sin[n r d°] Cos[r d°], Sin[n r d°] Sin[r d°]},
      {Sin[n (r + 1) d°] Cos[(r + 1) d°], Sin[n (r + 1) d°] Sin[(r + 1) d°]}}]},
    {r, 0, k}]], AspectRatio → Automatic]

$$\text{daisy3MC}\left[\frac{4}{3}, 137°, 1080, .7, \frac{1}{3540}\right]$$

Here is a similar adaption of **pattern1**: (Color Fig 3.5)

pattern1MC[θ°_, n_, k_, a_, b_] := Show[Graphics[
  Table[{Hue[a + b r], Line[{{ Cos[r θ°], Sin[r θ°]}, {Cos[n r θ°], Sin[n r θ°]}}]},
    {r, 0, k}]], AspectRatio → Automatic];

$$\text{pattern1MC}\left[1°, \frac{3}{4}, 1440, 0, \frac{1}{480}\right]$$

Exercise:
Experiment with **pattern1MC**. Some suggested values for the parameters $θ°$, n, k, a, and b are given in the following table:

1) 1°, 1.5, 1080, 0, 1 − 1/720.

2) 1°, 0.25, 480, 0, 1/480.

3) 1°, 0.5, 720, 0, 1/360.

4) 0.5°, 2, 720, 0.2, 1/1440.

5) 1°, 1.5, 1080, 0, 1 − 1/720.

6) 0.5°, 180, 720, 0, 1/360.

7) 0.5, 180, 720, 0, 1/1080.

Here is another example: (Color Fig 3.6)

pattern3MC[θ°_, n_, k_, a_, b_] := Show[Graphics[Table[
  {Hue[a + b r], Line[{{ (2 Cos[r θ°] − 1) (Cos[r θ°]), (2 Cos[r θ°] − 1) (Sin[r θ°])},
      {(2 Cos[n r θ°] − 1) (Cos[n r θ°]), (2 Cos[n r θ°] − 1) Sin[n r θ°]}}]},
    {r, 0, k}]], AspectRatio → Automatic];

$$\text{pattern3MC}\left[0.5°, \frac{4}{3}, 720, 0.6, \frac{1}{360}\right]$$

Exercise:
Experiment with **pattern3MC**. Some suggested values for the parameters $θ°$, n, k, a, and b are given in the following table:

1) 0.5°, 0.5, 1440, 1/360.

2) 1°, 1.5, 720, 0, 1/360.

3) 0.5°, 1.5, 1440, 0, 1/180.

Chapter 4

Orbits of Points Under a $\mathbb{C} \to \mathbb{C}$ Mapping

Introduction

In this Chapter, we define and construct orbits of points under the action of a complex function. Let $f : D \to D$ be a complex function, with D a subset of $\mathbb{C}$. The iterates of f are the functions f, $f \circ f$, $f \circ f \circ f$,..., which are denoted f^1, f^2, f^3,... . If $z \in \mathbb{C}$, then the orbit of z under f is the sequence $(z, f[z], f[f[z]], ...)$. We will discuss and illustrate some of the different ways in which such a sequence can behave. We start with definitions and examples of convergent sequences and sequences which tend to infinity in $\mathbb{C}$. Some of the tests for the behaviour of such sequences involve the notion of the derivative of a complex function, so we provide a definition of this notion, together with a brief discussion. We will also state and illustrate the contraction mapping theorem for $\mathbb{C}$. To do this, we need first to explain notions such as boundaries of subsets of $\mathbb{C}$, closed subsets, closure of a subset and so forth. This chapter provides a preparation for Chapter 7 in which Julia and Mandelbrot sets are constructed.

4.1 Limits, Continuity, Differentiability

4.1.1 Limits of Sequences in $\mathbb{C}$

The sequence (z_n) of elements of $\mathbb{C}$ is said to converge to the element α of $\mathbb{C}$ if the real sequence $(d[z_n, \alpha]) = (|z_n - \alpha|)$ converges to 0, and is said to tend to infinity if the real sequence $(|z_n|)$ tends to infinity as n tends to infinity. For example:

$$\mathbf{Limit}\left[\frac{1}{n} + \frac{n\,I}{n+1}, n \to \infty\right] \qquad i$$

4.1.2 Limits, Continuity, Differentiability of Complex Functions

A function f is said to be a complex function if its domain, D, and range R are subsets of $\mathbb{C}$. Let f be a complex function.

We say that $f[z]$ tends to the limit m as z tends to α, and we write $\lim_{z \to \alpha} f[z] = m$ or $f[z] \to m$ as $z \to \alpha$ if, for every $\epsilon > 0$, there exists $\delta > 0$ such that $0 < |z - \alpha| < \delta \Rightarrow |f[z] - m| < \epsilon$. Note that in this case $f[z]$ must be defined for all z in an open disk centre α, except possibly at α. We say that $f[z]$ tends to infinity as z tends to infinity if, for every $K > 0$, there exists $M > 0$ such that $|z| > M \Rightarrow |f[z]| > K$. For example, if f is a non-constant polynomial, then $f[z]$ tends to infinity as n tends to infinity. There are similar rules for limits of complex functions as there are for real functions.

We say that f is continuous at α if $f[z] \to f[\alpha]$ as $z \to \alpha$ and that f is analytic or differentiable at α with derivative $f'[\alpha]$ if $\frac{f[z] - f[\alpha]}{z - \alpha}$ tends to a limit called $f'[\alpha]$ as z tends to α. All these definitions are exact analogues of the corresponding definitions for real functions. The rules for differentiation exactly model those for real functions.

4.2 Constructing and Plotting the Orbit of a Point

4.2.1 Iterating a Function

Iteration of functions can be carried out with the command **Nest.**

> **? Nest**
>
> Nest[f, expr, n] gives an expression with f applied n times to expr.

Example:

$$g[z\_] := z^2 - 1;$$

Nest[g, z, 3]
$$-1 + \left(-1 + \left(-1 + z^2\right)^2\right)^2$$

Nest[g, 0.2 − 0.3 I, 3] $-1.05574 + 0.0444024\,i$

Exercise:
Find the 10th iterate of $\frac{\pi}{4}$ under the action of the Cosine function.

4.2.2 Calculating the Orbit of a Point

We can use the command **NestList** to calculate the orbit of a point under the action of a particular complex function.

> **? NestList**
>
> NestList[f, expr, n] gives a list of the results of applying f to expr 0 through n times.

NestList[Sin, 1.0 + I, 5]

{1. + i, 1.29846 + 0.634964 i, 1.16392 + 0.182506 i,
    0.933697 + 0.0726277 i, 0.805946 + 0.0432418 i, 0.72216 + 0.0299512 i}

We can replace the 'f' in the **NestList** command by a pure function. The following command exhibits the first 6 points in the orbit of the complex number 1 + I under the action of the function defined by: $z \to z^2 - 1$.

NestList[(#² − 1) &, 1.0 + I, 5]

{1. + i, −1. + 2. i, −4. − 4. i, −1. + 32. i, −1024. − 64. i, 1.04448 × 10^6 + 131072. i}

4.2.3 Plotting the Orbit of a Point

In Chapter 1, we showed how to plot a sequence of complex numbers using the pure function **{Re[#], Im[#]}&** .

We can apply the above function to a list of complex numbers using the command **Map** or **/@**.

> **{Re[#], Im[#]} & /@ NestList[(#² − 0.39 − 0.59 I) &, −0.3, 10]**

> {{−0.3, 0}, {−0.3, −0.59}, {−0.6481, −0.236},
> {−0.0256624, −0.284097}, {−0.470052, −0.575419},
> {−0.500157, −0.049046}, {−0.142248, −0.540939}, {−0.66238, −0.436105},
> {−0.14144, −0.0122653}, {−0.370145, −0.58653}, {−0.597011, −0.155797}}

In order to plot this orbit we apply the command **ListPlot** to the above list of points together with suitable options. The option **PlotJoined→True** is included so that the order of the points in the orbit can be seen.

> **ListPlot[{Re[#], Im[#]} & /@ NestList[(#² − 0.39 − 0.59 I) &, −0.3, 10],**
> **PlotJoined → True, AspectRatio → Automatic, PlotRange → All]**

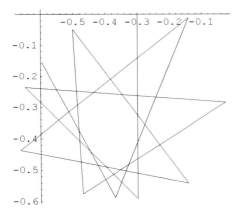

The above is an example of a bounded orbit.

The following program can be used to plot the orbit of length n of the complex number s under the action of the function f. We have included a variable t with a default value, as we wish to change the tick marks in some cases.

> **complexOrbit[f_, s_, n_, t_: Automatic] := ListPlot[{Re[#], Im[#]} & /@ NestList[f, s, n],**
> **PlotJoined → True, AspectRatio → Automatic, PlotRange → All, Ticks → t];**

Plotting the orbit of length 6 of the point -0.2+0.5I under the action of the function $f[z] = z^2 - 1 + I$ (or the pure function $(\#^2 - 1 + I)\, \&$) we get:

complexOrbit[($\#^2 - 1 + I$) &, $-0.2 + 0.5 I$, 6]

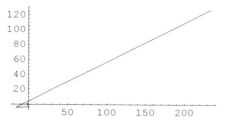

The above is an example of an unbounded orbit.

Here is an example of the orbit of a point under a rational function.

complexOrbit$\left[\left(\dfrac{\# - 4}{2 \#^2 + 2 I}\right)$ &, $-0.06 + 0.9 I$, 15$\right]$

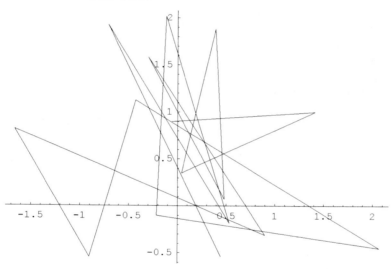

4.3 Types of Orbits

4.3.1 Bounded and Unbounded Orbits

The orbit of the point z under the action of the function f is said to be bounded if there exists $M \in \mathbb{R}$ such that $|f^n[z]| \leq M$ for all $n \in \mathbb{N}$. If the orbit is not bounded, it is said to be unbounded.

When plotting the orbit of a point, it is a good idea to start by plotting a few points, in order to obtain an idea if the orbit is bounded or unbounded. An example of a bounded orbit and of an unbounded orbit are shown in 4.2.3 above.

4.3.2 Fixed Points and Periodic Orbits

Let $f : X \to X$, $X \subseteq \mathbb{C}$.

The point z is said to be a fixed point of f if $f[z] = z$. We also say that ∞ is a fixed point of f if $|f[z]| \to \infty$ as $|z| \to \infty$.

For example, if $f[z] = z^2$ then 1 and ∞ are fixed points of f, as $f[1] = 1$ and $|f[z]| \to \infty$ as $|z| \to \infty$.

The point z is said to be a periodic point of period n of f if $f^n[z] = z$.

In this case the orbit of z under f is $(z, f^1[z], f^2[z], ..., f^{n-1}[z], z, f^1[z], ...)$.

If n is the smallest integer such that $f^n[z] = z$, then z is said to have prime period n. If z has period k, then it is a fixed point of f^k, and $f[z]$ is a fixed point of f^{k+1} etc. The orbit of a point with period n is called a cycle of period n.

For example, if $f[z] = z^2 - 1$, the points -1 and 0 both have period 2 as shown below:

NestList[(#^{2} − 1) &, −1, 4] $\{-1, 0, -1, 0, -1\}$

The diagram below shows a plot of a period 3 orbit:

complexOrbit[(#^{2} − I) &, −1.2904912332417333 − 0.7792817182359892 I, 15]

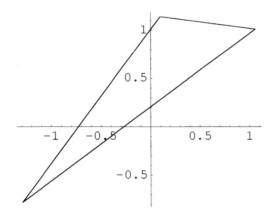

4.3.3 Convergent Orbits

Consider the following example:

h[z_] := z^2 + 0.33 + 0.35 I;

complexOrbit[h, −0.35 − 0.25 I, 100, {{−0.3, −0.1, 0.1, 0.3}, Automatic}]

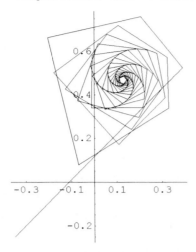

From the plot of the orbit, it seems as though the orbit is convergent.

Suppose that the function f is continuous and the orbit of the point z converges to α so $f^n[z] \to \alpha$ as $n \to \infty$.

Since f is continuous, $f[f^n[z]] = f^{n+1}[z] \to f[\alpha]$ as $n \to \infty$. But $f^{n+1}[z] \to \alpha$ as $n \to \infty$, so $f[\alpha] = \alpha$. This means that if an orbit of a point under f is convergent, it must converge to a fixed point of f.

We verify this fact, in the case of the example above. We first find fixed points of h using **NSolve**:

NSolve[h[z] == z, z]

$\{\{z \to 0.126485 + 0.468522\,i\}, \{z \to 0.873515 − 0.468522\,i\}\}$

We now calculate part of orbit of the point -0.35-0.25I. To save space we display the last 10 points in the orbit from 0 to 350.

Drop[NestList[h, −0.35 − 0.25 I, 350], 340]

$\{0.126474 + 0.468518\,i, 0.126487 + 0.46851\,i, 0.126497 + 0.468521\,i, 0.12649 + 0.468533\,i,$
$\quad 0.126477 + 0.468529\,i, 0.126477 + 0.468516\,i, 0.126489 + 0.468513\,i, 0.126495 + 0.468524\,i,$
$\quad 0.126487 + 0.468532\,i, 0.126476 + 0.468526\,i, 0.126479 + 0.468515\,i\}$

It seems that the orbit of -0.35-0.25I converges to the fixed point $\alpha = 0.126485+0.468522$ I

We now plot the orbit under h of a different point. We also show the fixed point in gray:

ListPlot[{Re[#], Im[#]} & /@ NestList[(#² + 0.33 + .35 I) &, 0.25 + 0.25 I, 100],
   PlotJoined → True, AspectRatio → Automatic,
   PlotRange → All, Epilog → {PointSize[0.02], Hue[0, 0, 0.5],
      Point[{0.12648530502056088, 0.4685223898047904}]}]

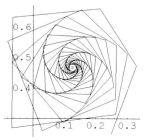

We see that the orbit of a different point converges to the same fixed point. This can be explained by the Contraction Mapping Theorem for $\mathbb{C}$.

4.4 The Contraction Mapping Theorem for $\mathbb{C}$

4.4.1 Contraction Mappings on $\mathbb{C}$

Let f be a complex function and A a subset of its domain. Then f is said to be a contraction mapping on A if and only if f maps A into A and there exists c with $0 \le c < 1$ such that $|f[z] - f[w]| \le c|z - w|$ for all z, w ∈ A. If such a c exists, it is called a contractivity factor for f on A.

Note that a function f may be a contraction mapping on one subset, A, of its domain, but not a contraction mapping on a different subset, B, of its domain.

Example: Let $f[z] = z^2$, z ∈ $\mathbb{C}$, and let A be the disk centre the origin radius r, $0 < r < \frac{1}{2}$.

If z, w ∈ A, then $|f[z]| < |z|$, and so f[z] ∈ A, i.e. f : A → A.

Also $|f[z] - f[w]| = |z + w| . |z - w| \le 2r|z - w|$, so f is a contraction mapping on A. However $|f[1.1] - f[1]| = 0.21 > |1.1 - 1|$, so f is not a contraction mapping on any set containing the points 1 and 1.1.

4.4.2 Boundary of a Subset of $\mathbb{C}$

Let A be a subset of $\mathbb{C}$. The point x is a boundary point of A if and only if every disk center x contains an element of A and an element of the complement of A. The set of all boundary points of A is called the boundary of A. For example, the boundary of the set $C = \{z \in \mathbb{C} \mid |z| < 1\}$ is the set $B = \{z \in \mathbb{C} \mid |z| = 1\}$, or the boundary of the interior of a circle is its circumference.

4.4.3 Closed Subsets of $\mathbb{C}$, Closure

A subset of $\mathbb{C}$ is closed if and only if it contains its boundary. For example the disk $D = \{z \in \mathbb{C} \mid |z| \leq 1\}$ is closed. The closure of A, denoted $\overline{A}$, is the union of A and its boundary. For example, the closure of the set $C = \{z \in \mathbb{C} \mid |z| < 1\}$ is $C \cup B$ where $B = \{z \in \mathbb{C} \mid |z| = 1\}$.

4.4.4 Compact Subsets of $\mathbb{C}$

A subset A of $\mathbb{C}$ is compact if and only if it is closed and bounded, where the set A is bounded if and only if it is a subset of a disk with radius r. If we regard the computer screen or the sheet of paper on which we print as part of the x-y plane (identified with $\mathbb{R}^2$ or $\mathbb{C}$) then all the diagrams we construct are compact, as they consist of finitely many 'points' or pixels. Our 'points' are not points in the mathematical sense as they are small disks or rectangles.

4.4.5 The Contraction Mapping Theorem for $\mathbb{C}$

Let A be a closed, bounded subset of $\mathbb{C}$ and let $f : A \to A$ be a contraction mapping. Then the contraction mapping theorem states that f has exactly one fixed point, a, in A. Further if z is any element of A then the sequence $(f^n[z])$ converges to a.

The following result provides a sufficient condition for a function to be a contraction mapping on a subset of its domain.
If f has a fixed point at a, and is analytic in a neighbourhood of a, and if $|f'[a]| < 1$, then there exists a disk, D, centre a such that f is a contraction mapping on D.

For example recall the function h considered in 4.3.3: $h[z] = z^2 + 0.33 + 0.35\,I$. We found that h has a fixed point $a = 0.126485 + 0.468522I$. We check the absolute value of the derivative of h at this point:

Abs[2 (0.126485 + 0.468522 *i*)]   0.97059

So there is a neighbourhood of the above fixed point, a, such that the orbit of every point in the neighbourhood converges to a.

4.5 Attracting and Repelling Cycles

4.5.1 Attracting and Repelling Fixed Points

Suppose that the complex function f is analytic in a region A of $\mathbb{C}$, and f has a fixed point at $a \in A$. Then a is said to be:

an attracting fixed point if $|f'[a]| < 1$;

a repelling fixed point if $|f'[a]| > 1$;

a neutral fixed point if $|f'[a]| = 1$.

As stated in 4.4.5, it can be proved that if a is an attracting fixed point of f, then there exists a neighbourhood D of a such that if $b \in D$ the orbit of b converges to a. In 4.3.3, we illustrated an orbit of the function h converging to an attracting fixed point.

If a is a repelling periodic point of f, then there is a neighbourhood N of a such that if $b \in N$ there are points in the orbit of b which are not in N. (The orbit of b leaves N, although it may return to N.)

In the case of polynomials of degree greater than 0 and some rational functions, ∞ is also called an attracting fixed point, as, for each such function, f, there exists $R > 0$ such that if $|z| > R$ then $f^n[z] \to \infty$ as $n \to \infty$. (For more details, see 7.1.2)

Here is an example of the orbit of a point under a trigonometric function converging to a fixed point of that function.

complexOrbit[(2 + 0.5 I) Sin[#] &, 2, 65]

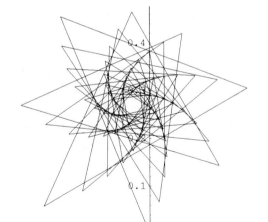

We check, by using the command **FindRoot** to find a fixed point of the above function.

FindRoot[$\{z \to 1.96563 + 0.268928\,i\}$
  (2 + 0.5 I) Sin[z] == z,
  {z, 1.96 + 0.273 I}]

We now show the orbit of a point being repelled from a neighbourhood of a repelling fixed point of the function h defined in 4.3.3.

  ListPlot[{Re[#], Im[#]} & /@ NestList[(#² + 0.33 + 0.35 I) &, 1 − 0.5, 8],
    PlotJoined → True, AspectRatio → Automatic,
    PlotRange → All, Epilog → {PointSize[0.02],
      Point[{0.873514699456611, 0.46852239083117725}]}]

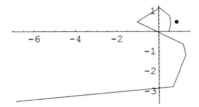

Exercise:
1) Use the command **Solve** to find fixed points of the function g defined by $g\,[z] = z^2 + \frac{2}{9}$.
2) Use **NSolve** to find fixed points of the function k defined by $k\,[z] = z^5 - 2\,z + 2$.
3) Use **FindRoot**, with starting point 0.75, to find a fixed point of the function Cos.
   In each of the above 3 cases decide if the fixed point is repelling, attracting or neutral.
4) Classify the fixed points 0 and 1 of the function $z \to z^2$.

4.5.2 Attracting and Repelling Cycles of Prime Period Greater than One

If f is analytic on a subset A of $\mathbb{C}$, so is f^k, and if f^k has a fixed point at $a \in A$, then, by definition, this fixed point is attracting, repelling, or neutral according as $|(f^{k})\,'[a]|$ is less than, greater than or equal to 1. The corresponding k-cycle $\{a, f\,[a], \dots f^{k-1}[a]\}$ is called attracting, repelling or neutral, respectively. The derivative of f^k is the same at each point of the above cycle. The derivative of f^k at its fixed point a can be calculated using the following formula:

$$(f^k)\,'[a] = f\,'[a]\,f\,'[f\,[a]] \dots f\,'[f^{k-1}[a]].$$

So the formula for $(f^k)\,'[a]$ does not involve the calculation of f^k, it is the product of the derivatives of f at each point of the cycle.

Here is an example of an attracting 5-cycle, and an example of the orbit of a point attracted to it:

a = complexOrbit[(#^{2} − 0.53 − 0.55 I) &,
0.07978143067846499 − 0.05170047012496698 I, 10]

b = complexOrbit[(#^{2} − 0.53 − 0.55 I) &, 0, 100]

Show[GraphicsArray[{a, b}, GraphicsSpacing → 0.1]]

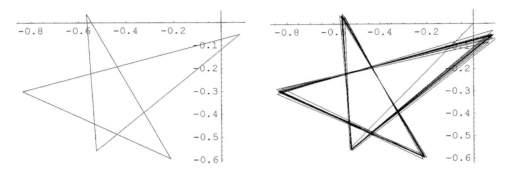

The following is an example of the orbit of a point under a trigonometric function attracted to a 4-cycle:

complexOrbit[I Sin[#] &, 0.5 + 0.4 I, 100]

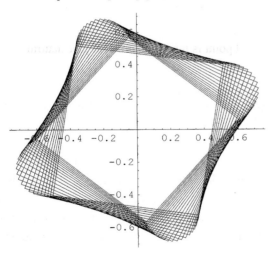

Here is an example of a repelling 3-cycle, and an example of the orbit of a point being repelled from it.

c = complexOrbit[(#^{2} − I) &, −1.2904912332417333 − 0.7792817182359892 I, 15]

d = complexOrbit[(#^{2} − I) &, −1.2904 − 0.7792 I, 13]

Show[GraphicsArray[{c, d}, GraphicsSpacing → 0.1]]

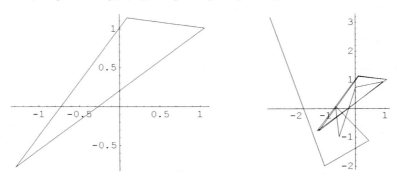

Exercise:

1) In the case of each of the above orbit diagrams, construct the orbits of different points, and see if you can find an orbit with different properties to the one portrayed.

2) Use the following programs to plot each orbit, and use the image to classify the orbits according to the specifications given above. Plot more points in the orbit if necessary.

$$\text{complexOrbit}[(\#^2 - 0.18 + 0.573\,I)\ \&,\ 0,\ 50]$$

$$\text{complexOrbit}[(\#^2 - 0.128 + 0.773\,I)\ \&,\ 0.2\,I,\ 100]$$

$$\text{complexOrbit}[(\#^2 - 1.18 + 1.573\,I)\ \&,\ 0,\ 5]$$

$$\text{complexOrbit}[(\#^2 - I)\ \&,\ -1 - I,\ 20]$$

$$\text{complexOrbit}[(\#^2 - 0.6 - 0.5\,I)\ \&,\ -0.12310562561766059`,\ 13]$$

$$\text{complexOrbit}[(1 - 0.5\,I)\ \text{Cos}[\#]\ \&,\ 0.25,\ 50]$$

$$\text{complexOrbit}[(1 + 0.5\,I)\ \text{Cos}[\#]\ \&,\ 1,\ 25]$$

$$\text{complexOrbit}[(0.8 + 0.5\,I)\ \text{Sin}[\#]\ \&,\ 0.1 + 0.1\,I,\ 155]$$

$$\text{complexOrbit}[(I)\ \text{Sin}[\#]\ \&,\ 0.5 + 0.4\,I,\ 25]$$

$$\text{complexOrbit}[(I)\ \text{Sin}[\#]\ \&,\ 0.9 + 0.9\,I,\ 105]$$

$$\text{complexOrbit}\left[\frac{2\,\#}{\#^4 + \#^2 + 1}\ \&,\ 0.9 - I,\ 45\right]$$

$$\text{complexOrbit}\left[\left(\frac{\# - 2}{3\,\#^2 + 2\,I}\right)\ \&,\ -0.06 + 0.9\,I,\ 35\right]$$

$$\text{complexOrbit}\left[\left(\frac{2\,\# - 4}{\#^2 + 1}\right)\ \&,\ 1 - 0.2\,I,\ 45\right]$$

$$e[z\_] := \frac{2\,z}{z^4 + z^2 + 1};$$

complexOrbit[e, 0.3 − 0.8 I, 10]

Check some of your conclusions, using **NestList** or by finding periodic points and checking if they are attracting, repelling or neutral.

4.6 Basins of Attraction

4.6.1 Basin of Attraction of a Fixed Point

Let f be a complex function with attracting fixed point a. The basin of attraction of a under f is defined to be the set:
$\{z \in \mathbb{C} \mid f^n[z] \to a \text{ as } n \to \infty\}$.

For example, consider the following function, h:

$$h[z\_] := z^2 + 0.33 + 0.35\,I;$$

We find fixed points of h using **NSolve**, and check if they are attracting:

Replace[z, NSolve[h[z] == z, z]] $\{0.126485 + 0.468522\,i, 0.873515 − 0.468522\,i\}$

Abs[D[h[z], z]] /. z −> % $\{0.970591, 1.98246\}$

The pont x = 0.126485+0.468522 I is an attracting fixed point of h. We wish to find some points in the basin of attraction of x. We choose a point 'near' x, and calculate part of its orbit. A bit of trial and error is needed in choosing the point, as we do not know the extent of the basin of attraction. To save space, we drop the first 190 terms of the orbit.

Drop[NestList[h, −0.35 − 0.25 I, 200], 190]

$\{0.125799 + 0.469392\,i, 0.125496 + 0.468098\,i, 0.126633 + 0.467489\,i, 0.12749 + 0.468399\,i,$
$0.126856 + 0.469432\,i, 0.125725 + 0.4691\,i, 0.125752 + 0.467956\,i, 0.126831 + 0.467693\,i,$
$0.12735 + 0.468636\,i, 0.126598 + 0.469361\,i, 0.125727 + 0.468841\,i\}$

It seems as though the point we chose is in the basin of attraction of x. All the other points in its orbit are also in the basin of attraction of x. In the following program, we omit the command **PlotJoined→True** from **complexOrbit**, in order to display points in the basin of attraction of x.

complexOrbit2[f_, s_, n_] := ListPlot[{Re[#], Im[#]} & /@ NestList[f, s, n],
　　　AspectRatio → Automatic, PlotRange → All, PlotStyle → PointSize[0.011]];

b1 = complexOrbit2[(#^{2} + 0.33 + 0.35 I) &, −0.35 − 0.25 I, 200]

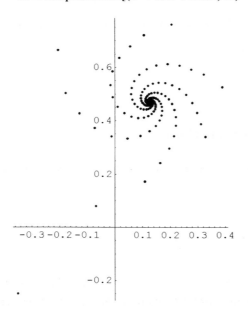

All the points in the above orbit belong to the basin of attraction of x. We find more points in the basin of attraction of x, and then plot the union of the two sets.

b2 = complexOrbit2[(#^{2} + 0.33 + 0.35 I) &, 0.2 + 0.25 I, 200]

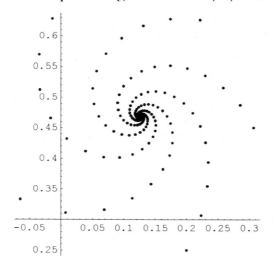

Show[b1, b2]

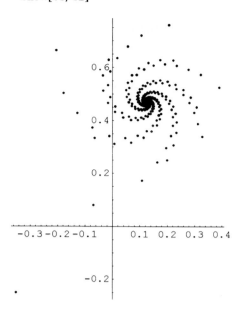

We have found 401 points in the basin of attraction of x, but, of course, since the basin of attraction contains a disk, center x, there are infinitely many (uncountably many, in fact) points in the basin.

Exercise:

Find points not in the basin of attraction of x.

4.6.2 Basin of Attraction of an Attracting Cycle of Period p > 1

Let f be a function with an attracting cycle, $c = \{z_1, z_2, \ldots z_p\}$, of period $p > 1$. Then the points of the cycle are attracting fixed points of f^p and so each z_k has a basin of attraction, B_k, consisting of all points in $\mathbb{C}$ whose orbits under the action of f^p converge to z_k. The union of all these basins of attraction is defined to be the basin of attraction of the attracting cycle c under the action of f. We illustrate this notion below with an example of a function with an attracting cycle of period 2.

j[z_] := z^2 – 1 – 0.2 I;

We find fixed points of j^2, and check if they are attracting:

Replace[z, NSolve[j[j[z]] == z, z]]

$\{-1.03393 - 0.187291\,i,\ -0.621583 - 0.0891597\,i,$
$\quad 0.0339271 + 0.187291\,i,\ 1.62158 + 0.0891597\,i\}$

Abs[D[j[j[z]], z]] /. z → %   {0.8, 1.57726, 0.8, 10.5499}

We see that j^2 has 2 attracting fixed points, and so j has an attracting 2-cycle $\{a_1, a_2\}$, say. We plot some points in each of the basins of attraction of these 2 fixed points under the action of j^2. So first we choose the point $-0.9 - 0.1\,I$, which is 'near' the point a_1, and plot part of its orbit; we then plot part of the orbit of $j\,[-0.9 - 0.1\,I]$ which is 'near' to a_2 ($= j\,[a_1]$).

complexOrbit2[Composition[j, j], −0.9 − 0.1 I, 100]

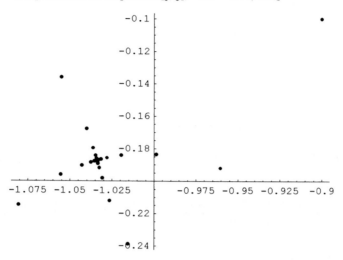

complexOrbit2[Composition[j, j], j[−0.9 − 0.1 I], 100]

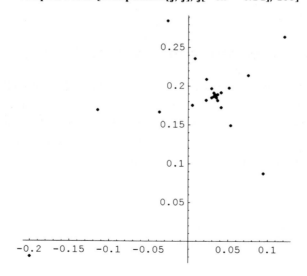

All the points in the above 2 diagrams form part of the basin of attraction of the attracting cycle $\{a_1, a_2\}$ of j.

We now plot the first 200 points in the orbit of -0.9-0.1I under j. The points in this orbit are alternately in the basins of attraction of a_1 and a_2 under j^2.

complexOrbit2[j, −0.9 − 0.1 I, 200]

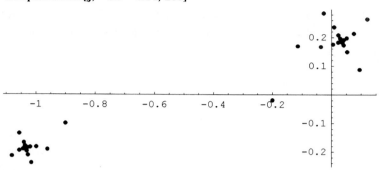

4.6.3 The Basin of Attraction of Infinity

If infinity is an attracting fixed point of f, then the basin of attraction of infinity is defined to be the set:
$\{z \in \mathbb{C} \mid f^n[z] \to \infty \text{ as } n \to \infty\}$.

In the following example, we first find a point whose orbit tends to infinity under the function $z \to z^2 - 1$.

NestList[$\#^2$ − 1 &, .28 + .4 I, 19]

$\{0.28 + 0.4\,i, -1.0816 + 0.224\,i, 0.119683 - 0.484557\,i, -1.22047 - 0.115986\,i,$
$\quad 0.476098 + 0.283115\,i, -0.853485 + 0.269581\,i, -0.344237 - 0.460167\,i,$
$\quad -1.09325 + 0.316813\,i, 0.094835 - 0.692714\,i, -1.47086 - 0.131387\,i,$
$\quad 1.14616 + 0.386503\,i, 0.164303 + 0.885991\,i, -1.75798 + 0.291142\,i,$
$\quad 2.00574 - 1.02364\,i, 1.97516 - 4.10634\,i, -13.9608 - 16.2214\,i, -69.2296 + 452.925\,i,$
$\quad -200349. - 62711.6\,i, 3.6207 \times 10^{10} + 2.51284 \times 10^{10}\,i, 6.79506 \times 10^{20} + 1.81965 \times 10^{21}\,i\}$

All the points in the above orbit are in the basin of attraction of infinity. We plot a few of them:

complexOrbit2[$\#^2$ − 1 &, 0.28 + 0.4 I, 13]

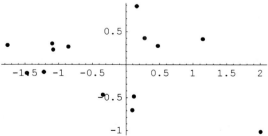

4.7 The 'Symmetric Mappings' of Michael Field and Martin Golubitsky

Chossat and Golubitsky (1988) define symmetric mappings on the complex plane. These mappings have the property that orbits of certain points under the action of these mappings display a degree of symmetry. Examples of such mappings are given by the family defined by:

$$F[z] = (\lambda + \alpha \, |z|^2 + \beta \, \text{Re} \, (z^n)) \, z + \gamma \, (\bar{z})^{n-1}$$

where λ, α, β, γ are real numbers and n is a natural number.

Here is a program for plotting the orbit of a point under the action of a member of the family F. The number of plot-points must be very large and the points are not joined by straight lines. As there are very many points in the image, we have included the directive **PointSize[0.0001].**

```
g[λ_, α_, β_, γ_, n_, z_] := ListPlot[{Re[#], Im[#]} & /@
    NestList[((λ + α (Abs[#])² + β Re[#ⁿ]) # + γ (Conjugate[#])ⁿ⁻¹) &, z, 100000],
    AspectRatio → Automatic, PlotRange → All,
    PlotStyle → {PointSize[0.0001]}, Axes → False];
```

Here is an example:

```
g[1.52, −1, 0.1, −0.8, 3, 0.1 − 0.1 I]
```

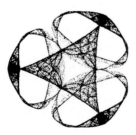

Programs to try:

```
g[−2.7, 5, 2, 1, 6, −0.35 − 0.25 I],  g[−2.6, 4, 2, 1, 8, −0.2 + 0.1 I],
g[−2.585, 5, 2, 1, 6, −0.2 − 0.1 I],  g2[−1.75, 2, −0.2, 1, 3, 0.2 + 0.1 I]
```

Exercise:

1) Experiment with the above programs by varying one or more of the parameters slightly.

2) Write a program similar to the above which includes a color directive, x, say, and a variable k, say, for the number of plot points.

3) Look in Field and Golubitsky (1995) to find other suitable values for the parameters of the family F.

Chapter 5

Using Roman Maeder's Packages AffineMaps, Iterated Function Systems and Chaos Game to Construct Affine Fractals

Introduction

An affine transformation is one that preserves linearity and parallelism but permits some type of distortion such as shearing. A similarity transformation is an affine transformation which preserves shape but may alter size. A geometric object is self-affine (self-similar) if it can be partitioned into sub-objects, all of which are affine (similar) copies of the parent object, or if a proper subset of it is an affine (similar) copy of the whole.

So, if a set S in the plane is self-similar then S has a proper subset, S_1, which is similar to (has the same shape as, but not the same size as) S. This means that S_1 is also self-similar, and a part of it, S_2, say, is similar to S_1 and hence also to S and so on ad infinitum. This means that often we can not actually construct a self-similar set, but only an approximation to it.

Although there is no generally accepted mathematical definition of a fractal, a property that characterizes many fractals is some type of self-affinity or self-similarity.

The Sierpinski triangle depicted below is an example of a representation of a self-similar fractal.

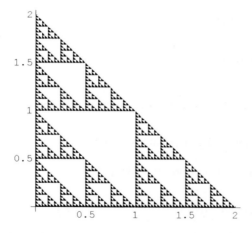

The following graphic is a representation of a fractal which is self-affine, but not self similar:

Notice that the the top left branch of the above tree-like graphic resembles the complete graphic, but the ratio of height to width is different.

In this Chapter, we shall construct self-affine fractals using Roman Maeder's commands **AffineMap**, **IFS**, and **ChaosGame** which, in *Mathematica* 4.2 can be found in: Help - Add-ons - Extras - Programming-in-*Mathematica*. These packages are included with *Mathematica* 5, but are not documented.

The packages were developed for Roman Maeder's Book: 'Programming in *Mathematica*', Maeder (1996).

Do not forget to load these packages in each of your *Mathematica* sessions involving the commands of this chapter, and remember to load these packages only once per *Mathematica* session.
We first define the notion of an affine map on $\mathbb{R}^2$.

5.1 Affine Maps from $\mathbb{R}^2$ to $\mathbb{R}^2$

5.1.1 Definitions

The map $g : \mathbb{R}^2 \to \mathbb{R}^2$ is linear if and only if $g\left[\begin{pmatrix} x \\ y \end{pmatrix}\right] = \begin{pmatrix} a\,x + b\,y \\ c\,x + d\,y \end{pmatrix} = \begin{pmatrix} a & b \\ c & d \end{pmatrix}\begin{pmatrix} x \\ y \end{pmatrix}$ for some $a, b, c, d \in \mathbb{R}$.

The map $f : \mathbb{R}^2 \to \mathbb{R}^2$ is affine if and only if $f\left[\begin{pmatrix} x \\ y \end{pmatrix}\right] = \begin{pmatrix} ax + by \\ cx + dy \end{pmatrix} + \begin{pmatrix} p \\ q \end{pmatrix} = \begin{pmatrix} a & b \\ c & d \end{pmatrix}\begin{pmatrix} x \\ y \end{pmatrix} + \begin{pmatrix} p \\ q \end{pmatrix}$

for some $a, b, c, d, p, q \in \mathbb{R}$.

So, an affine map is a linear map followed by a translation.

The matrix $\begin{pmatrix} a & b \\ c & d \end{pmatrix}$ can be written in the form $\begin{pmatrix} r\,\mathrm{Cos}[\theta] & -s\,\mathrm{Sin}[\phi] \\ r\,\mathrm{Sin}[\theta] & s\,\mathrm{Cos}[\phi] \end{pmatrix}$ for some $r, s, \theta, \phi \in \mathbb{R}$.

In Roman Maeder's package **ProgrammingInMathematica`AffineMaps`**, the command **AffineMap** can be applied to 2D graphics primitives such as points, lines, polygons as well as to other 2D graphics. The following command loads the package:

> **<< ProgrammingInMathematica`AffineMaps`**

> **? AffineMap**

> AffineMap[ϕ, ψ, r, s, e, f] generates an affine map with rotation angles ϕ, ψ, scale factors r, s, and translation
> components e, f. AffineMap[{x, y}, {fxy, gxy}] generates an affine map with the two components
> given as expressions in x and y. AffineMap[matrix] uses the 2x3 matrix for the affine map.

For example, we apply an affine map to a point:

AffineMap[30°, 30°, 2, 3, 1, 4][{3.0, −1}]   {7.69615, 4.40192}

5.1.2 Affine Maps which are Similarities

Let r > 0 and consider the command **AffineMap[θ, θ, r, r, e, f]**. To understand its effect on a point P in the plane, we first consider its effect on the line segment OP. When applied to OP, the above affine map scales OP, by the factor r, rotates it through θ radians about the origin and then translates it by the vector (e, f). The affine map **AffineMap[θ, θ, r, -r, e, f]** scales the line segment OP, by the factor r, reflects it in the x-axis, rotates it through θ radians about the origin and then translates it by the vector (e, f). If we denote the resulting line segment as QR, where Q is the point {e, f }, then the effect of applying the above affine map to P is to move it to R. When applied to a 2D graphic G, the above procedure is applied to each point in G.

The following example shows the 4 affine maps:

AffineMap$\left[0, 0, \frac{1}{2}, \frac{1}{2}, 0, 0\right]$, Affinemap$\left[\frac{2\pi}{3}, \frac{2\pi}{3}, \frac{1}{2}, \frac{1}{2}, 0, 0\right]$,
AffineMap$\left[\frac{2\pi}{3}, \frac{2\pi}{3}, \frac{1}{2}, \frac{1}{2}, -1, 1\right]$, Affinemap[0, 0, 1, −1, 0, 0]

applied to the triangle with vertices {0, 0}, {3, 2}, {2.5, 0.5}.

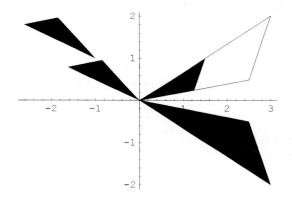

In the above diagram, the triangle with vertices {0, 0}, {3, 2}, {2.5, 0.5} has been contracted by the factor $\frac{1}{2}$, then rotated through $2\frac{\pi}{3}$ and then translated by the vector (−1, 1). Also shown is the reflection of the triangle in the x-axis. Notice that each of the 4 images of the original triangle is similar in shape to the original triangle. The above affine maps are examples of similarities.

5.1.3 Sheared Affine Transformations

If r ≥ 0, s ≥ 0 and r ≠ s, the affine transformation **AffineMap[θ, θ, r, s, e, f]** is a shear transformation. It scales the graphics primitive P by the factor r in the x direction and by the factor s in the y direction, rotates through θ radians, and translates by the vector (e, f).

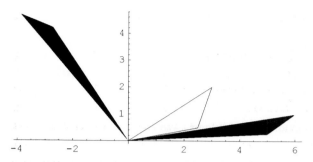

In the above diagram, the triangle with vertices {0, 0}, {3, 2}, {2.5, 0.5} has been scaled in the x direction by the factor 2 and in the y direction by the factor $\frac{1}{2}$. It has then been rotated through $2\frac{\pi}{3}$.

5.1.4 Definition of the Sierpinski Triangle

We define 3 affine contraction mappings using the command **AffineMap** as follows:

$$w_1 = \textbf{AffineMap}\left[0, 0, \frac{1}{2}, \frac{1}{2}, 0, 0\right];$$

$$w_2 = \textbf{AffineMap}\left[0, 0, \frac{1}{2}, \frac{1}{2}, 1, 0\right]; \quad w_3 = \textbf{AffineMap}\left[0, 0, \frac{1}{2}, \frac{1}{2}, 0, 1\right];$$

Each of the maps is a similarity which contracts by $\frac{1}{2}$. The second and third maps translate to the points {1, 0} and {0, 1} respectively. In the diagram below, each of these 3 maps has been applied to the triangle, D_1, and the 3 images of these maps form the set D_2 consisting of 3 triangles. The 3 maps are then applied to D_2 generating D_3, a set of 9 triangles.

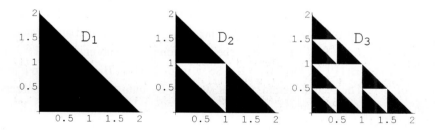

Proceeding in this way, we obtain an infinite sequence (D_n) of closed bounded subsets of $\mathbb{R}^2$. The Sierpinski triangle is defined to be $T = \bigcap_{n=1}^{\infty} D_n$.

In order to construct an approximation to the Sierpinski triangle and other affine fractals, we need to apply a set of affine maps repeatedly to a 2D graphic.

5.2 Iterated Function Systems

5.2.1 Contraction Mappings on Subsets of $\mathbb{R}^2$ and Compact Subsets of $\mathbb{R}^2$

In Chapter 4, we discussed contraction mappings on $\mathbb{C}$. We now discuss the analogous situation for $\mathbb{R}^2$.

There is a one-to-one correspondence between complex numbers in $\mathbb{C}$ of the form $a + Ib$ and points in $\mathbb{R}^2$ with co-ordinates $\{a, b\}$.

If $z = a + I b$, $w = c + I d$ are complex numbers then $|z - w| = \sqrt{(a - c)^2 + (b - d)^2}$ is equal to the distance between the points $\{a, b\}$ and $\{c, d\}$ corresponding to the complex numbers z and w respectively.

Let A be a subset of $\mathbb{R}^2$ and let $f : A \to A$.

Then f is said to be a contraction mapping on A if and only if there exists s with $0 \le s < 1$ such that the distance between $f[\{a, b\}]$ and $f[\{c, d\}]$ is less than or equal to $s \sqrt{(a - c)^2 + (b - d)^2}$, for $\{a, b\}, \{c, d\} \in A$. In particular if $A = \mathbb{R}^2$ then f is said to be a contraction mapping on $\mathbb{R}^2$.

For example, the command **AffineMap[θ, ϕ, r, s, e, f]** defines a contraction mapping on $\mathbb{R}^2$ if $0 \le |r| < 1$ and $0 \le |s| < 1$. Such a mapping is called an affine contraction mapping.

Compact subsets of $\mathbb{C}$ were defined in Chapter 4. A subset B of $\mathbb{R}^2$ is compact if and only if the corresponding subset, $C = \{x + Iy \in \mathbb{C} \mid \{x, y\} \in \mathbb{R}^2\}$ of $\mathbb{C}$ is compact.

5.2.2 Definition of an IFS

Let $f : \mathbb{R}^2 \to \mathbb{R}^2$, and let $B \subseteq \mathbb{R}^2$, then $f[B] = \{f[x] \mid x \in B\}$. So f can be regarded as a function which maps subsets of $\mathbb{R}^2$ onto subsets of $\mathbb{R}^2$. Let $w_1, w_2, \dots w_k : \mathbb{R}^2 \to \mathbb{R}^2$ be contraction mappings on $\mathbb{R}^2$. Then $\{w_1, w_2, \dots w_k\}$ or $\{\mathbb{R}^2 \mid w_1, w_2, \dots w_k\}$ is called an Iterated Function System or IFS for $\mathbb{R}^2$. Let B_1 be a compact subset of $\mathbb{R}^2$ and let $w[B_1] = w_1[B_1] \bigcup w_2[B_1] \bigcup \dots \bigcup w_k[B_1] = B_2$. So w maps a subset B_1 of $\mathbb{R}^2$ onto a union, B_2, of k subsets of $\mathbb{R}^2$. The second iterate, w^2, of w maps B_2 onto a union of k^2 subsets of $\mathbb{R}^2$, and so forth.

Sometimes the name 'Iterated Function System' is used to denote a finite set of maps acting on $\mathbb{R}^2$ with no conditions imposed on the maps, Barnsley (1988). We shall follow this convention. Barnsley calls the IFS 'hyperbolic' if each of its constituent mappings is a contraction.

If each constituent map is an affine map, then the IFS is called an affine IFS.

We shall initially be concerned with a set of affine maps each of which is a contraction mapping, and we shall be applying each one of them to a plane set of points, B, called the *initial* set, and then re-applying them to the resulting plane sets of points, etc. In order to do this, we load another package of Roman Maeder's: **ProgrammingInMathematica`IFS`**.

Needs["ProgrammingInMathematica`IFS`"]

? IFS

IFS[{maps..}, {options..}] generates an iterated function system (IFS).

The following is the IFS we will use to construct the Sierpinski triangle:

ifs1 = IFS[{AffineMap[0, 0, 0.5, 0.5, 0, 0],
AffineMap[0, 0, 0.5, 0.5, 1, 0], AffineMap[0, 0, 0.5, 0.5, 0, 1]}];

Roman Maeder's IFS command allows us to apply an IFS repeatedly to certain graphics primitives such as lines, polygons, circles, points and also to other 2D plots.

If P is the name of a graphic, then the command: **Show[Nest[ifs1, P, 0]]** returns P itself, while **Show[Nest[ifs1, P, 1]]** applies **ifs1** once to P and **Show[Nest[ifs1, P, 2]]** twice, etc.

5.2.3 Constructing the Sierpinski Triangle Using an Affine IFS

We now apply the maps w_1, w_2, and w_3 defined in 5.1.4 to triangle D_1 :

D_1 = Polygon[{{0, 0}, {0, 2}, {2, 0}}];

The command **t[x_,n_]** defined below applies the iterated function system **ifs1** to the graphics primitive x, n times. As we wish to show the **ifs1** applied to D_1 firstly 3 times and then 6 times in the same diagram, we use the command **GraphicsArray**.

t[x_, n_] := Graphics[Nest[ifs1, x, n], Axes → True,
AspectRatio –> 1, AxesOrigin → {0, 0}, Ticks → {{1, 2}, {1, 2}}];

Show[GraphicsArray[{t[D_1, 3], t[D_1, 6]}]]

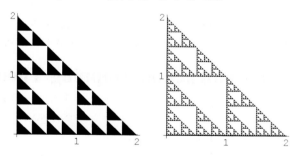

If you wish to display a single image, **t[x_, n_]**, use the command **Show[t[x_, n_]]**.

Notice that the Sierpinski triangle is self-similar. The top triangle is a reduced copy of the whole triangle, and it, in turn, is divided into 3 triangles each similar to the whole triangle etc.

We shall now apply the above command **t[x_, n_]** to a different graphics primitive, **x,** and see what happens.

We replace x by a line. An array of successive applications of **IFS** is shown:

L1 = Line[{{0, 0}, {1, 1}}];

Show[GraphicsArray[Table[t[L1, n], {n, 1, 6}]]]

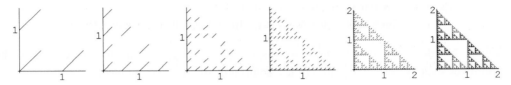

So, we obtained an approximation to the Sierpinski triangle again - although we applied the **IFS** to a very different initial set!

Let w_1, w_2, ..., w_n be contraction maps on $\mathbb{R}^2$ and let $w[B] = w_1[B] \cup w_2[B] \cup ... \cup w_n[B]$ for each compact subset B of $\mathbb{R}^2$. It can be proved that there exists a unique compact subset, A, of $\mathbb{R}^2$ such that if D is *any* compact subset of $\mathbb{R}^2$ then, the sequence of iterations of D, $(D, w[D], w^2[D], ...)$ 'approaches' A.

The Contraction Mapping Theorem for $\mathbb{R}^2$, in 5.3 below, explains this phenomenon.

Exercise:

1) Apply **ifs1** as defined above to a square, or to a disk, or to a point, or...

2) Construct the following fractal. Start with the square of side-length 3, sides parallel to the co-ordinate axes, and apply 5 contraction maps to it.

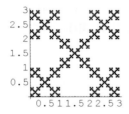

5.3 Introduction to the Contraction Mapping Theorem for $\mathcal{H}\,[\mathbb{R}^2]$

5.3.1 $\mathcal{H}\,[\mathbb{R}^2]$

The set of all non-empty compact subsets of $\mathbb{R}^2$ is denoted $\mathcal{H}[\mathbb{R}^2]$.

■ **Distance Between a Pair of Compact Sets in $\mathbb{R}^2$**

Suppose S_1, S_2,.... is a sequence of subsets of $\mathbb{R}^2$. We wish to define the notion of the sequence converging to a subset, S, of $\mathbb{R}^2$. In order to do this we need to define the distance between a pair of subsets of $\mathbb{R}^2$. The distance definition we use is called the Hausdorff distance. We shall not give a mathematical definition of the Hausdorff distance here, but shall try to give an intuitive idea. (The mathematical definition can be found in Barnsley (1988)). If 2 objects touch each other, we think of them as being 'close' to each other. The Hausdorff distance between 2 sets does not measure closeness in this sense, but it measures 'coincidence' or how nearly the sets coincide. The 2 subsets of the plane in the diagram below are 'nearly the same' or 'almost coincide'. The Hausdorff distance between them is small.

Show[Graphics[{Circle[{0, 0}, 1], Circle[{0, 0}, 1.05]},
Axes → True, AspectRatio → Automatic]]

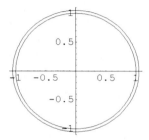

Now consider the following pair of subsets. Although they are contiguous, they do not almost coincide. The Hausdorff distance between them is large.

Show[Graphics[{Disk[{2, 0}, 2], Disk[{6, 0}, 2]}], AspectRatio → Automatic, Axes → True]

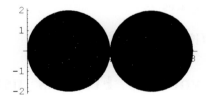

The Hausdorff distance between the subsets A, B of $\mathbb{R}^2$ is denoted h [A, B] and is a non-negative real number.

5.3.2 The Contraction Mapping Theorem for $\mathcal{H}$ [$\mathbb{R}^2$]

■ **Convergence of a Sequence of Elements of $\mathcal{H}[\mathbb{R}^2]$**

The sequence (S_n) of elements of $\mathcal{H}[\mathbb{R}^2]$ is said to converge to the element A of $\mathcal{H}[\mathbb{R}^2]$ if the real sequence (h [S_n, A]) tends to 0 as n tends to infinity.

■ **The Contraction Mapping Theorem for $\mathcal{H}[\mathbb{R}^2]$**

The mapping w on $\mathcal{H}[\mathbb{R}^2]$ is said to be a contraction mapping if there exists α, with $0 \le \alpha < 1$, such that for all A, B in $\mathcal{H}[\mathbb{R}^2]$, $h[w[A], w[B]] \le h[A, B]$.

Let w be a contraction mapping on $\mathcal{H}[\mathbb{R}^2]$, then w has a unique fixed point, $A \in \mathcal{H}[\mathbb{R}^2]$, i.e. there exists a unique compact subset A of $\mathbb{R}^2$ such that w[A]=A. Also, if B is any element of $\mathcal{H}[\mathbb{R}^2]$, the sequence $(w^n [B]) = (B, w[B], w^2 [B], ...)$ converges to A.

■ **Contraction Mappings on $\mathcal{H}[\mathbb{R}^2]$ defined by Iterated Function Systems**

Let $w_1, w_2, ..., w_n$ be contraction mappings on $\mathbb{R}^2$, with contractivity factors $\alpha_1, \alpha_2, .., \alpha_n$ respectively, and let $w[B] = w_1[B] \bigcup w_2[B] \bigcup ... \bigcup w_n[B]$, for each $B \in \mathcal{H}[\mathbb{R}^2]$. It can be proved that w is a contraction mapping on $\mathcal{H}[\mathbb{R}^2]$ with contractivity factor $\alpha = \text{Max}\{\alpha_1, \alpha_2, .., \alpha_n\}$. So, by the Contraction Mapping Theorem for $\mathcal{H}[\mathbb{R}^2]$, there exists an element A of $\mathcal{H}[\mathbb{R}^2]$ such that if w is applied repeatedly to any compact subset, B, of $\mathbb{R}^2$, the sequence of iterates $(B, w[B], w^2[B], ...)$ converges to A. The set A is called the *attractor* of the IFS $\{w_1, w_2, ... w_n\}$, and often has fractal-like qualities. The process of constructing an approximation to the attractor, A, of an IFS consisting of contraction mappings, by applying the IFS repeatedly to a compact subset of $\mathbb{R}^2$ is called the 'Deterministic Algorithm'. The reason that this process is given a special name is that it can be proved that an approximation to A can be constructed by other processes. For example, in this chapter, we shall construct approximations to certain attractors using the 'Random Algorithm'. Since an IFS consisting of affine contraction mappings from $\mathbb{R}^2$ to $\mathbb{R}^2$ is a contraction mapping on $\mathcal{H}[\mathbb{R}^2]$, Roman Maeder's command **IFS** can be used to construct such an attractor provided that all constituent maps are contraction mappings.

■ **Contraction Mapping Theorems for Other Spaces**

There are analogous definitions and theorems for $\mathcal{H}[\mathbb{C}]$, $\mathcal{H}[\mathbb{R}]$, $\mathcal{H}[\mathbb{R}^3]$.

5.4 Constructing Various Types of Fractals using Roman Maeder's Commands

5.4.1 Relatives of the Sierpinski Triangle

Many fractals may be constructed in the following way. Let S be the square with sides of length 4, sides parallel to the co-ordinate axes, center the origin. Divide S into 4 congruent squares, and choose 3 of these, S_1, S_2 and S_3, say, as shown in the diagram below:

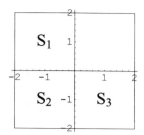

Three affine maps of the form **AffineMap$\left[\theta,\ \theta,\ \pm\frac{1}{2},\ \pm\frac{1}{2},\ \mathbf{p},\ \mathbf{q}\right]$** where θ is a multiple of $\frac{\pi}{2}$, and $\{p, q\} = \{-1, 1\}$, for S_1, $\{-1, -1\}$ for S_2 and $\{1, -1\}$ for S_3, are used to map S onto S_1, S_2 and S_3 respectively. An IFS is formed from these maps, and Peitgen calls the images resulting from applying such an IFS a Sierpinski relative, Peitgen (1992).

Note that

$$\textbf{AffineMap}\left[\theta,\ \theta,\ \frac{1}{2},\ -\frac{1}{2},\ \mathbf{p},\ \mathbf{q}\right]$$

represents a contraction by $\frac{1}{2}$ followed by a rotation through θ followed by a reflection in the x-axis followed by a translation.

Here is an example:

```
sr1[x_, n_] :=
    Show[Graphics[Nest[IFS[{AffineMap[180°, 180°, 0.5, −0.5, −1, 1], AffineMap[90°,
            90°, 0.5, 0.5, −1, −1], AffineMap[90°, 90°, 0.5, −0.5, 1, −1]}], x, n]],
        Axes → False, AspectRatio → Automatic, AxesOrigin → {0, 0}];

sr1[Point[{0, 0}], 8]
```

Exercise:
Construct other relatives of the Sierpinski Triangle. (Use the above program, and in each

command **AffineMap[θ, θ, r, s, e, f]** leave e and f unchanged, replace θ by a multiple of $\frac{\pi}{2}$, and each of r, s by 0.5 or -0.5).

5.4.2 Iterated Function Systems which Include the Identity Map

Suppose that w_1, w_2, ..., w_n are each contraction maps on $\mathbb{R}^2$, and w_0 is the identity map on $\mathbb{R}^2$. We consider the IFS $\{\mathbb{R}^2 \mid w_0, w_1, w_2, ... w_n\}$.

It can be shown that if $w[B] = w_0[B] \bigcup w_1[B] \bigcup ... \bigcup w_n[B]$, where B is any compact subset of $\mathbb{R}^2$, then the sequence of iterations of B, (B, w[B], w^2[B], ...), approaches (converges in the Hausdorff metric to) a subset A of $\mathbb{R}^2$. (For proof, see Appendix to 5.4.2.) This subset A is not unique, but depends on the choice of the initial set, B. Such an IFS is equivalent to Barnsley's definition, Barnsley (1988), of a (hyperbolic) IFS with condensation. Two illustrative examples are shown. In each case, the identity map has been added to an IFS for a Sierpinski relative. In the first example, the initial set is a (filled-in) triangle, and in the second example, the initial set is a circle. Notice that the resulting fractals are self-similar.

```
fract1[x_, n_] := Show[Graphics[
        Nest[IFS[{AffineMap[0, 0, 1, 1, 0, 0], AffineMap[−90°, −90°, 0.5, 0.5, −1, 1],
            AffineMap[0°, 0°, 0.5, 0.5, −1, −1], AffineMap[90°, 90°, 0.5, 0.5, 1, −1]}],
        x, n]], Axes → False, AspectRatio → Automatic, AxesOrigin → {0, 0}];

fract1[Polygon[{{1, 0}, {0, 1}, {2, 2}, {1, 0}}], 7]
```

Images such as the above can be assigned a color, using the command **ColorOutput**.
In the following example, a gray scale color has been chosen.

```
fract2[x_, n_] := Show[Graphics[
    Nest[IFS[{AffineMap[0, 0, 1, 1, 0, 0], AffineMap[−90°, −90°, 0.5, 0.5, −1, 1],
        AffineMap[0°, 0°, 0.5, 0.5, −1, −1], AffineMap[90°, 90°, 0.5, 0.5, 1, −1]}],
    x, n]], Axes → False, AspectRatio → Automatic,
    AxesOrigin → {0, 0}, ColorOutput → (GrayLevel[0.2] &)];

fract2[Circle[{1, 1}, 2], 6]
```

Exercise:

1) Add the identity map to an IFS you have defined, and apply the resulting IFS to a graphics primitive.

2) Adapt the program **fract2** to construct other fractals, using different rotation angles, or contractions or translations.

3) Construct fractals using iterated function systems in which one of the mappings is the identity map, and apply the IFS to different graphics primitives.

5.4.3 The Collage Theorem

As illustrated in the diagram below, the Sierpinski triangle, T, can be covered by 3 reduced copies of itself T_1, T_2 and T_3, say.

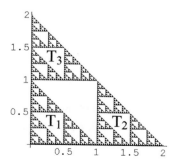

When we constructed the Sierpinski triangle in 5.1.4, we used 3 affine maps each of which mapped the original triangle, D_1, onto a sub-triangle of itself. It turns out that each of these 3 maps also maps the Sierpinski triangle, T, onto either T_1 or T_2 or T_3. The Collage theorem of Barnsley (1988) states that this result can be extended as follows:

Let S be a black 'picture' on a white background. Suppose that we can approximately cover S with a finite number of images of S under contraction mappings from S into itself, i.e. S is a collage made up of smaller images of itself. The Collage theorem of Barnsley states that the attractor of the IFS consisting of these contraction mappings forms an approximation to S.

For example, in the plant-like fractal below, we show, by using 4 rectangles, how the fractal can be covered by a collage of 4 affine copies of itself. By measuring the lengths and inclination of the sides of the rectangles, we can calculate the 4 affine maps required to map the largest rectangle onto each of the smaller rectangles.

Here is a routine for constructing the above fractal in green:

```
collage1[x_, n_] := Graphics[Nest[IFS[{AffineMap[-2°, -2°, 0.02, 0.6, -0.14, -0.8],
        AffineMap[0, 0, 0.6, 0.4, 0, 1.2], AffineMap[-30°, -30°, 0.4, 0.7, 0.6, -0.35],
        AffineMap[30°, 30°, 0.4, 0.65, -0.7, -0.5]}], x, n],
    Axes → False, AspectRatio → Automatic, AxesOrigin → {0, 0},
    ColorOutput → (RGBColor[0.316411, 0.699229, 0.0585946] &)];

Show[collage1[Point[{0, 0}], 8]]
```

The Collage theorem can be used to make images of natural objects, Barnsley (1988). For example, in making an image of a fern Barnsley uses the following affine maps:

AffineMap[0°, 0°, 0, 0, 0.16, 0], AffineMap[-2.5°, -2.5°, 0.85, 0.85, 0, 1.6],
AffineMap[49°, 49°, 0.3, 0.34, 0, 1.6], AffineMap[120°, -50°, 0.3, 0.37, 0.0, 0.37].

Exercise:

1) In each of the following cases, use the collage theorem to construct the following fractals. In each case the fractal may be constructed using an IFS consisting of 3 similarity mappings.

2) See if you can construct the following fractal which requires one similarity and two sheared affine transformation for its implementation.

3) Find the 4 affine maps which are required to construct the following fractal:

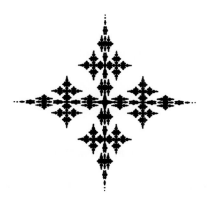

5.4.4 Constructing Your Own Fractals

One method of constructing a fractal by trial and error is as follows:

Start with a rectangle, and make a rough plan of a few affine copies of it on a piece of paper or use Roman Maeder's packages to make your plan, as shown below.

Start with an iterated function system consisting of the identity map, and apply it to a rectangle:

```
myfractal1[x_, n_] := Show[Graphics[Nest[IFS[{AffineMap[0, 0, 1, 1, 0, 0]}], x, n]],
    Axes → False, AspectRatio → Automatic];
```

myfractal1[Line[{{−2, −2}, {2, −2}, {2, 2}, {−2, 2}, {−2, −2}}], 1]

Now add some affine contraction maps to the above IFS, and apply the result to the rectangle using one iteration:

myfractal2[x_, n_] := Show[Graphics[Nest[
    IFS[{AffineMap[0, 0, 0.7, 0.7, 0, 0], AffineMap[−60°, −60°, 0.5, 0.5, −1.5, 1.5],
    AffineMap[60°, 60°, 0.5, 0.5, 1.5, 1.5], AffineMap[0, 0, 1, 1, 0, 0]}], x, n]],
  Axes → False, AspectRatio → Automatic];

myfractal2[Line[{{−2, −2}, {2, −2}, {2, 2}, {−2, 2}, {−2, −2}}], 1]

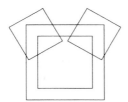

Now, either first deleting the identity map, or retaining it, apply the resulting IFS repeatedly to a graphics primitive.

Here is the case with no identity map:

myfractal3[x_, n_] :=
  Show[Graphics[Nest[IFS[{AffineMap[0, 0, 0.7, 0.7, 0, 0], AffineMap[−60°, −60°,
    0.5, 0.5, −1.5, 1.5], AffineMap[60°, 60°, 0.5, .5, 1.5, 1.5]}], x, n]],
  Axes → False, AspectRatio → Automatic];

myfractal3[Point[{0, 0}], 8]

In this example, the identity map is retained and he resulting IFS is applied to a circle:

myfractal4[x_, n_] :=
　　Show[Graphics[Nest[IFS[{AffineMap[0, 0, 1, 1, 0, 0], AffineMap[0, 0, 0.7, 0.7, 0, 0],
　　　　AffineMap[−60°, −60°, 0.5, 0.5, −1.5, 1.5], AffineMap[60°, 60°, 0.5,
　　　　0.5, 1.5, 1.5]}], x, n]], Axes → False, AspectRatio → Automatic];

myfractal4[Circle[{0, 0}, 0.7], 6]

Another example:

myfractal5[x_, n_] :=
　　Show[Graphics[Nest[IFS[{AffineMap[0, 0, 0.5, 0.5, 0, 0], AffineMap[−60°, −60°,
　　　　0.5, −0.5, −0.6, 1.8], AffineMap[150°, 150°, 0.5, 0.5, 1, −0.5],
　　　　AffineMap[15°, 15°, 0.5, 0.5, −1, −0.5]}], x, n]],
　　Axes → False, AspectRatio → Automatic, AxesOrigin → {0, 0}];

This shows the first iteration applied to a rectangle:

myfractal5[Line[{{−2, −2}, {2, −2}, {2, 2}, {−2, 2}, {−2, −2}}], 1]

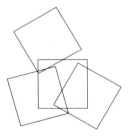

We now apply the IFS to a point:

myfractal5[Point[{0, 0}], 6]

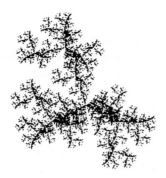

Sometimes, if one chooses the affine maps in an IFS and applies the IFS repeatedly to a graphics primitive, one may obtain an image which exhibits 'apparent disorder'. It may look as though the image has been formed in a random manner. However, the image is *determined* by a set of affine maps, and there is nothing random about it. Here is an example:

myfractal6[x_, n_] :=
      Show[Graphics[Nest[IFS[{AffineMap[−50 °, 0, 0.4, 0.6, 1, 0.5], AffineMap[
                0, 0, −0.7, −0.3, 0, 0], AffineMap[−30 °, −30 °, 0.7, 0.5, 0, 0.5],
              AffineMap[120 °, −60 °, 0.3, 0.5, 0.5, 0]}], x, n]],
        Axes → False, AspectRatio → Automatic];

myfractal6[Point[{0, 0}], 7]

Have fun !

5.4.5 Constructing Fractals with Initial Set a Collection of Graphics Primitives

Roman Maeder's package which we have been using allows us to apply an IFS to an initial set consisting of a collection of graphics primitives, which may have attached graphics directives such as color and thickness. A background color can also be introduced. This is particularly useful when constructing fractals with an IFS which includes the identity map. The set of graphics primitives must first be defined and named. Here is an example:

We start with an initial set consisting of a pair of concentric circles with different colors and a blue background: (Color Fig 5.1)

> circles = Show[Graphics[{{RGBColor[0, 0, 0], Thickness[0.01], Circle[{0, 0}, 0.8]},
> {RGBColor[0.792981, 0.679698, 0.226566], Thickness[0.03], Circle[{0, 0}, 0.7]}}],
> AspectRatio → Automatic, PlotRange → All, Background → RGBColor[0, 0, 0.9]]

We now apply an IFS to the above initial set: (Color Fig 5.2)

> myfractal7[x_, n_] :=
> Show[Graphics[Nest[IFS[{AffineMap[0, 0, 1, 1, 0, 0], AffineMap[0, 0, 0.7, 0.7, 0, 0],
> AffineMap[−30°, −30°, 0.5, 0.5, −1.5, 1.5], AffineMap[30°, 30°, 0.5,
> 0.5, 1.5, 1.5]}], x, n]], Axes → False, AspectRatio → Automatic];

> myfractal7[circles, 5]

In the following example, we omit the command **Show** from the routine for the initial set, as it is not necessary to display it. We have also introduced some randomness into the IFS. (Color Fig 5.3)

> n = Graphics[{RGBColor[1, 1, 1], PointSize[0.1], Point[{0, 0}]},
> AspectRatio → Automatic, Background → RGBColor[0.55079, 0.742199, 0.83595]];

> myfractal8[x_, n_] :=
> Show[Graphics[Nest[IFS[{AffineMap[0°, 0°, 0.2, 0.2, −0.6, Random[]],
> AffineMap[0°, 00.5, 0.3, 0.2, 0.1, 0.2], AffineMap[0°, 0°, 0.2, 0.2, 1.1, 0.3],
> AffineMap[0°, 0°, 0.3, 0.3, 0.7, 0], AffineMap[0°, 0°, 0.2, 0.3, 0.4, 0.6],
> AffineMap[0°, 0°, 0.5, 0.4, −0.4, 0.1], AffineMap[0°, 0°, 0.6, 0.4, 0.5, 0.3],
> AffineMap[0°, 0°, 0.5, 0.3, Random[], 0.5]}], x, n]],
> Axes → False, AspectRatio → Automatic];

> myfractal8[n, 6]

Exercise:

1) Apply **myfractal7** to other initial sets.

2) Experiment with other iterated function systems and various initial sets.

5.4.6 Constructing Tree-like Fractals

We use an IFS consisting of the identity map and 2 or more affine contraction mappings applied to a suitable polygon (or line) which will represent the trunk of the tree. In the following example, the IFS consists of the identity map and 4 affine maps, which are not all similarities. These are applied to the initial set, the polygon with vertices {{-0.2, 0}, {0.2, 0}, {0.35, 2.5}, {0.2, 3}, {-0.2, 3}, {0, 2.5}}. We omit the command **Show**, as we wish to display an array of images.

```
tree1[x_, n_] := Graphics[
    Nest[IFS[{AffineMap[0°, 0°, 1, 1, 0, 0], AffineMap[20°, 20°, 0.65, 2/3, -0.1, 3],
        AffineMap[45°, 45°, 0.5, 0.5, 0, 1.3], AffineMap[-30°, -30°, 0.6, 2/3, 0, 3],
        AffineMap[-55°, -55°, -0.6, 2/3, 0, 2]}], x, n],
    AspectRatio → Automatic, PlotRange → All];

p = Polygon[{{-0.2, 0}, {0.2, 0}, {0.35, 2.5}, {0.2, 3}, {-0.2, 3}, {0, 2.5}}]];
```

The initial set:

```
Show[tree1[p, 0]]
```

The first iteration:

```
Show[tree1[p, 1]]
```

The second to fifth iterations:

Show[GraphicsArray[{tree1[p, 2], tree1[p, 3], tree1[p, 4], tree1[p, 5]}]]

We now show how to construct a colored version of the above fractal, using the method described in 5.4.5 Our initial set will consist of a colored version of the first iteration of the above fractal, which consists of 5 polygons. We know the co-ordinates of the vertices of the largest polygon, P1. We find the co-ordinates of the vertices of the other 4 polygons by applying the affine maps of the IFS **tree1** to P1. We apply one of the affine maps to P1:

> **AffineMap[45°, 45°, 0.5, 0.5, 0, 1.3][**
> **Polygon[{{−0.2, 0}, {0.2, 0}, {0.35, 2.5}, {0.2, 3}, {−0.2, 3}, {0, 2.5}}]]**

Polygon[{{−0.0707107, 1.22929}, {0.0707107, 1.37071}, {−0.76014, 2.30763},
{−0.989949, 2.43137}, {−1.13137, 2.28995}, {−0.883883, 2.18388}}]]

Continuing thus, and assigning colors to the resulting polygons, we obtain the following routine for the initial set: (Color Fig 5.4)

> **initSet2 = Show⎡Graphics⎡{{RGBColor[0.6, 0.52, 0.344],**
> **Polygon[{{−0.2, 0}, {0.2, 0}, {0.35, 2.5}, {0.2, 3}, {−0.2, 3}, {0, 2.5}}]},**
> **{RGBColor[0.5, 0.61, 0.34], AffineMap[45°, 45°, 0.5, 0.5, 0, 1.3][**
> **Polygon[{{−0.2, 0}, {0.2, 0}, {0.35, 2.5}, {0.2, 3}, {−0.2, 3}, {0, 2.5}}]]},**
> **{RGBColor[0.5, 0.61, 0.34], AffineMap[−55°, −55°, −0.6, $\frac{2}{3}$, 0, 2][**
> **Polygon[{{−0.2, 0}, {0.2, 0}, {0.35, 2.5}, {0.2, 3}, {−0.2, 3}, {0, 2.5}}]]},**
> **{RGBColor[0.457, 0.855, 0.015], AffineMap[20°, 20°, 0.65, $\frac{2}{3}$, −0.1, 3][**
> **Polygon[{{−0.2, 0}, {0.2, 0}, {0.35, 2.5}, {0.2, 3}, {−0.2, 3}, {0, 2.5}}]]},**
> **{RGBColor[0.457, 0.855, 0.015], AffineMap[−30°, −30°, 0.6, $\frac{2}{3}$, 0, 3][**
> **Polygon[{{−0.2, 0}, {0.2, 0}, {0.35, 2.5}, {0.2, 3}, {−0.2, 3}, {0, 2.5}}]]}},**
> **AspectRatio → Automatic, PlotRange → All⎤**

We now apply **tree1** to the above initial set: (Color Fig 5.5)

      Show[tree1[initSet2, 6]]

Here is a program to try:

      tree2[x_, n_] :=
        Show[Graphics[Nest[IFS[{AffineMap[0 °, 0 °, 1, 1, 0, 0], AffineMap[45 °, 45 °, 0.65,

            0.5, −0.1, 3], AffineMap[55 °, 55 °, 0.7, 0.7, 0.1, 1.3], AffineMap[−15 °,

            −15 °, 0.6, $\dfrac{2}{3}$, 0, 3], AffineMap[−55 °, −55 °, −0.6, $\dfrac{2}{3}$, 0.13, 3]}], x, n]],

        AspectRatio → Automatic, PlotRange → All];

      tree2[Polygon[{{−0.25, 0}, {0.2, 0}, {0.2, 3}, {−0.2, 3}}], 5]

Another example, resembling a young Baobab tree of Africa. Here the trunk of the tree is formed from a rectangle.

      tree3[x_, n_] :=
        Show[Graphics[Nest[IFS[{AffineMap[0 °, 0 °, 1, 1, 0, 0], AffineMap[20 °, 20 °, 0.5,
            0.5, 0, 0.9], AffineMap[−50 °, −50 °, 0.4, 0.4, 0, 0.5], AffineMap[35 °,
            35 °, 0.5, 0.5, 0, 0.1], AffineMap[−45 °, −45 °, 0.5, .5, 0, 0.9]}], x, n]],
        AspectRatio → Automatic, PlotRange → All];

      tree3[Polygon[{{−0.1, 0}, {0.1, 0}, {0.1, 1}, {−0.1, 1}}], 5]

Exercise:

1) Construct other fractal-like trees and bushes, using different initial sets, and different branching arrangements. Start with a plan consisting of a polygon and 2 or more affine copies of it. Iterate a few times, and then, if needed, make alterations to the plan, by changing the trunk, or changing the rotation angles, scaling factors or positions of the branches, or adding extra branches. Possibly, use a real tree as a model.

2) Construct fractals described in 5.4.1, the section on Sierpinski relatives, using an initial set consisting of more than one graphics primitive.

5.4.7 Fractals Constructed Using Regular Polygons

Vertices of regular polygons may be obtained by loading the package **<<Geometry`Polytopes`**. Go to the Master Index in *Mathematica*'s Help and look up Polytopes to find out how to obtain vertices of regular polygons. We start with a regular hexagon, H, with vertices:

$$\left\{\left\{\tfrac{\sqrt{3}}{2}, \tfrac{1}{2}\right\}, \{0, 1\}, \left\{-\tfrac{\sqrt{3}}{2}, \tfrac{1}{2}\right\}, \left\{-\tfrac{\sqrt{3}}{2}, -\tfrac{1}{2}\right\}, \{0, -1\}, \left\{\tfrac{\sqrt{3}}{2}, -\tfrac{1}{2}\right\}\right\}.$$

We construct affine maps to make 6 affine copies of a scaled version of H which rotate, contract and translate the copies in the direction of the mid-points of the sides, as shown in the following example:

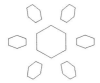

As we shall vary the contractivity factors and translation distances, we define the following iterated function system which has 6 parameters, contains the identity map and 6 affine maps. The arguments a and b are the scaling factors of the affine maps, the argument c determines the length of the translation, the argument x represents the starting graphics primitive while n is the number of iterations. The command **ColorFunction** has also been included, with the default color black (see 1.8.4).

$$\textbf{hexfractal1[a\_, b\_, c\_, x\_, n\_, col\_: RGBColor[0, 0, 0]]} :=$$
$$\textbf{Show}\Big[\textbf{Graphics}\Big[\textbf{Nest}\Big[\textbf{IFS}\Big[\Big\{\textbf{AffineMap[0°, 0°, 1, 1, 0, 0], AffineMap[0, 0, a, b, c, 0],}$$

$$\textbf{AffineMap}\Big[60°, 60°, a, b, c\,\frac{1}{2}, c\,\frac{\sqrt{3}}{2}\Big], \textbf{AffineMap}\Big[120°, 120°, a, b, -c\,\frac{1}{2},$$

$$c\,\frac{\sqrt{3}}{2}\Big], \textbf{AffineMap[180°, 180°, a, b, -c, 0], AffineMap}\Big[240°, 240°, a,$$

$$b, -c\,\frac{1}{2}, -c\,\frac{\sqrt{3}}{2}\Big], \textbf{AffineMap}\Big[300°, 300°, a, b, c\,\frac{1}{2}, -c\,\frac{\sqrt{3}}{2}\Big]\Big\}\Big], x, n\Big]\Big],$$

$$\textbf{AspectRatio} \rightarrow \textbf{Automatic, PlotRange} \rightarrow \textbf{All, ColorOutput} \rightarrow \textbf{(col \&)}\Big];$$

We shall apply the above IFS, for particular values of the parameters to scaled copies of the hexagon H defined above. We define a function **hex1**:

$$\textbf{hex1[k\_]} := \textbf{Line}\Big[\Big\{\Big\{\frac{k\sqrt{3}}{2}, \frac{k}{2}\Big\}, \{0, k\}, \Big\{-\frac{k\sqrt{3}}{2}, \frac{k}{2}\Big\},$$

$$\Big\{-\frac{k\sqrt{3}}{2}, -\frac{k}{2}\Big\}, \{0, -k\}, \Big\{\frac{k\sqrt{3}}{2}, -\frac{k}{2}\Big\}, \Big\{\frac{k\sqrt{3}}{2}, \frac{k}{2}\Big\}\Big\}\Big];$$

We choose values for a, b and c and apply **hexfractal1** to **hex1[0.9]**, 4 times:

hexfractal1[0.8, 0.4, 5, hex1[0.9], 4]

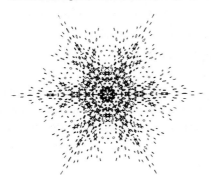

Exercise:

1) Try out the IFS for various values of the parameters.

2) Construct a variation on the above IFS by eliminating the identity map.

We now construct an IFS **hexfractal2** by adding to **hexfractal1** 6 more affine maps which contract, rotate and translate **hex1** in the direction of its vertices. When the resulting IFS is applied once to **hex1**, the resulting figure will resemble the following:

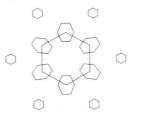

We use a technique described in a previous section to apply the above iterated function systems to an initial set consisting of one or more graphics primitives, with attached graphics directives and a background color. We first define and name the initial set, which consists of 2 hexagons. We do not use the command **Show** as it is not necessary to display the initial set. As the color will be determined by the initial set, we omit the color option when defining **hexfractal2**.

$$hex2[k\_, l\_, t\_, c\_, d\_, b\_] :=$$
$$\textbf{Graphics}[\{\{\textbf{Thickness}[t], c, \textbf{Line}[\{\{0.5\sqrt{3}\,k, 0.5\,k\}, \{0, k\}, \{-0.5\sqrt{3}\,k, 0.5\,k\},$$
$$\{-0.5\sqrt{3}\,k, -0.5\,k\}, \{0, -k\}, \{0.5\sqrt{3}\,k, -0.5\,k\}, \{0.5\sqrt{3}\,k, 0.5\,k\}\}]\},$$
$$\{\textbf{Thickness}[t], d, \textbf{Line}[\{\{0.5\sqrt{3}\,l, 0.5\,l\}, \{0, l\}, \{-0.5\sqrt{3}\,l, 0.5\,l\},$$
$$\{-0.5\sqrt{3}\,l, -0.5\,l\}, \{0, -l\}, \{0.5\sqrt{3}\,l, -0.5\,l\}, \{0.5\sqrt{3}\,l, 0.5\,l\}\}]\}\},$$
$$\textbf{AspectRatio} \rightarrow \textbf{Automatic, PlotRange} \rightarrow \textbf{All, Background} \rightarrow b];$$

Here is another initial set, consisting of 2 filled hexagons:

```
hex3[k_, l_, c_, d_, b_] :=
    Graphics[{{c, Polygon[{{0.5 √3 k, 0.5 k}, {0, k}, {−0.5 √3 k, 0.5 k},
            {−0.5 √3 k, −0.5 k}, {0, −k}, {0.5 √3 k, −0.5 k}}]},
        {d, Polygon[{{0.5 √3 l, 0.5 l}, {0, l}, {−0.5 √3 l, 0.5 l},
            {−0.5 √3 l, −0.5 l}, {0, −l}, {0.5 √3 l, −0.5 l}}]}},
        AspectRatio → Automatic, PlotRange → All, Background → b];
```

The following IFS consists of 13 affine maps, the first map is the identity map, the following 6 maps are the same as those of the previous IFS, while the last 6 maps contract, rotate and translate the initial set in the direction of its vertices:

```
hexfractal2[a_, b_, c_, d_, e_, f_, x_, n_] :=
    Show[Graphics[Nest[IFS[{AffineMap[0, 0, 1, 1, 0, 0],
            AffineMap[0, 0, a, b, c, 0], AffineMap[60°, 60°, a, b, 0.5 c, 0.5 √3 c],
            AffineMap[120°, 120°, a, b, −0.5 c, 0.5 √3 c], AffineMap[180°,
                180°, a, b, −c, 0], AffineMap[240°, 240°, a, b, −0.5 c, −0.5 √3 c],
            AffineMap[300°, 300°, a, b, 0.5 c, −0.5 √3 c],
            AffineMap[30°, 30°, d, e, 0.5 √3 f, 0.5 f], AffineMap[90°, 90°, d, e, 0, f],
            AffineMap[150°, 150°, d, e, −0.5 √3 f, 0.5 f], AffineMap[210°,
                210°, d, e, −0.5 √3 f, −0.5 f], AffineMap[270°, 270°, d, e, 0, −f],
            AffineMap[330°, 330°, d, e, 0.5 √3 f, −0.5 f]}], x, n]],
        AspectRatio → Automatic, PlotRange → All];
```

The above command can be used to construct a single color image with a colored background. (We have used 4 iterations to obtain a good printout, however, on the screen, we obtain a better image with 3 iterations.) (Color Fig 5.6)

```
hexfractal2[0.6, 0.3, 4, 0.2, 0.2, 4,
    hex2[2, 3, 0.01, RGBColor[1, 1, 1], RGBColor[1, 1, 1], RGBColor[0, 0.5, 0.7]], 4]
```

An example of a bi-colored fractal with a white background: (Color Fig 5.7)

```
hexfractal2[0.7, 0.5, 2, 0.2, 0.2, 5,
    hex2[2, 1, 0.01, RGBColor[0, 0, 1], RGBColor[1, 0, 1], RGBColor[1, 1, 1]], 3]
```

Here are some other programs:

```
hexfractal2[0.6, 0.3, 4, 0.3, 0.2, 4, hex3[4, 3, RGBColor[0, 0.8, 0.4],
    RGBColor[1, 1, 1], RGBColor[0.699229, 0.183597, 0.464851]], 2]
```

> **hexfractal2[0.7, 0.7, 7, 0.4, 0.4, 2, hex2[2, 1, 0.01, RGBColor[1, 1, 1],**
>   **RGBColor[0, 1, 0.7], RGBColor[0.699229, 0.183597, 0.464851]], 3]**

> **hexfractal2[0.6, 0.3, 4, 0.2, 0.2, 4,**
>   **hex2[2, 1, 0.018, RGBColor[1, 1, 1], RGBColor[0, 1, 0.5], RGBColor[0, 0.5, 0]], 3]**

In this example, we use **hex3** as initial set: (Color Fig 5.8)

> **hexfractal2[0.3, 0.3, 2.2, 0.2, 0.2, 1.1,**
>   **hex3[1, 0.5, RGBColor[0, 0, 1], RGBColor[0, 1, 0], RGBColor[1, 1, 1]], 4]**

Exercise:

1) Experiment with the above IFS.

2) Omit the identity map and experiment with the resulting IFS.

3) Adapt the programs to other regular polygons.

Examples with a square initial set:

> **square1[k_, l_, t_, c_, d_, b_] := Show[**
>   **Graphics[{{Thickness[t], c, Line[k {{−1, −1}, {1, −1}, {1, 1}, {−1, 1}, {−1, −1}}]},**
>   **{Thickness[t], d, Line[l {{−1, −1}, {1, −1}, {1, 1}, {−1, 1}, {−1, −1}}]}}],**
>   **AspectRatio → Automatic, PlotRange → All, Background → b,**
>   **DisplayFunction → Identity];**

> **squareFractal1[a_, b_, c_, d_, e_, f_, x_, n_] := Show[**
>   **Graphics[Nest[IFS[{AffineMap[0°, 0°, 1, 1, 0, 0], AffineMap[45°, 45°, a, b, c, c],**
>   **AffineMap[135°, 135°, a, b, −c, c], AffineMap[225°, 225°, a, b, −c, −c],**
>   **AffineMap[315°, 315°, a, b, c, −c], AffineMap[0°, 0°, d, e, f, 0],**
>   **AffineMap[90°, 90°, d, e, 0, f], AffineMap[180°, 180°, d, e, −f, 0],**
>   **AffineMap[270°, 270°, d, e, 0, −f]}], x, n]], AspectRatio → Automatic,**
>   **PlotRange → All, DisplayFunction → $DisplayFunction];**

> **squareFractal1[0.6, 0.6, 2, 0.3, 0.3, 2,**
>   **square1[1.2, 0.2, 0.03, RGBColor[1, 1, 1], RGBColor[1, 1, 1], RGBColor[0, 0, 0]], 3]**

Here are some programs to try:

squareFractal1[0.5, 0.5, 1, 0.5, 0.5, 2,
    square1[0.5, 0.3, 0.025, RGBColor[1, 0, 1], RGBColor[0, 1, 1], RGBColor[0, 0, 0]], 3]

squareFractal1[0.7, 0.5, 1, 0.5, 0.3, 4,
    square1[1, 0.8, .004, RGBColor[1, 0, 0], RGBColor[0, 0, 1], RGBColor[1, 1, 1]], 3]

squareFractal1[0.5, 0.4, 2, 0.6, 0.3, 4, square1[1.5, 0.6, .02, RGBColor[1, 1, 1],
    RGBColor[0.184314, 0.937255, 0.6], RGBColor[0, 0, 0.8]], 3]

5.4.8 Constructing Affine Fractals Using Parametric Plots

Roman Maeder's command IFS can also be applied to an initial set generated by the the command **ParametricPlot**, which may include the **PlotStyle** option, determining colors or thickness. We must first define and name the initial set. In the following example, our initial set will consist of a curve, C, and several differently colored affine transformations of C. These same affine transformations, together with the identity map, will form the IFS which we shall apply to the initial set to generate a fractal. The reason for starting with the above initial set instead of C itself, is that we can obtain a multiply colored fractal in this way.

Consider a curve in $\mathbb{R}^2$ with parametric equations: $x = f[t]$, $y = g[t]$. We can apply an affine transformation to this curve to obtain the curve with parametric equations: $x = a f[t] + b g[t] + c$; $y = \alpha f[t] + \beta g[t] + \gamma$.

In effect, for each separate value of t we apply the above affine transformation to the point $\{x, y\} = \{f[t], g[t]\}$.

By choosing a set of values for t and plotting the resultant curve by using the command **ParametricPlot**, we are applying the affine transformations to the whole curve in $\mathbb{R}^2$.

Consider the curve C given in parametric form by: $x = 0.5 t^2 \sin[5 t]$; $y = t$.
We plot this curve using the command **ParametricPlot**. (Color Fig 5.9)

> **ParametricPlot[{0.5 t^2 * Sin[5 t], t}, {t, 0, 0.5}, AspectRatio → Automatic,**
>   **PlotStyle –> {RGBColor[0.59, 0.41, 0.246], Thickness[0.08]}, Ticks → {False, True}]**

We apply an affine map to the point with parameter t on the above curve:

$$\textbf{AffineMap}\left[60°, 60°, \frac{2}{5}, \frac{2}{5}, 0.006, 0.14\right][\{0.5 \, t^2 * \text{Sin}[5 \, t], t\}]$$

$$\left\{0.006 - \frac{\sqrt{3} \, t}{5} + 0.1 \, t^2 \, \text{Sin}[5 \, t], \, 0.14 + \frac{t}{5} + 0.173205 \, t^2 \, \text{Sin}[5 \, t]\right\}$$

The first curve has been contracted by the factor $\frac{2}{5}$, rotated through 60°, and translated by the vector (0.006, 0.4). If we look at the two parametric curves on the same diagram we see that we have the beginning of a 'fern': (Color Fig 5.10)

ParametricPlot$\Big[\{\{0.5\,t^2 * Sin[5\,t],\ t\},$

$$\Big\{0.006` - \frac{\sqrt{3}\,t}{5} + 0.1`\,t^2\,Sin[5\,t],\ 0.14` + \frac{t}{5} + 0.17320508075688773`\,t^2\,Sin[5\,t]\Big\}\Big\},$$

$\{t,\ 0,\ 0.5\},\ AspectRatio \rightarrow Automatic,$
$\quad PlotStyle \rightarrow \{\{RGBColor[0.59,\ 0.41,\ 0.246],\ Thickness[.02]\},$
$\qquad \{RGBColor[0.414,\ 0.6,\ 0.226],\ Thickness[.0165]\}\},\ Ticks \rightarrow \{False,\ True\}\Big]$

We now apply an additional 6 affine transformations to the original curve, together with color specifications for each, and plot all 8 curves on the same diagram. The resulting image, which we call **fernStart** will be our initial set for constructing a fern-like fractal. (Color Fig 5.11)

fernStart = ParametricPlot$\Big[$

$\quad \Big\{\{0.5\,t^2 * Sin[5\,t],\ t\},\ AffineMap\Big[-60°,\ -60°,\ \dfrac{1}{6},\ \dfrac{1}{6},\ 0.0752,\ 0.5\Big][\{0.5\,t^2 * Sin[5\,t],\ t\}],$

$\qquad AffineMap\Big[60°,\ 60°,\ \dfrac{1}{6},\ \dfrac{1}{6},\ 0.0752,\ 0.5\Big][\{0.5\,t^2 * Sin[5\,t],\ t\}],$

$\qquad AffineMap\Big[-60°,\ -60°,\ \dfrac{1}{3},\ \dfrac{1}{3},\ 0.045,\ 0.3151\Big][\{0.5\,t^2 * Sin[5\,t],\ t\}],$

$\qquad AffineMap\Big[60°,\ 60°,\ \dfrac{1}{3},\ \dfrac{1}{3},\ 0.045,\ 0.3151\Big][\{0.5\,t^2 * Sin[5\,t],\ t\}],$

$\qquad AffineMap\Big[-60°,\ -60°,\ \dfrac{2}{5},\ \dfrac{2}{5},\ 0.006,\ 0.14\Big][\{0.5\,t^2 * Sin[5\,t],\ t\}],$

$\qquad AffineMap\Big[60°,\ 60°,\ \dfrac{2}{5},\ \dfrac{2}{5},\ 0.006,\ 0.14\Big][\{0.5\,t^2 * Sin[5\,t],\ t\}],$

$\qquad AffineMap\Big[0°,\ 0°,\ \dfrac{1}{6},\ \dfrac{1}{3},\ .0752,\ 0.5\Big][\{0.5\,t^2 * Sin[5\,t],\ t\}]\Big\},$

$\quad \{t,\ 0,\ 0.5\},\ AspectRatio \rightarrow Automatic,$
$\quad PlotStyle \rightarrow \{\{RGBColor[0.59,\ 0.41,\ 0.246],\ Thickness[0.02]\},$
$\qquad \{RGBColor[0.6,\ 0.82,\ 0.08]\},\ \{RGBColor[0.6,\ 0.82,\ 0.08]\},$
$\qquad \{RGBColor[0.45,\ 0.676,\ 0.15]\},\ \{RGBColor[0.45,\ 0.675,\ 0.15]\},$
$\qquad \{RGBColor[0.414,\ 0.6,\ 0.226],\ Thickness[0.015]\},\ \{RGBColor[0.41,\ 0.6,\ 0.226],$
$\qquad Thickness[0.015]\},\ \{RGBColor[0.53,\ 0.85,\ 0.1]\}\},\ Axes \rightarrow False\Big]$

The above image, **fernStart**, was generated by applying the identity map and 7 affine contraction mappings to the curve C. These 8 maps constitute an IFS. We now apply this IFS to **fernStart**: (Color Fig 5.12) fern[x_, n_] :=

Show$\Big[$Graphics$\Big[$Nest$\Big[$IFS$\Big[\Big\{$AffineMap[0, 0, 1, 1, 0, 0], AffineMap$\Big[-60°, -60°, \dfrac{1}{6}, \dfrac{1}{6},$

0.0752, 0.5$\Big]$, AffineMap$\Big[60°, 60°, \dfrac{1}{6}, \dfrac{1}{6},$ 0.0752, 0.5$\Big]$, AffineMap$\Big[-60°,$

$-60°, \dfrac{1}{3}, \dfrac{1}{3},$ 0.045, 0.3151$\Big]$, AffineMap$\Big[60°, 60°, \dfrac{1}{3}, \dfrac{1}{3},$ 0.045, 0.3151$\Big]$,

AffineMap$\Big[-60°, -60°, \dfrac{2}{5}, \dfrac{2}{5},$ 0.006, 0.14$\Big]$, AffineMap$\Big[60°, 60°,$

$\dfrac{2}{5}, \dfrac{2}{5},$ 0.006, 0.14$\Big]$, AffineMap$\Big[0°, 0°, \dfrac{1}{6}, \dfrac{1}{3},$ .0752, 0.5$\Big]\Big\}\Big]$, x, n$\Big]\Big]$,

Axes → False, AspectRatio → Automatic, AxesOrigin → {0, 0}$\Big]$;

fern[fernStart, 3]

One can use the fact that a thickness directive can be added to a parametric plot to construct a tree-like fractal.

First we obtain the initial set, which forms the trunk:

t4 = ParametricPlot[{x, Abs[x^3 − 1]}, {x, 0.2, 1.26},
    PlotRange → All, AspectRatio → Automatic, PlotStyle → Thickness[0.06]]

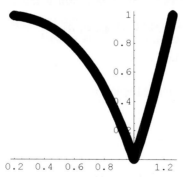

The following command was worked out one affine map at a time and applied to the initial set.

tree4[x_, n_] :=
    Show[Graphics[Nest[IFS[{AffineMap[0, 0, 1, 1, 0, 0], AffineMap[−45°, −45°, −0.6,
        0.6, 1.65, 0.6], AffineMap[35°, 35°, 0.5, 0.5, −0.3, 0.68], AffineMap[−10°,
        −10°, −0.5, 0.5, 0.9, 0.9], AffineMap[40°, 40°, 0.2, 0.3, 1, 0.7]}], x, n]],
    Axes → False, AspectRatio → Automatic, AxesOrigin → {0, 0}];

tree4[t4, 5]

Exercise:
Construct other trees, bushes, ferns by experimenting with the above technique.

5.4.9 Constructing Fractals from Polygonal Arcs

Here is another method of constructing a large collection of interesting affine fractals:
Choose points $\{x_r, y_r\}$, $1 \leq r \leq n$ in the x-y plane with $x_1 = 0$, $x_n = 1$ and $x_r \neq x_s$, if $|r - s| \neq 1$ and such that the distance between each point and its successor is less than 1. For each r, let f_r be a similarity which maps $\{0, 0\}$ onto $\{x_r, y_r\}$ and $\{1, 0\}$ onto $\{x_{r+1}, y_{r+1}\}$. The x-co-ordinates of the points do not have to be in increasing order. A set of such maps, with or without the identity map can be used to generate a fractal.

In order to save the work of calculating similarities, we make the following definition:

$$
\begin{aligned}
&\textbf{similarity1[p\_, q\_, s\_, t\_] :=} \\
&\quad \textbf{AffineMap}\left[\textbf{ArcTan}\left[\frac{(t-q)}{s-p}\right],\right. \\
&\quad \left.\textbf{ArcTan}\left[\frac{(t-q)}{s-p}\right], \sqrt{(s-p)^2 + (t-q)^2}, \sqrt{(s-p)^2 + (t-q)^2}, \textbf{p, q}\right];
\end{aligned}
$$

Check that the above map is a similarity which maps $\{0, 0\}$ onto $\{p, q\}$ and $\{1, 0\}$ onto $\{s, t\}$.

Here is an example using just 3 points:

```
simfractal1[x_, n_] :=
    Graphics[Nest[IFS[{similarity1[0, 0, 0.5, 0.45], similarity1[0.5, 0.45, 1, 0]}], x, n],
    AspectRatio → Automatic, AxesOrigin → {0, 0}];
```

Show[simfractal1[Line[{{0, 0}, {1, 0}}], 1], Axes → True]

Iterating 10 times we get:

Show[simfractal1[Line[{{0, 0}, {1, 0}}], 10]]

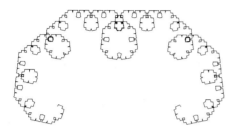

Here is a variation of the above including the identity map. We have used a line as the initial set.

simfractal2[x_, n_] :=
    Show[Graphics[Nest[IFS[{AffineMap[0, 0, 1, 1, 0, 0], similarity1[0, 0, 0.5, 0.45],
        similarity1[0.5, 0.45, 1, 0]}], x, n]],
      Axes → False, AspectRatio → Automatic, AxesOrigin → {0, 0}];

simfractal2[Line[{{0, 0}, {1, 0}}], 10]

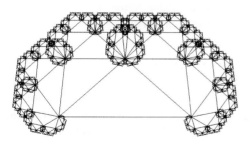

Another example:

simfractal3[x_, n_] :=
    Graphics[Nest[IFS[{AffineMap[0, 0, 1, 1, 0, 0], similarity1[0, 0, 0.250, 0.297],
        similarity1[0.250, 0.297, 0.750, −0.297], similarity1[0.750, −0.297, 1, 0]}],
      x, n], AspectRatio → Automatic];

Show[simfractal3[Circle[{0.5, 0}, 0.3], 1], Axes → True]

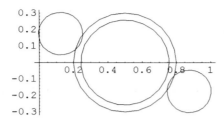

Show[simfractal3[Circle[{0.5, 0}, 0.3], 6]]

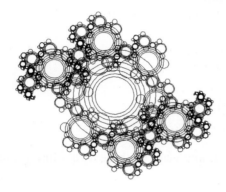

Try the next 2 programs in which the option **ColorOutput** has been included.

```
simfractal4[x_, n_] := Show[
    Graphics[Nest[IFS[{similarity1[0, 0, 0.4, −0.08], similarity1[0.4, −0.08, 0.5, 0.25],
            similarity1[0.25, 0.5, 0.5, 0.25], similarity1[0.25, 0.5, 0.6, 0.45],
            similarity1[0.6, 0.45, 0.7, 0.1], similarity1[0.7, 0.1, 1, 0]}], x, n]],
    Axes → False, AspectRatio → Automatic, ColorOutput →
        (RGBColor[0.343755, 0.609384, 0.0781262] &)];
```

simfractal4[Circle[{0, 0}, 0.5], 4]

```
simfractal5[x_, n_] := Show[Graphics[
    Nest[IFS[{similarity1[0, 0, 0.250, 0.297], similarity1[0.250, 0.297, 0.350, −0.297],
            similarity1[0.350, −0.297, 1, 0]}], x, n]], AspectRatio → Automatic,
    ColorOutput → (RGBColor[0.10547, 0.800793, 0.332036] &)];
```

simfractal5[Line[{{0, 0}, {1, 0}}], 10]

Exercise:

1) Try the above examples with different starting sets, colors or number of iterations.

2) Use the above methods to construct other fractals.

5.5 Construction of 2D Fractals Using the Random Algorithm

5.5.1 Introduction

In 5.1 to 5.4, we constructed fractals by repeatedly applying iterated function systems to plane sets of points. The deterministic algorithm was used. In this section, we construct plane fractals by applying iterated function systems to a single point using the random algorithm.

Consider the IFS $\{X \mid w_1, w_2, w_3\}$ where $X \subseteq \mathbb{R}^2$ and where w_1, w_2 and w_3 are contraction mappings on $\mathbb{R}^2$.

Let A be the fractal determined by repeatedly applying the function $w = w_1 \bigcup w_2 \bigcup w_3$ to any compact subset of X, so A is the attractor of the IFS $\{X \mid w_1, w_2, w_3\}$.

Choose a point $a \in X$. Let $S = (b_1, b_2, ..., b_n)$ be a random sequence of the numbers 1, 2 and 3. For each i = 1, 2, ... n, the probability that $b_i = 1$ or 2 or 3 is $\frac{1}{3}$.

Now consider the sequence: $(a, w_{b_1}[a], w_{b_2}[w_{b_1}[a]], w_{b_3}[w_{b_2}[w_{b_1}[a]]], ...)$. It can be proved that for suitable choices of m and n, if the first m points of the sequence are dropped and the remaining n − m points are plotted, the resulting set has a high probability of approximating A. This method of construction is called 'The Random Iteration Algorithm'. The algorithm can be extended to an IFS consisting of any number of contraction maps on X.

5.5.2 Roman Maeder's Package: The ChaosGame

In this section we load the following packages of Roman Maeder:

> << **ProgrammingInMathematica`AffineMaps`**

> << **ProgrammingInMathematica`IFS`**

> << **ProgrammingInMathematica`ChaosGame`**

In 5.4 we constructed, with the deterministic algorithm, a Sierpinski relative using the following IFS which we denote **ifs2**:

> **ifs2 = IFS[{AffineMap[180°, 180°, 0.5, −0.5, −1, 1],**
> **AffineMap[90°, 90°, 0.5, 0.5, −1, −1], AffineMap[90°, 90°, 0.5, −0.5, 1, −1]}]**

We shall construct the same fractal using the Random Algorithm. The package of Roman Maeder's entitled: **"ProgrammingInMathematica`ChaosGame"** can be used to construct affine fractals with the random algorithm.

In the above package, the command **AffineMap** can be applied to 2D graphics primitives such as points, lines, polygons etc.

> **? ChaosGame**
>
> ChaosGame[−ifs−, n, opts..] iterates random maps applied to a point n times and plots the result.

We simply apply the command **ChaosGame** to **ifs2** a large number of times:

> **ChaosGame[ifs2, 50000];**

The option **Coloring** can be used to assign 3 colors, c_1, c_2, and c_3 to sections of the above image as follows. Let w_1, w_2 and w_3 denote the 3 maps of **ifs2**. Images of points under w_i are assigned color c_i.

> **? Coloring**
>
> Coloring −> val is an option of ChaosGame. Possible values are None, Automatic, a list of color
>   directives, or a function of two arguments so that val[i, n] is the color of the ith out of n objects.

We choose 3 colors with the color selector. (Color Fig 5.13)

> **ChaosGame[ifs2, 50000,**
> **Coloring->{RGBColor[0.902358, 0.0937514, 0.820325],RGBColor[0.203128, 0.68751, 0.14844],**
> **RGBColor[0.500008, 0, 0.996109]}];**

Exercise:

Apply the above procedure to other Sierpinski Relatives.

■ Using the command AverageContraction

Notice that the 3 affine maps of **ifs2** have the same contractivity factor. Each of the maps would map a square of side-length 1 onto a square of side-length 0.25. The affine contraction map defined by: **AffineMap[$\theta°$, $\theta°$, r, s, e, f]** maps a square of side-length 1 onto a parallelogram of area rs. Suppose we use the random algorithm to generate a fractal, **frac**, using two contraction mappings w_1 and w_2 with contractivity factors c_1 and c_2 respectively such that $c_1 < c_2$, and we denote the images of **frac** under w_1, w_2 by A_1, A_2 respectively, then A_2 will be sparser than A_1. For example, in 5.4.3, we constructed a fractal using the following IFS:

> **ifs3 = IFS[{AffineMap[$-2°$, $-2°$, 0.02, 0.6, -0.14, -0.8],**
> **AffineMap[0, 0, 0.6, 0.4, 0, 1.2], AffineMap[$-30°$, $-30°$, 0.4, 0.7, 0.6, -0.35],**
> **AffineMap[$30°$, $30°$, 0.4, 0.65, -0.7, -0.5]}]**

The command **Probabilities** can be used to distribute the points more evenly as follows. We first use the command **AverageContraction**:

> **? AverageContraction**

> AverageContraction[map] gives the average area contraction factor (the determinant) of an affine map.

So an affine map has average contraction k if it maps a square of side-length 1 onto a parallelogram of area k. We calculate the **AverageContraction** for each of the affine maps in **ifs3**:

> **AverageContraction/@ {AffineMap[$-2°$, $-2°$, 0.02, 0.6, -0.14, -0.8],**
> **AffineMap[0, 0, 0.6, 0.4, 0, 1.2], AffineMap[$-30°$, $-30°$, 0.4, 0.7, 0.6, -0.35],**
> **AffineMap[$30°$, $30°$, 0.4, 0.65, -0.7, -0.5]}**

> {0.012, 0.24, 0.28, 0.26}

We now normalise the above average contractions, so that their sum is 1.

> **ifs3probs = $\dfrac{\%}{\textbf{Apply[Plus, \%]}}$**

> {0.0151515, 0.30303, 0.353535, 0.328283}

We now use the option **Probabilities** applied to **ifs3probs** when applying **ChaosGame** to **ifs3**. We also use the option **Coloring**.

> **? Probabilities**

> Probabilities $->$ {pr..} is an option of IFS that gives the probabilities of the maps for the chaos game.

> **ChaosGame[ifs3, 100000, Probabilities -> ifs3probs,**
> **Coloring->Automatic];**

Color Fig 5.1.4 illustrates the Collage Theorem (5.4.3) very clearly.

Exercise:

Apply the above procedures to fractals constructed in 5.4.3 and 5.4.4.

Chapter 6

Constructing Non-Affine and 3D Fractals Using the Deterministic and Random Algorithms

Introduction

In the previous Chapter, we defined an Iterated Function System (IFS), $\{\mathbb{C} \mid w_1, w_2, ..., w_n\}$, on $\mathbb{C}$. Now, for each r, w_r was defined to be a 2D affine map and so we were able to use Roman Maeder's commands **AffineMap** and **IFS** to construct certain fractals.

In this chapter we will be doing two types of constructions. Firstly, we will construct 2D fractals, but we will be using Iterated Function Systems whose constituent maps are not affine.

Secondly, we will be using Iterated Function Systems whose constituent maps are affine but are applied to 3D subsets of $\mathbb{R}^3$.

6.1 Construction of Julia Sets of Quadratic Functions as Attractors of Non-Affine Iterated Function Systems

6.1.1 Julia Sets and Filled Julia Sets

Let $\mathbb{C}$ denote the set of complex numbers. Let f: $\mathbb{C} \to \mathbb{C}$, be a polynomial in z of degree ≥ 2 where $z \in \mathbb{C}$.

The orbit of z under f is bounded if there exists $K > 0$ such that $|f^n[z]| < K$ for all $n \in \mathbb{N}$.

The filled Julia set of f is the set:

$K_f = \{z \in \mathbb{C} \mid$ the orbit of z under f is bounded$\}$.

The Julia set of f, J_f, is the boundary of the filled Julia set of f.

Examples of Julia sets and filled Julia sets for the family of complex quadratic functions $\{Q_c : z \to z^2 + c \mid c \in \mathbb{C}\}$ are shown below:

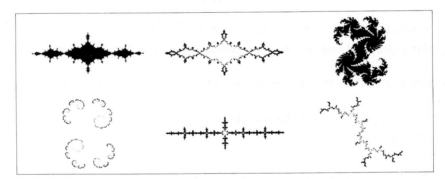

6.1.2 The Quadratic Family Q_c

Consider the family of functions $\{Q_c : z \to z^2 + c \mid c \in \mathbb{C}\}$, where c is a parameter.

Barnsley (1988) shows that the the Julia set of Q_c is the attractor of the IFS $\{A \mid w_1, w_2\}$, where A is a subset of $\mathbb{C}$ and is chosen in a way that depends on the parameter c, and, for $z \in \mathbb{C}$, $w_1(z) = \sqrt{z - c}$, $w_2(z) = -\sqrt{z - c}$.

The maps w_1 and w_2 are not contraction maps on the whole of $\mathbb{C}$ but on the subset A of $\mathbb{C}$. For this reason it seems that the starting point to be chosen for the iterative process which will be used to generate the Julia set might present a problem in that we would have to ensure that the starting point is chosen to lie in A. However it can be shown that the starting point can be chosen to be any point in $\mathbb{C}$ except in the case $c = 0$ in which case the point 0 must be avoided (Keen (1994)).

■ **Choice of c**

Julia sets can be divided into 2 classes. They are either connected or totally disconnected. Roughly speaking, a set which is connected is all in one piece (no breaks) while a set which is totally disconnected is like a cloud of dust particles, its only connected components being points. If $|c| \leq 2$ then both types of Julia sets occur and both types are generally aesthetically pleasing. If $|c| > 2$ then the Julia set of Q_c is totally disconnected. This fact, together with the largeness of c, produces Julia sets which are rather dispersed and spread out and so not very pleasing. We show below: a) The Julia set of $Q_{-0.757}$ which is connected; b) The Julia set of $Q_{0.306-0.0214I}$ which is totally disconnected; c) The Julia set of $Q_{2.1I}$ which is totally disconnected.

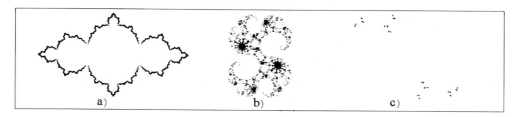

6.1.3 Construction of Julia Sets Using the Deterministic Algorithm

The *Mathematica* commands required to generate the Julia sets for the quadratic function Q_c using the deterministic algorithm on the IFS $\{A \,|\, w_1, w_2\}$ are given below. Each command will be discussed separately. Note that we cannot use Roman Maeder's add on command **IFS** as it applies only to affine maps.

Let $w = w_1 \bigcup w_2$ be the contraction mapping on A as defined in 5.2.2.

Using the deterministic algorithm to generate the attractor of w and starting with the single point set $\{z\}$ we apply the function w, n times. The first application is as follows:

STEP 1:
Consider the function w operating on $\{z\}$:
$$w[\{z\}] = \{w_1(z), \ w_2(z)\}$$
$$= \left\{ \sqrt{z-c}, \ -\sqrt{z-c} \right\}.$$

Theoretically w is a function of the set $\{z\}$, however, when using *Mathematica* commands we omit the curly braces and speak loosely of w as a function of z i.e. we write 'w[z]'.
We define w as a *Mathematica* function. (For reasons which will appear later, we replace c with k.)

$$w[z\_] := \left\{ \sqrt{z - k}, \, -\sqrt{z - k} \right\};$$

To find w[1-2I], when c = −1, we use the replacement command /. to replace k by -1.

$$w[1 - 2\,I] \, / . \, k \to -1$$

$$\left\{ \sqrt{2 - 2\,i}, \, -\sqrt{2 - 2\,i} \right\}$$

As the construction of a Julia set involves many calculations, it is advisable to speed the process by working with approximations. This can be done in the above case by writing the argument of the function as an approximation:

$$w[1.0 - 2\,I] \, / . \, k \to -1.0$$

{ 1.55377-0.643594 i,-1.55377+0.643594 i}

STEP 2:
We must now perform the second iterative step by finding the image of the above set of points under w. So we need to evaluate w[w[z]]. The *Mathematica* command **Map** is used to generate the set of 4 points **w[w[1 − 2 I]]** as follows:

$$\textbf{Map}[\textbf{w}, \, \textbf{w}[1.0 - 2\,\textbf{I}]] \, / . \, \textbf{k} \to -1.$$

{{1.6105 − 0.199812 i, −1.6105 + 0.199812 i}, {0.384234 + 0.837502 i, −0.384234 − 0.837502 i}}

STEP 3:
Continuing in this manner we must now generate the set, w[w[w[1 − 2 I]]] which consists of 8 points. However w cannot be mapped onto the last output generated since arguments of w are complex numbers, and the last output generated is a list of pairs of complex numbers. Some of the brackets must be removed in order to get a list of co-ordinates of points. This is done by using the *Mathematica* command **Flatten**.

$$\textbf{h} = \textbf{Flatten}[\textbf{Map}[\textbf{w}, \, \textbf{w}[1.0 - 2\,\textbf{I}]] \, / . \, \textbf{k} \to -1, \, 1]$$

{1.6105 − 0.199812 i, −1.6105 + 0.199812 i, 0.384234 + 0.837502 i, −0.384234 − 0.837502 i}

Now consider the following pure function, g, in which # can be replaced by a list of the above type:

$$\textbf{g} = \textbf{Flatten}[(\textbf{Map}[\textbf{w}, \, \#] \, / . \, \textbf{k} \to -1), \, 1] \, \&;$$

When g is applied to a list of complex numbers, it applies w, with k = −1, to each one, replacing it by a pair of complex numbers and then flattens the resulting list of pairs, thus obtaining a list of single complex numbers. We apply the above function to the previous list:

Flatten[(Map[w, #] /. k → −1.0), 1] &[h]

{1.61688 − 0.0617893 *i*, −1.61688 + 0.0617893 *i*,
  0.126228 + 0.791473 *i*, −0.126228 − 0.791473 *i*, 1.22518 + 0.341789 *i*,
  −1.22518 − 0.341789 *i*, 0.909746 − 0.460295 *i*, −0.909746 + 0.460295 *i*}

In 4.2.1 we discussed the *Mathematica* command **Nest**. The pure function g defined above can be nested, or repeatedly applied, to a list of points:

Nest[Flatten[(Map[w, #] /. k → −1), 1] &, {1.0 − 2 I}, 4]

{1.61779 − 0.0190968 *i*, −1.61779 + 0.0190968 *i*, 0.0392861 + 0.786402 *i*,
  −0.0392861 − 0.786402 *i*, 1.11865 + 0.353763 *i*, −1.11865 − 0.353763 *i*,
  1.01309 − 0.390622 *i*, −1.01309 + 0.390622 *i*, 1.49607 + 0.114229 *i*, −1.49607 − 0.114229 *i*,
  0.303415 − 0.563237 *i*, −0.303415 + 0.563237 *i*, 1.39179 − 0.16536 *i*,
  −1.39179 + 0.16536 *i*, 0.528826 + 0.435204 *i*, −0.528826 − 0.435204 *i*}

We now use the command **Nest** to iterate g a sufficient number of times to obtain a list of complex numbers which, when plotted, form an approximation to the Julia set. We first apply the pure function (**{Re[#], Im[#]}**)**&** to the above list of complex numbers to obtain a list of co-ordinates of points (see 1.8.2) and plot these ponts, using **ListPlot**:

ListPlot[{Re[#], Im[#]} & /@ Nest[Flatten[(Map[w, #] /. k → −1.0), 1] &, {1 − 2 I}, 15],
  AspectRatio −> Automatic, Prolog −> PointSize[0.004]]

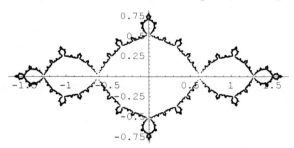

When constructing a particular Julia set, parameters c, z and m must be chosen. The following routine can be used to generate a Julia set for any suitable choice of the parameters. We use the function w defined above.

julia[c_, z_, m_] :=
  ListPlot[{Re[#], Im[#]} & /@ Nest[Flatten[(Map[w, #] /. k → c), 1] &, {z}, m],
    AspectRatio −> Automatic, Prolog −> PointSize[0.004]];

julia[0.4 + 0.7 I, 2 + I, 15]

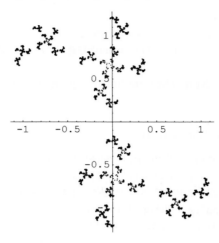

Exercise:

Construct the Julia set of Q_c for the following values of c :

1) c = -1.25.

2) c = I.

3) c = 0.4 + 0.07 I.

4) c = -0.597 - 0.7 I.

5) c = -1.58.

6) c = 0.0433 + 0.0545 I.

Notice that some Julia sets are 'all in one piece' or 'connected', while others are broken in bits or 'disconnected'.

6.1.4 Construction of Julia Sets Using the Random Algorithm

In 6.1.3, to construct the Julia set of the quadratic function Q_{-1}, using the deterministic algorithm, we used the IFS consisting of the maps:

$$w_1[z\_] := \sqrt{z+1} \, ; \, w_2[z\_] := -\sqrt{z+1} \, ;$$

We shall construct the same Julia set using the random algorithm. In this case we cannot use Roman Maeder's **ChaosGame** package as we did in 5.5 as it applies only to affine maps.

In order to obtain a random sequence of the maps w_1 and w_2, we apply the pure function $w_\# \, \&$ to a random sequence of the numbers 1 and 2 as illustrated below. We start with a short list, for demonstration purposes.

w$_\#$ & /@ Table[Random[Integer, {1, 2}], {n, 1, 15}]

{w$_1$, w$_1$, w$_2$, w$_1$, w$_2$, w$_1$, w$_2$, w$_2$, w$_1$, w$_1$, w$_1$, w$_2$, w$_2$, w$_2$, w$_2$}

We use the command **ComposeList** to generate a sequence similar to the sequence S described in 5.5.1

? ComposeList

ComposeList[{f1, f2, ... }, x] generates a list of the form {x, f1[x], f2[f1[x]], ... }.

We choose x = 2 I (the starting point we used in constructing the Julia set of Q_{-1} in 6.1.3).

ComposeList[w$_\#$ & /@ Table[Random[Integer, {1, 2}], {n, 1, 10}], 2.0 I]

{2. i, 1.27202 + 0.786151 i, 1.52909 + 0.257066 i, −1.59236 − 0.0807187 i,
  −0.052318 + 0.771424 i, −1.04155 − 0.370325 i, 0.406879 − 0.455081 i,
  1.20115 − 0.189435 i, −1.485 + 0.063783 i, −0.0456953 − 0.697916 i, −1.03358 + 0.33762 i}

We use the command **Drop** to omit the first few elements of the list:

Drop[ComposeList[w$_\#$ & /@ Table[Random[Integer, {1, 2}], {n, 1, 15}], 2.0 I], 5]

{0.0674778 + 0.717305 i, 1.0848 + 0.330617 i,
  −1.44839 − 0.114133 i, −0.0845512 + 0.674934 i, −1.01312 − 0.333098 i,
  −0.40015 + 0.416216 i, 0.815462 + 0.255203 i, 1.3507 + 0.0944706 i,
  1.53351 + 0.0308021 i, 1.59173 + 0.00967568 i, −1.60989 − 0.00300508 i}

We use the pure function **{Re[#], Im[#]}&** to convert the complex numbers to point co-ordinates:

{Re[#], Im[#]} & /@
  Drop[ComposeList[(w$_\#$) & /@ Table[Random[Integer, {1, 2}], {n, 1, 15}], 2.0 I], 5]

{{0.334455, 0.49363}, {−1.17416, −0.210206},
  {−0.222286, 0.472829}, {0.918663, 0.257346}, {−1.38826, −0.0926869},
  {0.0738584, −0.627464}, {1.07647, −0.291444}, {−1.44452, 0.100879},
  {0.0751759, 0.670952}, {−1.08225, −0.30998}, {−0.345295, 0.448863}}

We replace the number of points by 100000, drop the first 50 and plot the resulting list of points using **ListPlot**:

ListPlot[{Re[#], Im[#]} & /@ Drop[
CoposeList[(w#) & /@ Table[Random[Integer, {1, 2}], {n, 1, 100000}], 2.0 I], 50],
AspectRatio → Automatic, Prolog → PointSize[0.005]]

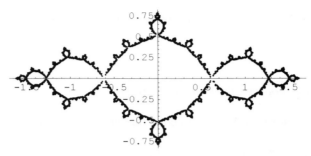

Exercise:
Use the above method to construct Julia sets of Q_c for other values of c.

6.2 Attractors of Iterated Function Systems whose Constituent Maps are not Injective

In the previous section we constructed Julia sets of the family of dynamical systems $\{Q_c \,|\, c \in \mathbb{C}\}$ where $Q_c : z \rightarrow z^2 + c$ is the quadratic function. Although Q_c is two-to-one on the complex plane, if Q_c is restricted to either the upper or lower half-plane, it is one-to-one there and so has an 'inverse' which is also one-to-one. We called these 2 inverses w_1 and w_2 respectively. Now w_1 and w_2 form an IFS whose attractor is the Julia set of the family of dynamical systems $\{Q_c \,|\, c \in \mathbb{C}\}$.

We now start with an IFS $\{X \,|\, w_1, w_2\}$ where w_1 and w_2 are not necessarily one-to-one, and we use the deterministic algorithm to generate the attractor of this IFS, if it exists.

Consider the IFS $\{\mathbb{R}^2 \,|\, w_1, w_2\}$ where
$$w_1[\{x, y\}] = \{Cos[\,x^2] + Cos[y^2], \; Sin[\,x] + Sin[\,y]\}$$
$$w_2[\{x, y\}] = \{-Cos[\,x^2] - Cos[y^2], \; -Sin[\,x] - Sin[\,y]\}$$

It is difficult to prove that these functions w_1 and w_2 are contraction mappings. However, if we use a deterministic algorithm program similar to the one developed in 6.1 on this IFS with starting point $\{0.5, 0.5\}$ and 15 iterations we see that we obtain an approximation to a limiting set. Applying the same IFS to a different starting point we get the same picture of a limiting set which seems to indicate that the functions w_1 and w_2 are contraction mappings.

f[{x_, y_}] :=
{{ (Cos[x^2] + Cos[y^2]), Sin[x] + Sin[y]}, {−(Cos[x^2] + Cos[y^2]), − Sin[x] − Sin[y]}};

ListPlot[Nest[Flatten[Map[f, #], 1] &, {{0.5, 0.5}}, 16],
    AspectRatio → Automatic, Axes → False, PlotStyle → {PointSize[0.001]}]

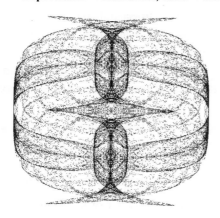

Exercise:

Try the above program with the following functions:

1) w[{x_, y_}] := {{ Sin[x + y], (Cos[x^3] + Cos[y^3])}, {− Sin[x + y], −(Cos[x ^{3}] + Cos[y^3])}}.

2) w[{x_, y_}] :=
    {{ Sin[x] + Cos[y], Sin[x^3] + Sin[y^3]}, {− (Sin[x] + Cos[y]), −(Sin[x ^{3}] + Sin[y^3])}}.

3) w[{x_, y_}] :=
    {{ Cos[x^2] + Cos[y^2], Sin[x] + Sin[y]}, {− Sin[x] − Sin[y], −(Cos[x^2] + Cos[y^2])}}.

4) w[{x_, y_}] := {{ Sin[x] + Sin[y], Cos[x^2 + y^2]}, {−(Sin[x] + Sin[y]), − Cos[x^2 + y^2]}}.

5) w[{x_, y_}] := {{ Cos[x + y], (Sin[x^3 + y^3])}, {− Cos[x + y], − Sin[x ^{3} + y^3]}}.

6) w[{x_, y_}] := {{Exp[Sin[x y]], Sin[x] + Sin[y]}, {−Exp[Sin[x y]], −(Sin[x] + Sin[y])}}.

Hint: When trying out your own IFS, in **ListPlot** choose a small number of iterations initially and choose large-sized points so that you can tell, without waiting too long, if the attractor is likely to be interesting.

6.3 Attractors of 3D Affine Iterated Function Systems Using Cuboids

In this section we construct 3D fractal cities using the **Cuboid,** a *Mathematica* primitive. A **Cuboid** is essentially a cube with sides parallel to the axes as explained in *Mathematica*:

? Cuboid

Cuboid[{xmin, ymin, zmin}] is a three−dimensional graphics primitive that
    represents a unit cuboid, oriented parallel to the axes. Cuboid[{xmin, ymin, zmin},
    {xmax, ymax, zmax}] specifies a cuboid by giving the coordinates of opposite corners.

Our first project will be to develop an IFS consisting of affine transformations which we will then apply repeatedly to a cuboid (using the deterministic algorithm defined in 5.3.2) to construct a fractal city. The technique is similar to that used to generate the Sierpinski triangle.

In order to construct the mappings of the IFS, we start with an initial rectangle and a few smaller rectangles inside it. With the ground plan in place we now construct an IFS which maps a cuboid based on the surrounding rectangle onto the union of cuboids based on the internal rectangles, and then iterate. We'll keep the bases of the cuboids on the same horizontal plane.

It's best to start with a ground plan in the x-y plane. This is is easily drawn on graph paper or one can construct it with *Mathematica* using the command **Line.**

Show$\Big[$**Graphics**$\Big[\Big\{$**Text["X", {−2.5, 10}], Text["y ➔", {12, 21}], Text**$\Big[$**"↓", {−2.5, 8}**$\Big]$,

  Text["A", {10, 3}], Text["B", {8, 8}], Table[Text[2 n, {−0.8, 18 − 2 n}], {n, 0, 9}],

  Table[Text[2 n, {2 n, 19}], {n, 0, 12}], Line[{{0, 0}, {24, 0}, {24, 18}, {0, 18}, {0, 0}}],

  Line[{{1, 0}, {6, 0}, {6, 5}, {1, 5}, {1, 0}}],

  Line[{{18, 0}, {23, 0}, {23, 5}, {18, 5}, {18, 0}}],

  Line[{{1, 6}, {17, 6}, {17, 11}, {1, 11}, {1, 6}}],

  Line[{{1, 12}, {6, 12}, {6, 17}, {1, 17}, {1, 12}}],

  Line[{{12, 12}, {23, 12}, {23, 17}, {12, 17}, {12, 12}}]$\Big\}\Big]$,

 AspectRatio → Automatic, PlotRegion → {{0.09, 0.99}, {0.08, 0.95}}$\Big]$

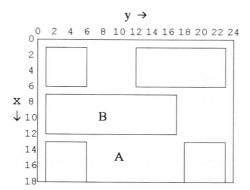

We now have a basis for 5 boxes. Notice that the origin is in the top left-hand corner, the x-axis is vertical and the y-axis horizontal. Let's choose 24 as the height of the bounding box A (with base dimensions 18 x 24), and 6 as the height of box B (with base dimensions 5 x 16). We need 5 affine maps each of which maps box A onto one of the smaller boxes. We show how to construct an affine map from box A onto box B.

A must be scaled in the x direction by the factor $\frac{5}{18}$, in the y direction by $\frac{16}{24} = \frac{2}{3}$ and in the z direction by $\frac{6}{24} = \frac{1}{4}$.

The resulting box is then translated by the vector (7, 1, 0).

So the required affine map is defined by: $\{x, y, z\} \rightarrow \left\{\frac{5}{18} x + 7, \frac{2}{3} y + 1, \frac{1}{4} z\right\}$.

The other affine maps are constructed in a similar way.

How do we apply an affine map to a cuboid? We simply apply it to each of the defining opposite vertices.

$$\textbf{city1}[\{\{x\_, y\_, z\_\}, \{p\_, q\_, r\_\}\}] :=$$
$$\left\{\left\{\left\{\frac{5}{18} x + 1, \frac{5}{24} y + 1, \frac{7}{12} z\right\}, \left\{\frac{5}{18} p + 1, \frac{5}{24} q + 1, \frac{7}{12} r\right\}\right\},\right.$$
$$\left\{\left\{\frac{5}{18} x + 1, \frac{2}{3} y + 7, \frac{5}{12} z\right\}, \left\{\frac{5}{18} p + 1, \frac{2}{3} q + 7, \frac{5}{12} r\right\}\right\},$$
$$\left\{\left\{\frac{5}{18} x + 7, \frac{2}{3} y + 1, \frac{1}{4} z\right\}, \left\{\frac{5}{18} p + 7, \frac{2}{3} q + 1, \frac{1}{4} r\right\}\right\},$$
$$\left\{\left\{\frac{5}{18} x + 13, \frac{7}{24} y + 1, \frac{1}{6} z\right\}, \left\{\frac{5}{18} p + 13, \frac{5}{24} q + 1, \frac{1}{6} r\right\}\right\},$$
$$\left.\left\{\left\{\frac{5}{18} x + 13, \frac{5}{24} y + 18, \frac{13}{24} z\right\}, \left\{\frac{5}{18} p + 13, \frac{5}{24} q + 18, \frac{13}{24} r\right\}\right\}\right\};$$

We now map the function **Cuboid** onto the above list of pairs of points, for particular values of the variables.

Show[Graphics3D[Apply[Cuboid, city1[{{0, 0, 0}, {18, 24, 24}}], {1}]],
ViewPoint -> {3.045, 1.475, 0.042}]

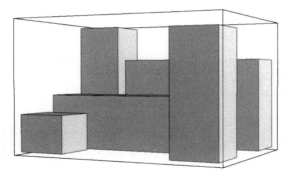

If you have *Mathematica* 4 or later you can use the command:

Show[Graphics3D[Cuboid @@@ city1[{{0, 0, 0}, {18, 24, 24}}]]],
ViewPoint -> {3.045, 1.475, 0.042}]

We now use a method similar to that used for constructing Julia sets. We define a pure function **Flatten[Map[city2, #], 1]&.**

We iterate the above function applied to the initial set, apply **Cuboid** to the result and then plot the resulting cuboids.

Second iteration:

> **Show[Graphics3D[**
> **Apply[Cuboid, Nest[Flatten[Map[city1, #], 1] &, {{{0, 0, 0}, {18, 24, 24}}}, 2], {1}]],**
> **ViewPoint −> {3.045, 1.475, 0.042}, Boxed → False, PlotRange → All]**

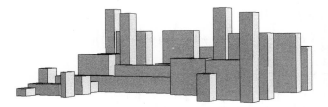

Third iteration:

> **Show[Graphics3D[**
> **Apply[Cuboid, Nest[Flatten[Map[city1, #], 1] &, {{{0, 0, 0}, {18, 24, 24}}}, 3], {1}]],**
> **ViewPoint −> {3.045, 1.475, 0.042}, Boxed → False, PlotRange → All]**

Fourth iteration:

> **Show[Graphics3D[**
> **Apply[Cuboid, Nest[Flatten[Map[city1, #], 1] &, {{{0, 0, 0}, {18, 24, 24}}}, 4], {1}]],**
> **ViewPoint −> {3.045, 1.475, 0.042}, Boxed → False,**
> **PlotRange → All, ColorOutput → CMYKColor]**

Here is an example of the application to a cuboid, of an IFS which includes the identity map.
We show the results of the first and fifth iterations:

> **s[{{x_, y_, z_}, {p_, q_, r_}}] := {{{x, y, z}, {p, q, r}},**
> **{{0.5 x − 1, 0.5 y − 1, 0.5 z + 1}, {0.5 p − 1, 0.5 q − 1, 0.5 r + 1}},**
> **{{0.5 x + 1, 0.5 y − 1, 0.5 z − 1}, {0.5 p + 1, 0.5 q − 1, 0.5 r − 1}},**
> **{{0.5 x − 1, 0.5 y + 1, 0.5 z − 1}, {0.5 p − 1, 0.5 q + 1, 0.5 r − 1}}};**

Show[Graphics3D[
    Apply[Cuboid, Nest[Flatten[Map[s, #], 1] &, {{{−2, −2, −2}, {0, 0, 0}}}, 1], {1}]],
    Boxed → True, Axes → True, PlotRange → All, ViewPoint −> {2, 2, 2}]

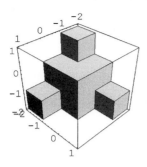

Show[Graphics3D[
    Apply[Cuboid, Nest[Flatten[Map[s, #], 1] &, {{{−2, −2, −2}, {0, 0, 0}}}, 5], {1}]],
    Boxed → False, PlotRange → All, ViewPoint −> {2, 2, 2}]

Exercise:

1) Apply the IFS above to the cuboid with opposite vertices {0, 0, 0} and {2, 2, 2}. Use the viewpoint {-2, -2, 2}.

2) Use the diagrams below as starting points for constructing 3D fractals:

a)

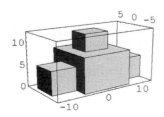

b)

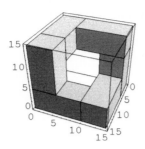

c)

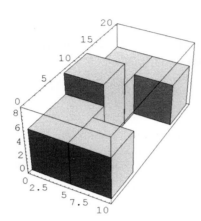

3) Include the identity map in some of the iterated function systems that you constructed above, and apply it to a cuboid of your choice.

4) In the formula for **city1** replace one or more of the coefficients by **Random[ ]**, and construct the resulting fractal.

5) Make a stereogram of one or more of the above constructions.

6.4 Construction of Affine Fractals Using 3D Graphics Shapes

6.4.1 Construction of Cylinders

In this section we shall apply 3D Iterated Function Systems to *Mathematica*'s cylinders to construct tree-like forms.

In order to use the *Mathematica* construct **Cylinder** we require a package:

```
<< Graphics`Shapes`
```

Cylinder is specified by its radius, height and 'smoothness':

> **? Cylinder**
>
> Cylinder[(r:1, h:1, (n:20r))] is a list of n polygons approximating
>   an open cylinder centered around the z−axis with radius r and half height h.

So a cylinder of radius 1, height 8 and consisting of 10 polygons is specified thus:

> **cylinder1 = Cylinder[1, 4, 10];**
>
> **Show[Graphics3D[cylinder1]]**

We shall be constructing the shape **'treeFirst'** below as a basis for our first tree.

However, prior to that, we demonstrate how to use the *Mathematica* commands which enable us to rotate, translate and scale.

6.4.2 Scaling, Rotating and Translating Cylinders

In the previous section we applied non-rotational affine transformations directly to the co-ordinates of the pair of diagonally opposite points on a **Cuboid**. We cannot use that method here. However in order to rotate, translate and scale, *Mathematica* has the commands **Rotate-Shape, TranslateShape** and **AffineShape** which can be applied to *Mathematica* 3D graphic objects such as the **Cylinder, Sphere** etc.

We start with rotation:

? RotateShape

RotateShape[graphics3D, phi, theta, psi] rotates
   the three–dimensional graphics object by the specified Euler angles.

From Help - Add-ons - Standard Packages - Geometry - Rotations:

The rotation given by the Euler angles phi, theta, and psi can be decomposed into a sequence of three successive rotations. The first by angle phi about the z-axis, the second by angle theta about the x-axis, and the third by angle psi about the z-axis (again). The angle theta is restricted to the range 0 to π.

The following command rotates **cylinder1** through $\frac{\pi}{4}$ radians about the x-axis.

$$\textbf{Show}\Big[\textbf{Graphics3D}\Big[\Big\{\textbf{cylinder1, RotateShape}\Big[\textbf{cylinder1, 0, }\frac{\pi}{4}\textbf{, 0}\Big]\Big\}\Big]\textbf{, Axes} \rightarrow \textbf{True,}$$

$$\textbf{Ticks} \rightarrow \textbf{\{\{1, 0\}, Automatic, Automatic\}, ViewPoint --> \{2.997, 1.014, 1.200\}}\Big]$$

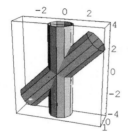

Note that **cylinder1** has been rotated in the z-y plane about the mid-point of its axis, not about the mid-point of the base of the cylinder.

Now a double rotation of **cylinder1**, first through $\frac{\pi}{4}$ about the x-axis and then through $\frac{\pi}{2}$ about the z-axis gives:

$$\textbf{Show}\Big[\textbf{Graphics3D}\Big[$$

$$\Big\{\textbf{cylinder1, RotateShape}\Big[\textbf{cylinder1, 0, }\frac{\pi}{4}\textbf{, 0}\Big]\textbf{, RotateShape}\Big[\textbf{cylinder1, 0, }\frac{\pi}{4}\textbf{, }\frac{\pi}{2}\Big]\Big\}\Big]\textbf{,}$$

$$\textbf{Axes} \rightarrow \textbf{True, ViewPoint --> \{2.997, 1.014, 1.200\}}\Big]$$

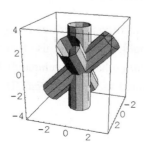

Next we look at translation.

? TranslateShape

TranslateShape[graphics3D, {x, y, z}] translates the three–dimensional graphics object by the specified vector.

So, translating **cylinder1** 2 units along the x-axis, 4 units along the y-axis and 5 units along the z-axis gives:

**Show[Graphics3D[{cylinder1, TranslateShape[cylinder1, {2, 4, 5}]}], Axes → True,
ViewPoint –> {2.997, 1.014, 1.200}, Ticks → {{0, 2}, Automatic, Automatic}]**

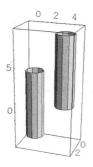

Finally, in order to scale we use the *Mathematica* command **AffineShape**:

? AffineShape

AffineShape[graphics3D, {x, y, z}] multiplies all coordinates of
    the three–dimensional graphics object by the respective scale factors x, y, and z.

Show[Graphics3D[

$\left\{\text{cylinder1, TranslateShape}\left[\text{AffineShape}\left[\text{cylinder1, } \left\{\frac{5}{4}, \frac{3}{2}, \frac{2}{3}\right\}\right], \{2, 4, 1\}\right]\right\},$

Axes → True, ViewPoint –> {2.997, 1.014, 1.200},

Ticks → {{2, 0}, Automatic, Automatic}]

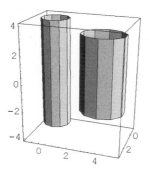

6.4.3 Constructing the Initial Branches of a Tree

We are now ready to construct **'treeFirst'**, as shown in 6.4.1. We start with the cylinder, **Cylinder**[1, 6, 10]. This will form the trunk of the tree. In order to construct the branches we will use the tree-trunk as a starting point, scale it by the factor $\frac{2}{3}$ and then rotate and translate the resultant cylinder. The first branch is scaled and then rotated through $\frac{\pi}{6}$ about the x-axis.

$$\text{Show}\Big[\text{Graphics3D}\Big[\Big\{\text{Cylinder}[1, 6, 10],$$

$$\text{RotateShape}\Big[\text{AffineShape}\Big[\text{Cylinder}[1, 6, 10], \Big\{\frac{2}{3}, \frac{2}{3}, \frac{2}{3}\Big\}\Big], 0, \frac{\pi}{6}, 0\Big]\Big\}\Big],$$

$$\text{ViewPoint} \rightarrow \{2.997, \ 1.014, \ 1.200\}\Big]$$

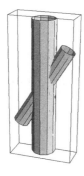

We must now work out what translation is required in order to move the branch to the top of the trunk. We shall do this more generally. Suppose cylinder X of radius r and half height h is scaled by the factor k where k < 1, and then rotated through the angle θ about the x-axis, to form cylinder Y. In the diagram below the axes of the 2 cylinders in the z-y plane are shown:

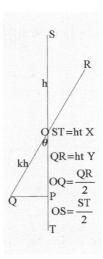

In triangle OQP, we have $QP = kh \, Sin[\theta]$; $OP = kh \, Cos[\theta]$. Translating by the vector $(0, -kh \, Sin[\theta], h + kh \, Cos[\theta])$, moves the point Q to S, the top of the trunk cylinder. Below is a right-half cross-sectional diagram showing the new position of the 2 cylinders in the z-y plane. Note that AS and CS are the axes of the 2 cylinders.

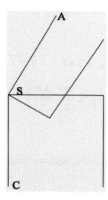

Finally, the smaller cylinder must be translated in order to move the 'branch' away from the interior of the trunk and obtain the following diagram:

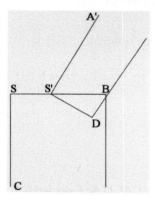

We need to find the length of SS'. Since $S'D = kr$, $S'B = kr \, Sec[\theta]$ so $SS' = r - kr \, Sec[\theta]$. So, in order to obtain the first branch, we start off with the original tree-trunk, scale and rotate it and then translate by the vector:

$(0, -kh \, Sin[\theta] + r - kr \, Sec[\theta], h + kh \, Cos[\theta])$.

The other branches can be constructed by suitable rotations and permuting and sign-changing of the x- and y- translations.

6.4.4 The Routine for Generating the Tree

In the following function the argument x can be replaced by any shape in the graphics shapes package.

$$\text{tree1}[x\_] := \Big\{ \text{TranslateShape}[\text{AffineShape}[x, \{1, 1, 1\}], \{0, 0, 0\}], \text{TranslateShape}\big[$$

$$\text{RotateShape}\big[\text{AffineShape}\big[x, \{\tfrac{2}{3}, \tfrac{2}{3}, \tfrac{2}{3}\}\big], 0, \tfrac{\pi}{6}, 0\big], \{0, 2.42, 6 + 2\sqrt{3}\}\big],$$

$$\text{TranslateShape}\big[\text{RotateShape}\big[\text{AffineShape}\big[x, \{\tfrac{2}{3}, \tfrac{2}{3}, \tfrac{2}{3}\}\big], 0, \tfrac{\pi}{6}, \tfrac{\pi}{2}\big],$$

$$\{2.42, 0, 6 + 2\sqrt{3}\}\big], \text{TranslateShape}\big[$$

$$\text{RotateShape}\big[\text{AffineShape}\big[x, \{\tfrac{2}{3}, \tfrac{2}{3}, \tfrac{2}{3}\}\big], 0, \tfrac{\pi}{6}, \pi\big], \{0, -2.42, 6 + 2\sqrt{3}\}\big],$$

$$\text{TranslateShape}\big[\text{RotateShape}\big[\text{AffineShape}\big[x, \{\tfrac{2}{3}, \tfrac{2}{3}, \tfrac{2}{3}\}\big], 0, \tfrac{\pi}{6}, \tfrac{3\pi}{2}\big],$$

$$\{-2.42, 0, 6 + 2\sqrt{3}\}\big]\Big\};$$

We now use a technique similar to that used in the construction of Julia sets to construct the tree.

```
treeFirst = Show[Graphics3D[Nest[Flatten[Map[tree1, #], 1] &, Cylinder[1, 6, 10], 1],
    PlotRegion → {{−0.1, 1.1}, {−0.4, 1.3}}],
    PlotRange → All, Boxed → False, ColorOutput → CMYKColor]
```

After 3 iteratons we get:

```
treeThird = Show[Graphics3D[Nest[Flatten[Map[tree1, #], 1] &, Cylinder[1, 6, 10], 3],
    PlotRegion → {{−0.25, 1.25}, {−0.55, 1.4}}],
    PlotRange → All, Boxed → False, ColorOutput → CMYKColor]
```

Foliage can be added to the ends of the branches of the tree by using the above IFS without the 'trunk' function (i.e. the identity function). We apply this IFS to one of the standard shapes, and iterate n+1 times where n is the number of iterations used to construct the branches. For example, using a helix we get:

$$\text{foliage1}[x\_] := \Big\{ \text{TranslateShape}\Big[$$
$$\text{RotateShape}\Big[\text{AffineShape}\Big[x, \Big\{\frac{2}{3}, \frac{2}{3}, \frac{2}{3}\Big\}\Big], 0, \frac{\pi}{6}, 0\Big], \{0, 2.42, 6+2\sqrt{3}\}\Big],$$
$$\text{TranslateShape}\Big[\text{RotateShape}\Big[\text{AffineShape}\Big[x, \Big\{\frac{2}{3}, \frac{2}{3}, \frac{2}{3}\Big\}\Big], 0, \frac{\pi}{6}, \frac{\pi}{2}\Big],$$
$$\{2.42, 0, 6+2\sqrt{3}\}\Big], \text{TranslateShape}\Big[$$
$$\text{RotateShape}\Big[\text{AffineShape}\Big[x, \Big\{\frac{2}{3}, \frac{2}{3}, \frac{2}{3}\Big\}\Big], 0, \frac{\pi}{6}, \pi\Big], \{0, -2.42, 6+2\sqrt{3}\}\Big],$$
$$\text{TranslateShape}\Big[\text{RotateShape}\Big[\text{AffineShape}\Big[x, \Big\{\frac{2}{3}, \frac{2}{3}, \frac{2}{3}\Big\}\Big], 0, \frac{\pi}{6}, \frac{3\pi}{2}\Big],$$
$$\{-2.42, 0, 6+2\sqrt{3}\}\Big]\Big\};$$

After 4 iterations we get:

Fol4 = Show[Graphics3D[Nest[Flatten[Map[foliage1, #], 1] &, Helix[2, 13, 5, 5], 4]],
PlotRange → All, ColorOutput → CMYKColor]

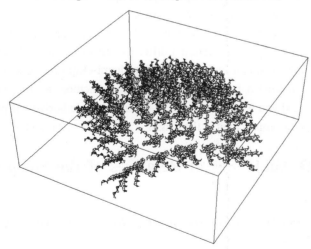

Adding the foliage to the trunk we get:

Show[treeThird, Fol4, Boxed → False, PlotRegion → {{−0.25, 1.25}, {−0.55, 1.4}}]

Exercise:

1) Try the above tree with foliage formed from cones or tori.

2) Construct other trees with different branching arrangements.

3) Construct another tree with seedpods made from **Sphere**[8, 5, 5].

4) Construct other fractals with cylinders, spheres, tori or cones.

5) Take the affine transformations used to construct the 'city' of the last section and make a function, say, 'city2', and apply it to a sphere to obtain a different version of city1. By choosing the plot range and scaling suitably, you can make spheres into hemi-ellipsoids. Use a similar function applied to a cone to construct a fractal-like 'landscape', remembering that you may need to apply vertical translations if you wish your cones to be on the same horizontal plane, as *Mathematica* places the centre of the cone's axis at the origin.

6.4.5 Constructing 3D Analogues of Relatives of the Sierpinski Triangle

In 5.4.1, 'Relatives of the Sierpinski Triangle' were constructed. We now show how to construct analogous 3D fractals.

Start with a cube centre the origin and side-length 4. Divide this cube into 8 congruent cubes and choose 4 of these sub-cubes whose union is not a cuboid.

Show[Graphics3D[{Cuboid[{2, −2, −2}, {0, 0, 0}], Cuboid[{−2, −2, −2}, {0, 0, 0}],
Cuboid[{−2, −2, 2}, {0, 0, 0}], Cuboid[{−2, 2, −2}, {0, 0, 0}]}],
Axes → True, ViewPoint −> {2.035, −0.834, 0.215}];

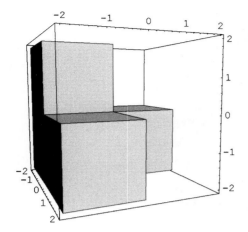

Now choose 4 affine contraction mappings, with contractivity $\frac{1}{2}$, which may include a rotation, each of which maps the larger cube onto one of the smaller cubes. A fractal can be constructed using these mappings as an IFS. Cuboids cannot be used to construct such fractals as they cannot be rotated. The easiest 3D graphics shape to use is the sphere, as a sphere with radius 1 fits nicely inside a cube with side-length 2.

? Sphere

Sphere[(r:1, (n:20r, m:15r))] is a list of n∗(m−2)+2 polygons approximating a sphere with radius r.

We use the following package which was loaded in 6.4.1:

<< Graphics`Shapes`

The commands **TranslateShape, RotateShape** and **AffineShape**, which are included in the above package, can be applied to *Mathematica*'s 3D Graphics Shape, **Sphere**. Here is an example:

```
sr1[x_] :=
    {TranslateShape[RotateShape[AffineShape[x, {0.5, 0.5, 0.5}], 0, 0, 0], {−1, −1, −1}],
      TranslateShape[RotateShape[AffineShape[x, {0.5, 0.5, 0.5}], 0, 0.5 π, 0],
        {−1, −1, 1}],
      TranslateShape[RotateShape[AffineShape[x, {0.5, 0.5, 0.5}], 0.5 π, 0, 0.5 π],
        {1, −1, −1}],
      TranslateShape[RotateShape[AffineShape[x, {0.5, 0.5, 0.5}], 0.5 π, 0, 0],
        {−1, 1, −1}]};
```

Show[Graphics3D[Nest[Flatten[Map[sr1, #], 1] &, Sphere[2, 5, 5], 5]],
    PlotRange → All, Axes → False, Boxed → False,
    ColorOutput → CMYKColor, ViewPoint –> {2.460, 2.240, 0.170}]

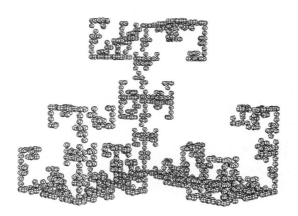

Exercise:

The above routine can be used to construct many such examples, merely by replacing the angles in the command **TranslateShape** by other multiples of $\frac{\pi}{2}$, and/or taking the negative of one or more of the numbers in the **AffineShape** command (thus producing 'flips' in one of the axes).

6.4.6 Constructing other 3D Fractals with Spheres

Other, irregular 3D fractals may be constructed with spheres. Here is an example:

fractal3D[x_] :=

$\Big\{$**TranslateShape**$\Big[$**RotateShape**$\Big[$**AffineShape**$\Big[$x, $\{\frac{1}{2}, \frac{1}{2}, \frac{3}{4}\}\Big]$, 0, 0, 0$\Big]$, {−1, −1, −1}$\Big]$,

    TranslateShape$\Big[$**RotateShape**$\Big[$**AffineShape**$\Big[$x, $\{\frac{2}{3}, \frac{1}{2}, \frac{1}{2}\}\Big]$, 0, $\frac{\pi}{3}$, 0$\Big]$, {−1, −1, 1}$\Big]$,

    TranslateShape$\Big[$

      RotateShape$\Big[$**AffineShape**$\Big[$x, $\{\frac{1}{2}, \frac{2}{3}, \frac{2}{3}\}\Big]$, $\frac{\pi}{2}, \frac{\pi}{6}, \frac{\pi}{2}\Big]$, {1, −1, −1}$\Big]$,

    TranslateShape$\Big[$**RotateShape**$\Big[$**AffineShape**$\Big[$x, $\{\frac{1}{2}, \frac{1}{2}, \frac{1}{2}\}\Big]$, $\frac{\pi}{2}, \frac{\pi}{6}, 0\Big]$,

      {−1, 0.5, −2}$\Big]\Big\}$;

Show[Graphics3D[Nest[Flatten[Map[fractal3D, #], 0] &, Sphere[2, 5, 5], 5]],
  PlotRange → All, Axes → False, Boxed → False,
  ColorOutput → CMYKColor, ViewPoint –> {2.460, 2.240, 0.170}]

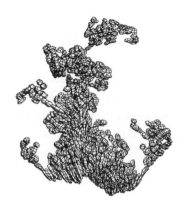

Exercise:

1) Experiment with constructing your own fractals with various 3D graphics primitives.
2) Use each starting set given in the exercise in 6.3 to construct an IFS which includes rotations as well as contractions and translations. Then apply the IFS to a sphere.

6.5   Construction of Affine Fractals Using 3D Parametric Curves

In this section, we construct a 3D tree using the function **ParametricPlot3D** applied to straight lines. An advantage of using this command is that the color and thickness of each plot is determined by a graphics directive which is part of the definition of the plot. We start with the plot of a straight line of thickness 0.2 which will form the trunk of the tree. (Color Fig 6.1)

ParametricPlot3D[{0, 0, t, {RGBColor[0.4, 0.4, 0.3], Thickness[0.2]}},
  {t, 0, 1}, BoxRatios → {1, 1, 4}, Ticks → {False, False, True}]

We shall construct an IFS consisting of 4 maps including the identity and apply it to the above graphic. Again Roman Maeder's commands **AffineMap** and **IFS** cannot be used as these apply only to 2D graphics.

We use **RGBColor** to specify colors of parts of the tree, as it is easy to choose suitable colors from the color selector. Each line that is plotted will have specifications of the form **{x, y, z, {RGBColor[r, g, b], Thickness[s]}},** where x, y, z are functions of t. When we apply an affine similarity scaling to a parametric plot each co-ordinate is multiplied by a constant factor.

In a similar way each of r, g, and b in the color specification **RGBColor[r, g, b]** is multiplied by a constant factor. How are the factors chosen?

We choose a color for the trunk, say, c_1 = **RGBColor[0.4, 0.4, 0.3]**, and a color for the final set of branches, say, c_2 = **RGBColor[0.34, 0.83, 0.06]**. We then choose the number of iterations, say 4.

We then calculate u = $(\frac{0.34}{0.4})^{\frac{1}{4}}$, v = $(\frac{0.83}{0.4})^{\frac{1}{4}}$ and w = $(\frac{0.06}{0.3})^{\frac{1}{4}}$. The color for the first set of branches is chosen to be **RGBColor[ur, vg, wb]**.

For the first 3 branches, we contract the trunk by 0.75, rotate about the x-axis through 0.5 radians, and then through 0, $2\frac{\pi}{3}$, $4\frac{\pi}{3}$ radians respectively about the z-axis and translate to the point {0, 0, 1}. We use the add-on **GeometryRotations** to calculate the rotations required. Let us reduce the trunk thickness by the factor 0.7 for the first 3 branches.

<< Geometry`Rotations`

Trunk specification:

{x, y, z, {RGBColor[r, g, b], Thickness[s]}}

First branch:

0.75 Rotate3D[{x, y, z}, 0, 0.5, 0] // Simplify

{0.75 x, 0.658187 y + 0.359569 z, −0.359569 y + 0.658187 z}

{0.75 x, 0.66 y + 0.36 z, −0.36 y + 0.66 z + 0.8,
{RGBColor[0.96 r, 1.2 g, 0.67 b], Thickness[0.7 s]}}

Second branch:

0.75 Rotate3D$\left[$ {x, y, z}, 0, 0.5, 2 $\dfrac{\pi}{3}$ $\right]$ // Simplify

{−0.375 x + 0.570007 y + 0.311396 z,
 −0.649519 x − 0.329093 y − 0.179785 z, −0.359569 y + 0.658187 z}

{−0.37 x + 0.58 y + 0.31 z, −0.65 x − 0.33 y − 0.18 z,
−0.36 y + 0.66 z + 1, {RGBColor[0.96 r, 1.2 g, 0.67 b], Thickness[0.7 s]}}

Third branch:

0.75 Rotate3D$\left[$ {x, y, z}, 0, 0.5, 4 $\dfrac{\pi}{3}$ $\right]$ // Simplify

{−0.375 x − 0.570007 y − 0.311396 z,
 0.649519 x − 0.329093 y − 0.179785 z, −0.359569 y + 0.658187 z}

$$\left\{-0.37\,x - 0.58\,y - 0.31\,z,\ 0.65\,x - 0.33\,y - 0.18\,z,\right.$$

$$\left.-0.36\,y + 0.66\,z + 1,\ \left\{\text{RGBColor}\left[\frac{r}{1.4},\ 1.04\,g,\ \frac{b}{1.4}\right],\ \text{Thickness}[.7\,s]\right\}\right\}$$

We now define a function 'tree':

```
tree[{x_, y_, z_, {RGBColor[r_, g_, b_], Thickness[s_]}}] :=
   {{x, y, z, {RGBColor[r, g, b], Thickness[s]}},
     {0.75 x, 0.66 y + 0.36 z, −0.36 y + 0.66 z + 0.8,
       {RGBColor[0.96 r, 1.2 g, 0.67 b], Thickness[0.7 s]}},
     {−0.37 x + 0.58 y + 0.31 z, −0.65 x − 0.33 y − 0.18 z, −0.36 y + 0.66 z + 1,
       {RGBColor[0.96 r, 1.2 g, 0.67 b], Thickness[0.7 s]}},
     {−0.37 x − 0.58 y − 0.31 z, 0.65 x − 0.33 y − 0.18 z, −0.36 y + 0.66 z + 1,
       {RGBColor[0.96 r, 1.2 g, 0.67 b], Thickness[.7 s]}}};
```

To construct the fractal-like tree, we use a nesting process similar to that used in previous parts of this Chapter.

```
tr[list_] := Flatten[Map[tree, list], 1];
```

```
ParametricPlot3D[
    Evaluate[Nest[tr, {{0, 0, t, {RGBColor[0.4, 0.4, .3], Thickness[0.03]}}}, 4]],
    {t, 0, 1}, ViewPoint -> {2.997, 1.014, 1.200}, PlotRange → All, Boxed → False,
    Axes → False, BoxRatios → {1, 1, 1.5}, Compiled → True] (Color Fig 6.2)
```

Another example: (Color Fig 6.3)

```
tree2[{x_, y_, z_, {RGBColor[r_, g_, b_], Thickness[s_]}}] :=
```

$$\left\{\{x, y, z, \{\text{RGBColor}[r, g, b], \text{Thickness}[s]\}\},\ \left\{0.5\,x,\ \frac{\sqrt{3}}{4}\,y - 0.25\,z,\right.\right.$$

$$\left.0.25\,y + \frac{\sqrt{3}}{4}\,z,\ \{\text{RGBColor}[0.96\,r,\ 1.2\,g,\ 0.67\,b],\ \text{Thickness}[0.7\,s]\}\right\},$$

$$\left\{-0.25\,x - 0.375\,y + \frac{\sqrt{3}}{8}\,z + 1,\ \frac{\sqrt{3}}{4}\,x - \frac{\sqrt{3}}{8}\,y + \frac{z}{8} - 1,\right.$$

$$\left.0.25\,y + \frac{\sqrt{3}}{4}\,z + 3\,\frac{\pi}{4},\ \{\text{RGBColor}[0.96\,r,\ 1.2\,g,\ 0.67\,b],\ \text{Thickness}[0.7\,s]\}\right\},$$

$$\left\{-0.25\,x + \frac{3}{8}\,y - \frac{\sqrt{3}}{8}\,z + 1,\ \frac{-\sqrt{3}}{4}\,x - \frac{\sqrt{3}}{8}\,y + \frac{z}{8} + -1,\right.$$

$$\left.0.25\,y + \frac{\sqrt{3}}{4}\,z + 3\,\frac{\pi}{4},\ \{\text{RGBColor}[0.96\,r,\ 1.2\,g,\ 0.67\,b],\ \text{Thickness}[0.7\,s]\}\right\}\right\};$$

```
tr2[list_] := Flatten[Map[tree2, list], 1];
```

ParametricPlot3D[Evaluate[

   Nest[tr2, {{Sin[2 t], Cos[2 t] − 1, 3 t, {RGBColor[0.4, 0.4, 0.3], Thickness[0.08]}}}, 4]],

   $\left\{ t, 0, \dfrac{\pi}{4} \right\}$, ViewPoint −> {3.583, −1.275, 0.845}, PlotRange → All,

   Boxed → False, Axes → False, BoxRatios → {1, 1, 1.5}, Compiled → True]

Exercise:
Construct different trees by varying the thickness of the lines, the color sequence, the branching arrangements or the number of branches or by starting with a curved instead of a straight line. Construct bushes, ferns, vines in a similar way.

6.6 Constructing Affine Fractals Using 3D Parametric Surfaces

In this section we construct a fractal-like vine using the function **ParametricPlot3D** applied to a surface.

We start with a shape constructed in 1.9.5, which will represent the trunk of a vine:

ParametricPlot3D[{(2 + Cos[y]) Cos[θ], (2 + Cos[y]) Sin[θ], 2 θ + Sin[y]},
   {y, 0, 2 π}, {θ, π, 3 π}, PlotRange → All, ViewPoint −> {−3.200, 1.099, 0.059}]

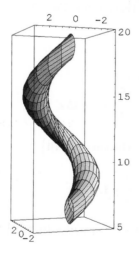

To form the first 3 branches, the trunk is scaled by the factor $\frac{2}{3}$, rotated about the x-axis through $\frac{\pi}{3}$ radians and then rotated about the z-axis through 0, $\frac{2\pi}{3}$, or $\frac{4\pi}{3}$ radians respectively. Finally these branches are translated to the top of the trunk.

As in the previous section we need to use the command **Rotate3D** in the add-on package **Geometry`Rotations`**.

We apply the above command to the vector {p, q, r}, scaled by the factor $\frac{2}{3}$.

<< Geometry`Rotations`

Rotate3D$\left[\frac{2}{3}\{p, q, r\}, 0, \frac{\pi}{3}, 0\right]$ $\left\{\frac{2p}{3}, \frac{q}{3}+\frac{r}{\sqrt{3}}, -\frac{q}{\sqrt{3}}+\frac{r}{3}\right\}$

Replacing p, q and r by the co-ordinates of a general point on the trunk, gives us our first branch, but it will not be in the right position. The branch must be moved from the bottom to the top of the trunk. To work out a suitable translation, we calculate the co-ordinates of a point at the top of the trunk, the co-ordinates of a point at the bottom of the branch and subtract the one from the other to obtain our translation vector.

Co-ordinates of a point at the top of the trunk are:

$\{(2 + \text{Cos}[y]) \text{ Cos}[\theta], (2 + \text{Cos}[y]) \text{ Sin}[\theta], 2\theta + \text{Sin}[y]\} \ /. \ \{y \to 2\pi, \theta \to 3\pi\}$

$\{-3, 0, 6\pi\}$

Co-ordinates of a point at the bottom of the branch are:

Rotate3D$\left[\frac{2}{3}\{(2 + \text{Cos}[y]) \text{ Cos}[\theta], (2 + \text{Cos}[y]) \text{ Sin}[\theta], 2\theta + \text{Sin}[y]\}, 0, \frac{\pi}{3}, 0\right] \ /.$

$\{y \to 0, \theta \to \pi\}$

$\left\{-2, \frac{2\pi}{\sqrt{3}}, \frac{2\pi}{3}\right\}$

Subtracting, we get our translation vector as:

$\{-3, 0, 6\pi\} - \left\{-2, \frac{2\pi}{\sqrt{3}}, \frac{2\pi}{3}\right\}$ $\left\{-1, -\frac{2\pi}{\sqrt{3}}, \frac{16\pi}{3}\right\}$

We now plot the 2 surfaces. The map **vine1** defined below applies 2 affine transformations to its argument {p, q, r}.

vine1$[\{p\_, q\_, r\_\}] := \left\{\{p, q, r\}, \left\{\frac{2p}{3} - 1, \frac{q}{3} + \frac{r}{\sqrt{3}} - 2\frac{\pi}{\sqrt{3}}, -\frac{q}{\sqrt{3}} + \frac{r}{3} + 16\frac{\pi}{3}\right\}\right\};$

We now map the function **vine1** onto the co-ordinates of a variable point on the trunk, and then plot:

ParametricPlot3D[
   Evaluate[vine1[{(2 + Cos[y]) Cos[θ], (2 + Cos[y]) Sin[θ], 2 θ + Sin[y]}]],
   {y, 0, 2 π}, {θ, π, 3 π}]

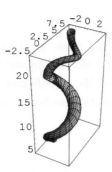

Making small adjustments to the translation vector in order to get a slightly better 'fit' and constructing the other 2 branches in a similar way, we obtain:

$$\text{vine2}[\{p\_, q\_, r\_\}] := \Big\{ \{p, q, r\}, \Big\{ \frac{2}{3}\, p - .4,\ \frac{1}{3}\, q - \frac{\sqrt{3}}{3}\, r + 4,\ \frac{\sqrt{3}}{3}\, q + \frac{1}{3}\, r + 17 \Big\},$$

$$\Big\{ \frac{-1}{3}\, p + \frac{1}{2}\, r - \frac{1}{2\sqrt{3}}\, q - 6,\ \frac{1}{\sqrt{3}}\, p + \frac{1}{2\sqrt{3}}\, r - \frac{1}{6}\, q - .2,\ \frac{1}{3}\, r + \frac{1}{\sqrt{3}}\, q + 17 \Big\},$$

$$\Big\{ \frac{-1}{3}\, p - \frac{1}{2}\, r + \frac{1}{2\sqrt{3}}\, q,\ \frac{-1}{\sqrt{3}}\, p + \frac{1}{2\sqrt{3}}\, r - \frac{1}{6}\, q - 2.5,\ \frac{1}{3}\, r + \frac{1}{\sqrt{3}}\, q + 17 \Big\} \Big\};$$

The map **vine2** defined above applies 3 affine transformations to the argument {p, q, r}.

ParametricPlot3D[
   Evaluate[vine2[{(2 + Cos[y]) Cos[θ], (2 + Cos[y]) Sin[θ], 2 θ + Sin[y]}]],
   {y, 0, 2 π}, {θ, π, 3 π}]

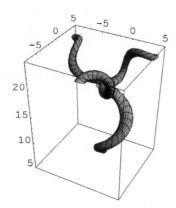

To construct the fractal-like vine, we use a nesting process similar to that used in previous parts of this Chapter.

> **v2[list_] := Flatten[Map[vine2, list], 1];**

After 4 iterations we get:

> **ParametricPlot3D[**
> **Evaluate[Nest[v2, {{(2 + Cos[y]) Cos[θ], (2 + Cos[y]) Sin[θ], 2 θ + Sin[y]}}, 4]],**
> **{y, 0, 2π}, {θ, π, 3π}, AspectRatio $\rightarrow$ Automatic, PlotRange $\rightarrow$ All,**
> **ViewPoint -> {3.263, -0.185, 0.878}, Boxed $\rightarrow$ False,**
> **Axes $\rightarrow$ False, AmbientLight -> RGBColor[0.443, 0.379, 0]]**

In the above example, we used the option **AmbientLight $\rightarrow$ RGBColor[r, g, b]**. With our particular choice of **RGBColor** parameters the screen image has a lighter color than the default color, obtained without using the **AmbientLight** option. As well the printed image in grayscale is clearer than that obtained without using the **AmbientLight** option.

Exercise:

1) Construct a fractal-like mountain range from the following plot, or another of your choice. (Hint: when defining affine transformations, it helps to view your trial copies of the plot from the z-axis.)

> **ParametricPlot3D[{x, y, (x^2 + y + 2y^4) E$^{1-x^2-y^2}$},**
> **{x, -3, 3}, {y, -3, 3}, ViewPoint -> {3.375, -0.191, 0.144}**
> **, PlotRange $\rightarrow$ {0, 4}]**

2) Here is an idea for constructing a fractal-like tree: start with the following plot (or a plot of your choice) as the cross-section of the trunk:

ParametricPlot[{4 Cos[θ] − Cos[4 θ], 4 Sin[θ] − Sin[θ] Sin[9 θ]},
  {θ, 0, 2π}, AspectRatio → Automatic]

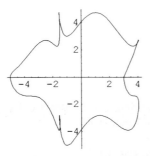

We refer you to the section in 1.9.6 on constructing a surface from a 2D parametric plot. In order to construct the trunk, we multiply the x- and y-co-ordinates of the plot by a scaling factor, to reduce the cross-sectional area with the height. We also add a translation term to make the trunk slightly curved.

ParametricPlot3D[{(1 − t / (9 π)) (4 Cos[θ] − Cos[4 θ]) + Cos[t],
  (1 − t / (9 π)) (4 Sin[θ] − Sin[θ] Sin[9 θ]) + Sin[t], 7 t}, {t, 0, 2π}, {θ, 0, 2π},
  PlotPoints → 70, ViewPoint −> {2.870, 1.385, 1.137}, BoxRatios → {1, 1, 5}]

Now carry on with the branches as in the previous examples.

Note: Some 2D affine fractals can be generated using the methods of this chapter.The tree-like image on the cover is an example of this.
Here are the codes for 1) the initial set, **is**; 2) an IFS, **tree2**, consisting of 4 affine maps; 3) a thickness table, **tt[n]**; 4) a color table, **c1[n]**; 5) the program, **p**, for generating the image:

is = ParametricPlot[{−0.5 x, Abs[x³ − 1]}, {x, 0.2, 1.2},
  AspectRatio → Automatic, PlotStyle → Thickness[0.04]]
tree2[{x_, y_}] := {{x, y}, {0.4596 x + 0.3858 y + 0.1, −0.3858 x + 0.4596 y + 0.75},
  {−0.52 x − 0.25 y − 0.845, −0.3 x + 0.433 y + 0.57},
  {−0.58 x − 0.13 y − 0.5, −0.155 x + 0.483 y + 0.9}}
tt[n_] := Which[n ≤ 0, 0.04, 0 < n && n ≤ 4, 0.6 ∗ 0.04, 4 < n && n ≤ 20,
  (0.55)² ∗ 0.04, 20 < n && n ≤ 84, (0.6)³ ∗ 0.04, 84 < n && n ≤ 340,
  (0.6)⁴ ∗ 0.04, 340 < n && n ≤ 1364, (0.6)⁵ ∗ 0.04, 1364 < n, (0.6)⁶ ∗ 0.04]
c1[n_] := Which[n ≤ 1364, RGBColor[0.470588, 0.411765, 0.305882],
  1364 < n && n ≤ 3600, RGBColor[0.5, 0.61, 0.34],
  n > 3600, RGBColor[0.457, 0.855, 0.015]]
p = ParametricPlot[Evaluate[Nest[Flatten[Map[tree2, #], 1] &, {{−0.5 t, Abs[t³ − 1]}}, 6]],
  {t, 0.2, 1.2}, PlotStyle → Table[{Thickness[tt[n]], c1[n]}, {n, 0, 5460}],
  AspectRatio → 0.8, Axes → False, PlotRange − All]

Chapter 7

Constructing Julia and Mandelbrot Sets with the Escape-Time Algorithm and Boundary Scanning Method

Introduction

In this Chapter, we use the notions discussed in Chapter 4 to define and construct Julia sets and the Mandelbrot set as well as other parameter sets. In each case we use a program which classifies points in a rectangular grid according to the behaviour of their orbits under the appropriate complex function, and color them according to this classification.

The finer the grid and the larger the number of iterations we take, the better the approximation we obtain to the fractal. However the image we obtain is always an *approximation*.

7.1 Julia Sets and Filled Julia Sets

Iteration of a complex analytic function f will decompose the complex plane into two disjoint sets: the stable or Fatou set on which the iterates are 'well-behaved' and the Julia set on which the map is chaotic.

We will consider 3 classes of analytic functions: polynomials, rational functions and certain entire transcendental functions. (An entire function is one which is analytic at every point in the complex plane. It can be proved that polynomials are entire. Examples of entire transcendental functions are Sin, Cos, Exp together with sums or products of these with each other and/or polynomials.) In all 3 cases, the Julia set is defined to be the closure of the set of repelling periodic points of the function. However, the criterion determining Julia sets of rational functions differs from that for Julia sets of entire transcendental functions, so different techniques are used to generate these different types of Julia sets.

Although polynomials form a subset of the set of rational functions, Julia and filled Julia sets of polynomials will be discussed separately first, as these are easier to generate and their generation provides an introduction to the escape time algorithm.

7.1.1 Julia Sets and Filled Julia Sets of Polynomials

Let P be a polynomial of degree ≥ 2, let J_P denote the Julia set of P and let K_P denote the filled Julia set of P. As defined in 6.1.1, $K_P = \{z \in \mathbb{C} \mid \text{the orbit of } z \text{ under P is bounded}\}$ and J_P is the boundary of K_P. For any polynomial P of degree ≥ 2, infinity is an attracting fixed point. The basin of attraction of infinity, $A(\infty)$, is the set of all points whose orbits are unbounded. For these polynomials, the Julia set of P is the boundary of the basin of attraction of infinity. But this boundary is also the boundary of the filled Julia set, K_P. Furthermore, the Julia set of P is also the boundary of the basin of attraction of any attracting cycle. This means that all attracting cycles have the same boundary.

■ Escape Criteria for Polynomials

To construct the filled Julia set for a polynomial, P, of degree ≥ 2, it is necessary to find the set of points in the plane whose orbits tend to infinity under P. The following result can be used: if P is a polynomial of degree ≥ 2, there exists $R > 0$, such that if $|z| > I$, then $|P[z]| > |z|$, and it follows that $|P^n[z]| \to \infty$ as $n \to \infty$ for all z satisfying $|z| > R$, Devaney (1989). The condition $|z| > R$ is called an escape criterion.

In some cases, an escape criterion is known for a whole class of polynomials. In the case of polynomials of degree 2, i.e. polynomials of the form $P[z] = az^2 + bz + d$ (a, b, d $\in \mathbb{C}$), it can be proved that the (filled) Julia set of P is geometrically similar to the (filled) Julia set of a quadratic poynomial of the form $Q_c[z] = z^2 + c$ for some $c \in \mathbb{C}$. For this reason, for polynomials of degree 2, we examine only the (filled) Julia sets of Q_c. (An example is given in Appendix 7.1.1.) Further, any polynomial of degree 3 can be written in the form $C_{a,b}[z] = z^3 + az + b$ after a suitable affine transformation, so we need only examine polynomials of the above form.

In view of this we shall discuss polynomials of the form $z^2 + c$, $z^3 + az + b$ and $z^n + c$, where a, b, c $\in \mathbb{C}$. For each such class of polynomials, an escape criterion is known. A method of finding an escape criterion for a polynomial not of one of the above forms, is discussed in section 7.1.2.

■ The Role of the Critical Points

The finite critical points of a polynomial P are the points a $\in \mathbb{C}$ which satisfy the equation $P'[a] = 0$.

The behaviour of the orbits of critical points tells us a lot about the structure of the (filled) Julia sets in that the (filled) Julia set of P is *connected* if and only if there does not exist a finite critical point of P in $A(\infty)$. So (K_P) J_P is connected if and only if the orbit of every finite critical point is bounded. Conversely, if $|P^n[a]| \to \infty$ as $n \to \infty$ for each finite critical point a, then (K_P) J_P is *totally disconnected* and $K_P = J_P$.

■ Polynomials of Degree Two

Let $P[z] = az^2 + bz + d$ (a, b, d $\in \mathbb{C}$) be a quadratic polynomial. As mentioned previously, it can be proved that the (filled) Julia set of P is geometrically similar to the (filled) Julia set of a quadratic poynomial of the form $Q_c[z] = z^2 + c$ for some c $\in \mathbb{C}$.

Since a polynomial of degree 2 has only one finite critical point, its (filled) Julia set is either connected or totally disconnected. In particular, the polynomial Q_c has the single critical point 0, so testing whether or not the orbit of 0 is bounded tells us whether or not (K_P) J_P is connected or totally disconnected. The orbit of 0 will prove to be of vital importance in the generation of the Mandelbrot set for polynomials of the form Q_c.

■ Computation of the Filled Julia Set for the Quadratic Polynomial $Q_c[z] = z^2 + c$

Let $Q_c[z] = z^2 + c$ (z, c $\in \mathbb{C}$). We describe an algorithmic process for obtaining an approximation to the image of the filled Julia set K_c of Q_c, where $K_c = \{z \in \mathbb{C} |$ the orbit of z under Q_c is bounded$\}$.

We use the following escape criterion:
Let $Q_c[z] = z^2 + c$ (z, c $\in \mathbb{C}$). Suppose that $|z| \geq 2$, then $|Q_c^n[z]| \to \infty$ as $n \to \infty$. So, for any $k \in \mathbb{N}$, if $|Q_c^k[z]| \geq 2$, then $z \notin K_c$, Devaney (1992).

Based on this fact, the escape time algorithm applies the following reasoning: if, after m iterations, the orbit of z is bounded by the number 2, then assume that the orbit of z is bounded and so lies in K_c. The choice of m depends on the accuracy required and the speed and resolution of the computer being used.

We use an adaptation of *Mathematica*'s programs for generating Julia and Mandelbrot sets. They can be found in Help by clicking along the route: Getting-started - Demos - Demos - Programming Sampler - Mandelbrot and Julia sets.

Our adaptation is simpler.

In order to obtain the iterates of z, we apply the pure function $Q_c = \#^2 + c \,\&$ repeatedly to the complex number z, at most 100 times. For example, choosing $c = 0.5 + I$, and $z = -1 - 0.4\,I$, and applying Q_c twice, we obtain:

$(\#^2 + 0.5 + I)\,\&[$ $\qquad\qquad\qquad -0.9444 + 5.824\,i$
$\quad (\#^2 + 0.5 + I)\,\&[-1 - 0.4\,I]]$

We next use the *Mathematica* command **FixedPointList** as follows:

$\qquad$ **FixedPointList$[\#^2 + c\,\&,\ x + I\,y,\ 100,\ $SameTest$ \to (Abs[\#] > 2.0\ \&)]$**

This command applies the function Q_c repeatedly to $x + I\,y$ until the the outcome, #, has absolute value greater than 2 or until 100 applications have been made. One can think of **SameTest** as a *'Stopping Test'*. The pure function **Abs[#] > 2&** returns **True** or **False** to the outcome of every application of Q_c to z. For example:

(Abs[#] > 2.0 &)[1] $\qquad\qquad$ False

(Abs[#] > 2.0 &)[3] $\qquad\qquad$ True

The first time that **True** is returned, the process stops and **FixedPointList** returns the orbit of z under Q_c up to this point. For example:

$\qquad$ **FixedPointList$[\#^2 + I\,\&,\ 1 + 0.3\,I,\ 100,\ $SameTest$ \to (Abs[\#] > 2.0\ \&)]$**

$\qquad \{1 + 0.3\,i,\ 0.91 + 1.6\,i,\ -1.7319 + 3.912\,i\}$

We now use the *Mathematica* command **Length**, which counts the number of points, n, say, in the orbit and returns the integer n. For example:

$\qquad$ **Length[FixedPointList$[\#^2 + I\,\&,\ 1 + 0.3\,I,\ 100,\ $SameTest$ \to (Abs[\#] > 2.0\ \&)]]$**

$\qquad$ 3

If the above command returns an integer $n \geq 100$, then $z \in K_c$, as z has not escaped i.e. we assume that the orbit of z is bounded, and so the point representing it must be colored black. If $n < 100$ then the point representing z has escaped and so must be colored white.

Now if the command **DensityPlot** is applied to the above command [**Length ....**] then each argument returned to **DensityPlot** will be a natural number lying between 1 and 100 and the output will be plotted in **GrayLevel** i.e. in different shades of gray within the two extremes of white and black. If it is desired to output in black and white only then we can use the **Color-Function** command which appears in the routine below.

We choose c = 0.377 − 0.248 I, and implement the following:

```
DensityPlot[Length[
       FixedPointList[#² + 0.377 − 0.248 I &, x + I y, 100, SameTest → (Abs[#] > 2.0 &)]],
       {x, −1.6, 1.6}, {y, −1.2, 1.2}, Mesh → False, Frame → False,
       Axes → False, PlotPoints → 400, AspectRatio → Automatic,
       ColorFunction → (If[# ≥ 1, RGBColor[0, 0, 0], RGBColor[1, 1, 1]] &)];
```

The above is an example of a connected filled Julia set.

In the command **DensityPlot**, one can specify the number of plot points in each direction. The default is 15. We generally choose between 300 and 400 to get a good image. However, we start with about 50 points to get an estimate of the size of the filled Julia set, and then we can possibly reduce the range of x- and/or y-values.

A good approximation to the Julia set of Q_c can sometimes be obtained by adapting the above program to the command **ContourPlot**, and using one contour and no contour shading. Below is a representation of the (connected) Julia set corresponding to the filled Julia set above.

ContourPlot[Length[
    FixedPointList[$\#^2 + 0.377 - 0.248\,I$ &, x + I y, 100, SameTest → (Abs[#] > 2.0 &)]],
{x, −1.6, 1.6}, {y, −1.2, 1.2}, ContourShading → False, Frame → False,
Axes → False, PlotPoints → 350, AspectRatio → Automatic, Contours → {100},
ColorFunction → (If[# ≥ 1, RGBColor[0, 0, 0], RGBColor[1, 1, 1]] &)]

An example of a totally disconnected Julia set is given by:

DensityPlot[
    Length[FixedPointList[$\#^2 + 0.66\,I$ &, x + I y, 100, SameTest → (Abs[#] > 2.0 &)]],
{x, −1.6, 1.6}, {y, −1.1, 1.1}, Mesh → False, Frame → False,
Axes → False, PlotPoints → 400, AspectRatio → Automatic,
ColorFunction → (If[# ≥ 1, RGBColor[0, 0, 0], RGBColor[1, 1, 1]] &)];

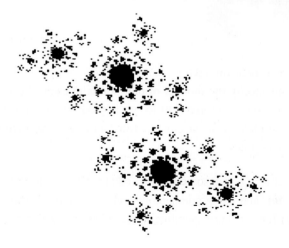

■ Variations on the Plot of the Filled Julia Set

Instead of colouring the complement of the filled Julia set white, we can use the command **DensityPlot** together with the option **Hue,** to shade or colour the complement i.e. the points whose orbits are unbounded, according to the number of iterations needed for a point in it to 'escape' from the disk $\{z \in \mathbb{C} \mid |z| \le 2\}$. This number is called the 'escape-time' of the point. We shall color the filled Julia set itself black. Recall that, in Chapter 2, we discussed the assignment of 'height numbers', between 0 and 1 to points in a density plot. The points in the filled Julia set have the largest 'height number', 1, while points not in the filled Julia set will have a 'height number' lying between 0 and 1. (Color Fig 7.1)

> **DensityPlot[Length[**
>   **FixedPointList[#^{2} + 0.308 − 0.0214 I &, x + I y, 100, SameTest → (Abs[#2] > 2 &)]],**
>   **{x, −1.1, 1.1}, {y, −1.1, 1.1}, Mesh → False, Frame → False,**
>   **Axes → False, PlotPoints → 400, AspectRatio → Automatic,**
>   **ColorFunction → (If[# ≥ 1, RGBColor[0, 0, 0], Hue[#]] &)];**

This color palette explains the coloring of the above image. For example, points which escape after 1 to 10 iterations are colored orange to yellow. Points which do not escape after 99 iterations, and so are taken to be in the filled Julia set, are colored black. (Color Fig 7.2)

> **DensityPlot[x, {x, 1, 100}, {y, 0, 1}, PlotPoints → {100, 2},**
>   **AspectRatio → 0.1, Axes → {True, False}, Frame → False,**
>   **ColorFunction → (If[# ≥ 1, RGBColor[0, 0, 0], Hue[#]] &)]**

In the following example, we use the fact that for **Hue[h]**, values of h outside the range (0, 1) are treated cyclically: (Color Fig 7.3)

> **DensityPlot[Length[FixedPointList[#^{2} − 1.8 &, x + I y, 100, SameTest → (Abs[#2] > 2 &)]],**
>   **{x, −2, 2}, {y, −0.8, 0.8}, Mesh → False, Frame → False,**
>   **Axes → False, PlotPoints → 240, AspectRatio → Automatic,**
>   **ColorFunction → (If[# ≥ 1, RGBColor[0, 0, 0], Hue[0.5 − 3 #]] &)];**

In the following example, we use coloring methods discussed in Chapter 2. We also demonstrate the use of the command **ContourPlot** to generate the graphic. If **Hue[h[#]]&** is chosen for the escaping points, then black or white can be chosen for the remaining points, as neither black nor white occurs in the above palette. If one wishes to choose a palette involving **RGB-Color, CMYKColor** or **Hue[h, s, b]** for the Fatou set then a contrasting color which does not appear in the palette should be chosen for the remaining points. The following example illustrates this.

We choose the following palette for the escaping points: (Color Fig 7.4)

ContourPlot[x, {x, 0, 20}, {y, 0, 2}, ColorFunction → (RGBColor[0, Sin[π#], #] &),
  AspectRatio → Automatic, Contours → 20]

The color **RGBColor[1, 0, 0]** (red) does not occur in the above sequence of colors, and it contrasts well with them. (Color Fig 7.5)

ContourPlot[−Length[FixedPointList[#² + 0.4245132 + 0.207530 I &,
  x + I y, 100, SameTest → (Abs[#] > 2 &)]], {x, −1, 1}, {y, −1.2, 1.2},
  ContourShading → True, Frame → False, Axes → False, PlotPoints → 200,
  Contours → 100, AspectRatio → Automatic, ContourLines → False,
  ColorFunction → (If[# ≤ 0, RGBColor[1, 0., 0.], RGBColor[0, Sin[π#], #]] &)];

Exercise:

Filled Julia sets for the following values of c are interesting and are worth trying out with various types of plots: -1.25-1.25I, 0.377+0.207530I, -0.744+0.148I, -0.765+0.109I, 0.66I, 0.32+0.043I, 0.5+0.5I, -0.745+0.11301I, 0.768+0.199I, -0.744+0.0971I, -0.759+0.108I, -0.735+0.194I.

■ Polynomials of Degree Three

For polynomials of degree > 2, the (filled) Julia set can be connected, disconnected or totally disconnected.

As mentioned previously any polynomial of degree 3 can be written in the form $C_{a,b}[z] = z^3 + a z + b$ after a suitable affine transformation, so we need only examine polynomials of the above form.

The finite critical points of $C_{a,b}$ are $z = \pm\sqrt{\frac{-a}{3}}$. So if the orbits of both critical points are bounded then the (filled) Julia set of $C_{a,b}$ is connected, if both orbits tend to infinity then the (filled) Julia set is totally disconnected. If one orbit is bounded and the other is not then the (filled) Julia set is disconnected but not totally.

An escape criterion for $C_{a,b}[z]$ is given by the following theorem:

Let $C_{a,b}[z] = z^3 + a z + b$. Suppose $|C_{a,b}^n[z]| > h = \max\left\{|b|, \sqrt{|a|+2}\right\}$, for some $n \in \mathbb{N}$, then the orbit of z escapes to infinity, Devaney (1992).

So for example if $a = -1$ and $b = 1$, then $h = \sqrt{2}$ and the following routine can be used to plot the filled Julia set of this cubic: (Color Fig 7.6)

DensityPlot[
  Length[FixedPointList[#³ − # + 1 &, x + I y, 50, SameTest → (Abs[#] > √2 &)]],
  {x, −1.6, 1.6}, {y, −1, 1}, Mesh → False, Frame → False,
  Axes → False, PlotPoints → 200, AspectRatio → Automatic,
  ColorFunction → (If[# ≥ 1, RGBColor[0, 0, 0], Hue[#]] &)];

Above is an example of a filled Julia set which is disconnected but not totally disconnected as the orbit of $\frac{1}{\sqrt{3}}$ is bounded but the orbit of $\frac{-1}{\sqrt{3}}$ is not.

In the following example, we use a = 0 and b = -1, so we can take h = $\sqrt{2}$. In this case, the Julia set is connected, as the orbits of both critical points are bounded. We use **ContourPlot** to construct the Julia set.

> **ContourPlot[**
> **Length[FixedPointList[$(\#^3 - I\,\#)$ &, x + I y, 100, SameTest → $\left(\text{Abs}[\#2] > \sqrt{2}\ \&\right)$]],**
> **{x, −1.5, 1.5}, {y, −1.2, 1.2}, ContourShading → False, Frame → False,**
> **Axes → False, PlotPoints → 350, AspectRatio → Automatic, Contours → {100},**
> **ColorFunction → (If[# ≥ 1, RGBColor[0, 0, 0], RGBColor[1, 1, 1]] &)]**

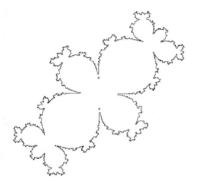

In the following case the Julia set is totally disconnected as the orbit of each of the critical points is unbounded.

> **DensityPlot[**
> **Length[FixedPointList[$\#^3 - \# + I$ &, x + I y, 15, SameTest → $\left(\text{Abs}[\#2] > \sqrt{3}\ \&\right)$]],**
> **{x, −2, 2}, {y, −1.2, 1.2}, Mesh → False, Frame → False,**
> **Axes → False, PlotPoints → 1000, AspectRatio → Automatic,**
> **ColorFunction → (If[# ≥ 1, RGBColor[0, 0, 0], RGBColor[1, 1, 1]] &)];**

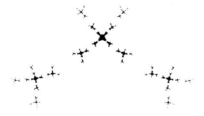

Exercise:

1) Construct Julia sets for the following values of the pair {a, b}:

{-0.48, 0.70626 + 0.502896I}, {-1.5, 0}, {-0.4, -1}, {-I, 0}.

2) Experiment with other values of a and b.

3) Devise a program for generating the above Julia sets in black and their complements in white.

■ Special Polynomials of Higher Degree

We use the folowing theorem:

Let $P_{m, c}[z] = z^m + c$. Suppose $|z| \geq |c|$ and $|z|^{m-1} > 2$, then the orbit of z escapes to infinity, Devaney (1992).

In the following example, $m = 5$, $c = -0.7 - 0.3\,I$ and the escape criterion is chosen accordingly. (Color Fig 7.7)

```
DensityPlot[
    Length[FixedPointList[#^5 + .7 - .3 I &, x + I y, 100, SameTest → (Abs[#2] > 1.2 &)]],
    {x, −1.5, 1.5}, {y, −1.2, 1.2}, Mesh → False, Frame → False,
    Axes → False, PlotPoints → 300, AspectRatio → Automatic,
    ColorFunction → (If[# ≥ 1, RGBColor[0, 0, 0], Hue[1 − #]] &)];
```

Here is a program for representing the above image using the command **Plot3D** instead of **DensityPlot**. The Julia set is colored white.

```
Plot3D[Length[
    FixedPointList[#^5 + 0.7 − 0.3 I &, x + I y, 100, SameTest → (Abs[#2] > 1.2 &)]],
    {x, −1.2, 1.2}, {y, −1.2, 1.2}, Mesh → False, PlotPoints → 250,
    BoxRatios → {1, 1, 10}, Boxed → False, Axes → False,
    ViewPoint → {0, 0, 3}, ColorFunction → (If[# ≥ 1, RGBColor[1, 1, 1],
        RGBColor[Abs[Cos[π #]], 1 − Abs[Sin[2 π #]], (4 #^2 − 4 # + 1)]] &)];
```

Another example: (Color Fig 7.8)

```
DensityPlot[Length[
    FixedPointList[#^8 + 0.2461 + 1.0651 I &, x + I y, 100, SameTest -> (Abs[#2] > 2.0 &)]],
    {x, −1.2, 1.2}, {y, −1.1, 1.15}, Mesh → False, Frame → False,
    Axes → False, PlotPoints → 350, AspectRatio → Automatic,
    ColorFunction → (If[# ≥ 1, RGBColor[0, 0, 0], Hue[2 #^3]] &)]
```

7.1.2 Notes on Julia Sets of Rational Functions

The mathematics involved in the construction of the Julia sets of rational functions is much more intricate than that of the other functions considered in this Chapter.

We consider rational functions of degree $d \geq 2$. If R is such a rational function then we can write $R[z] = \frac{P[z]}{Q[z]}$, where P and Q are polynomials with no common factor, and $d = \max \{\deg P, \deg Q\}$.

For rational functions as for polynomials, the Julia set turns out to be the boundary of the basin of attraction of any attracting cycle. So in order to generate the Julia sets of rational functions we need to know the basin of attraction of an attracting cycle.

The main difference between rational functions and polynomials is that infinity is not necessarily an attracting fixed point of the rational function R.

■ The Riemann Sphere

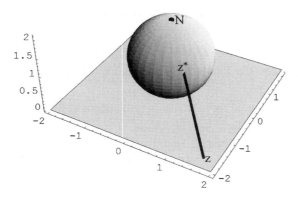

Let RS, the Riemann sphere, be the sphere centre $\{0, 0, 1\}$ and radius 1. Every point, z, in the complex plane can be represented by a unique point on RS in the following way:

The line joining the point z to $N\{0, 0, 2\}$, the north pole of the sphere, intersects the sphere RS in exactly one point, which we denote z^*. Also, to every point, P, on RS, *except* N, there corresponds a unique complex number z such that $z^* = P$.

If f is a complex function, the orbit, $(f^n[z])$ of a point z in $\mathbb{C}$ under f corresponds to the sequence $((f^n[z])^*)$ on the sphere. If $(f^n[z])$ tends to infinity, as n tends to infinity, then $((f^n[z])^*)$ tends to N as n tends to infinity. Investigations of Julia sets of rational functions are often done on the Riemann Sphere, or a model of it called $\mathbb{C} \cup \{\infty\}$, the extended complex plane. On the Riemann Sphere, there is nothing special about the point $N\{0, 0, 2\}$. It is simply called '∞'. So if r is a function on the Riemann sphere, then $r[\infty]$ will be some point, P, on the sphere.

We transfer these results to a complex rational function, R, on $\mathbb{C} \cup \{\infty\}$. If $R[z] = \frac{P[z]}{Q[z]}$ tends to infinity as z tends to a, then we write $R[a] = \infty$, while if $R[z]$ tends to β (or ∞) as z tends to infinity, we write $R[\infty] = \beta$ (or ∞). These definitions can be made mathematically precise, Milnor (1999). Also, continuity, and differentiability of R at infinity and at a pole can be defined. The derivative of R at infinity, if $Q[0] \neq 0$, is defined to be the derivative of $\frac{1}{R[\frac{1}{z}]}$ at 0, Devaney (1989).

So, in studying the behaviour of the orbit of a point in the complex plane, we can think of it as studying the behaviour of the orbit of a point on the Riemann Sphere, where there are no problems about infinity, and then map the orbit back onto the complex plane.

■ Critical Points and Critical Values of Rational Functions

Let $R[z] = \frac{P[z]}{Q[z]}$. If $Q[z] = (z-a)^k S[z]$ ($k \in \mathbb{N}$) and $S[a] \neq 0$, then R is said to have a pole of order k at a.

The complex number α is said to be a critical point of R if either $R'[\alpha] = 0$ or if $(z - \alpha)^k$, $k \geq 2$, is a factor of Q[z].

The image of α under R, $R[\alpha]$, is called a critical value of R; so in the case that $(z - \alpha)^k$ is a factor of Q, $R[\alpha] = \infty$.

The 'point' ∞ is a critical point of R (with critical value $R[\infty]$), if $R'[\infty] = 0$.

As mentioned above in order to generate the Julia set of a rational function we need to know the basin of attraction of an attracting cycle.

We use the following results in order to obtain these basins of attraction and thence to generate the Julia sets:

if R is a rational function of degree $d \geq 2$ then R has at most $2d - 2$ critical points and, further, the basin of attraction of every attracting cycle contains at least one critical point, Keen (1994), and, of course, a critical value. This tells us that there are at most $2d - 2$ attracting cycles.

Of course the orbit of a critical point is not necessarily attracted to an attracting cycle. For example:

Let $p[z] = z^2 - 2$, then the orbit of the critical point 0 is (0, -2, 2, 2, ...). Now 2 is a repelling fixed point of p so the orbit of 0 is not attracted to an attracting cycle. (We say that the above critical point is 'eventually fixed'.)

However, in some cases, by examining the orbits of the critical points, and testing if they do converge to attracting cycles, we are able to find the attracting cycles of the rational function.

■ The Role of Infinity

If degree P > degree Q, then infinity is a fixed point of R.

If infinity is an attracting fixed point for the rational function R, then the Julia set of R is the boundary of the basin of attraction of infinity. In order to locate the points whose orbits tend to infinity under R, we need an escape criterion.

The following result may be used:

If infinity is an attracting fixed point of R then there exists $S > 0$ such that $|z| > S$ implies $|R[z]| > z$ and $|R^n[z]| \to \infty$ as $n \to \infty$, for all z satisfying $|z| > S$. (For proof, see

Appendix to 7.1.2, Theorem 1.) The condition $|z| > S$ is called an 'escape criterion' for R.

If degreeP $\leq$ degreeQ then infinity may be a critical point or critical value.

In this case we may need to find part of the orbit of infinity in order to find out if the orbit converges to an attracting cycle.

Suppose $R[\infty] = k \in \mathbb{C}$, try calculating further terms in the orbit by hand, or try using *Mathematica*'s command **NestList**.

Examples:

If $f[z] = \frac{z-1}{z^2+1}$ then the orbit of infinity needs to be calculated by hand.

If $g[z] = \frac{z+1}{(z-1)^2}$ then there is a double pole at 1, so infinity is a critical value. The orbit of $g[\infty] = 0$ can be calculated using **NestList**.

One way of finding out if infinity is a critical point if $Q[0] \neq 0$, is to find the value of the derivative of $\frac{1}{R[\frac{1}{z}]}$ at $z = 0$. For example:

$$\mathbf{R[z\_]} := \frac{z^2 - 2}{z^2 + 1};$$

$$\mathbf{Together}\left[\frac{1}{\mathbf{R}[\frac{1}{z}]}\right] \qquad \frac{-1-z^2}{-1+2\,z^2}$$

$$\mathbf{D[\%, z] /. z \rightarrow 0} \qquad\qquad 0$$

So infinity is a critical point.

■ The Boundary Scanning Method for Generating the Julia Sets of Rational Functions

Let R be a rational function as defined above. If R has an attracting cycle, then the Julia set of R is the boundary of the basin of attraction of that cycle, Carleson (1993). The Julia set of R may therefore be represented by constructing the basin of attraction of one attracting cycle or by constructing the union of two or more basins of attraction of attracting cycles, since these basins must have a common boundary, the Julia set of R.

It can be proved that if R has no neutral cycles, then the Julia set of R is the set of points whose orbits do not converge to any attracting cycle. (See Appendix to 7.1.2, Theorem 2.) So, if R has no neutral cycles, and if one wishes to represent the actual Julia set, one must find all attracting cycles and construct the union of their basins of attraction. If every critical point converges to an attracting cycle, then there are no neutral cycles. If there are neutral cycles, one can still represent the Julia set as the boundary of a union of basins of attraction of attracting cycles. In each case, the union of the basins of attraction is colored in one way, and the complement in a different way.

7.1.3 Julia Sets of Rational Functions with Numerator not of Higher Degree than Denominator

In this case, infinity is not a fixed point.

■ Case1: The Attracting Cycles of R are all Fixed Points

Let R be a rational function as defined above. If R has no neutral cycles, the Julia set of R is the set of points whose orbits are NOT attracted to any of the attracting cycles.

Example 1

$$j[z\_] := \frac{z^2 + 0.5z}{z^2 + 0.2};$$

We find the finite critical points of j:

Replace[z,   {1., −0.2}
    NSolve[D[j[z], z] = 0, z]]

Since d = degree j = 2, giving 2d - 2 = 2, there are only 2 critical points and hence at most 2 attracting cycles. We calculate part of the orbits of the finite critical points:

    NestList[j, 1, 10]

    {1, 1.25, 1.24113, 1.24165, 1.24162, 1.24162, 1.24162, 1.24162, 1.24162, 1.24162, 1.24162}

    NestList[j, −0.2, 10]

    {−0.2, −0.25, −0.238095, −0.242933, −0.241105,
      −0.241818, −0.241543, −0.24165, −0.241608, −0.241624, −0.241618}

It seems as if each critical orbit converges to a fixed point of j. We check, by finding the fixed points of j and testing to find out if they are attracting:

Replace[z, NSolve[j[z] = z, z]]   {1.24162, −0.24162, 0.}

D[j[z], z] /. z → %   {−0.0574178, −0.387027, 2.5}

So we have found 2 finite attracting fixed points, and since j has degree 2, there are no other attracting and no neutral cycles. The orbits of all points in ℂ except those in the Julia set of j will converge to one of these fixed points. To generate the Julia set of j, we use the following program, which colors the points which converge to one of the 2 fixed points white and those which do not converge to either fixed point black. The black points constitute the Julia set of j.

$$\textbf{DensityPlot}\left[\textbf{Length}\left[\textbf{FixedPointList}\left[\frac{\textbf{0.5` \# + \#}^2}{\textbf{0.2` + \#}^2}\ \&,\right.\right.\right.$$

$$\textbf{x + I y, 35, SameTest} \rightarrow \textbf{(Abs[\#1 - 1.2416198487095664`] < 10}^{-4}\ \textbf{\textbar\textbar}$$

$$\textbf{Abs[\#1 - (-0.2416198487095662`)] < 10}^{-4}\ \&)]\Big],$$

$$\textbf{\{x, -0.55, .1\}, \{y, -.25, .2\}, Mesh} \rightarrow \textbf{False, Frame} \rightarrow \textbf{False, Axes} \rightarrow \textbf{True,}$$

$$\textbf{PlotPoints} \rightarrow \textbf{250, AspectRatio} \rightarrow \textbf{Automatic,}$$

$$\textbf{ColorFunction} -> \textbf{(If[\# $\geq$ 1, RGBColor[0, 0, 0], RGBColor[1, 1, 1]] \&)}\Big];$$

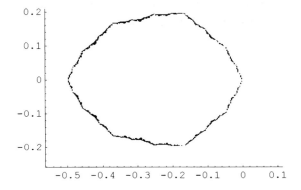

Here is another method of generating the above Julia set:

Test whether 2 successive iterates of the orbit of a point are within an ϵ distance of each other, where $\epsilon = 10^{-4}$, and assume that the orbit converges and color the point white if this is so. We color the remaining points black. The black points constitute the Julia set.

$$\textbf{DensityPlot}\Big[\textbf{Length}\Big[$$

$$\textbf{FixedPointList}\left[\frac{\textbf{0.5` \# + \#}^2}{\textbf{0.2` + \#}^2}\ \&,\ \textbf{x + I y, 35, SameTest} \rightarrow \textbf{(Abs[\#1 - \#2] < 10}^{-4}\ \&)\right]\Big],$$

$$\textbf{\{x, -0.55, .1\}, \{y, -.25, .2\}, Mesh} \rightarrow \textbf{False, Frame} \rightarrow \textbf{False,}$$

$$\textbf{Axes} \rightarrow \textbf{True, PlotPoints} \rightarrow \textbf{250, AspectRatio} \rightarrow \textbf{Automatic,}$$

$$\textbf{ColorFunction} -> \textbf{(If[\# $\geq$ 1, RGBColor[0, 0, 0], RGBColor[1, 1, 1]] \&)}\Big];$$

We now illustrate the fact that the Julia set of j is the common boundary of the basins of attraction of the 2 attracting fixed points, by plotting the basins of attraction of each of the 2 attracting fixed points. We first plot the basin of attraction of the fixed point -0.2416198487095662, colored in white, except for the fixed point, and then the basin of attraction of the other fixed point, colored in black except for the fixed point. The fixed points are shown in black.

k1 = DensityPlot$\Big[$**Length**$\Big[$**FixedPointList**$\Big[\dfrac{0.5\text{\textasciigrave }\,\#+\#^2}{0.2\text{\textasciigrave }+\#^2}$ **&, x + I y,**

15, **SameTest** $\to$ **(Abs[#1 − (−0.2416198487095662\`)] < 10^{-3} &)**$\Big]\Big]$,

{x, −0.55, 1.3}, {y, −.25, .2}, Mesh $\to$ **False, Frame** $\to$ **False,**

Axes $\to$ **True, PlotPoints** $\to$ **250, AspectRatio** $\to$ **Automatic,**

Epilog $\to$ **{PointSize[.02], Point[{−0.2416198487095662, 0}]},**

ColorFunction −> **(If[# ≥ 1, RGBColor[0, 0, 0], RGBColor[1, 1, 1]] &),**

DisplayFunction $\to$ **Identity**$\Big]$;

k2 = DensityPlot$\Big[$**Length**$\Big[$**FixedPointList**$\Big[\dfrac{0.5\text{\textasciigrave }\,\#+\#^2}{0.2\text{\textasciigrave }+\#^2}$ **&, x + I y,**

15, **SameTest** $\to$ **(Abs[#1 − 1.2416198487095664\`] < 10^{-3} &)**$\Big]\Big]$,

{x, −0.55, 1.3}, {y, −.25, .2}, Mesh $\to$ **False, Frame** $\to$ **False,**

Axes $\to$ **True, PlotPoints** $\to$ **250, AspectRatio** $\to$ **Automatic,**

Epilog $\to$ **{PointSize[.02], Point[{1.2416198487095664, 0}]},**

ColorFunction −> **(If[# ≥ 1, RGBColor[0, 0, 0], RGBColor[1, 1, 1]] &),**

DisplayFunction $\to$ **Identity**$\Big]$;

Show[GraphicsArray[{k1, k2}, GraphicsSpacing $\to$ **−0.05]]**

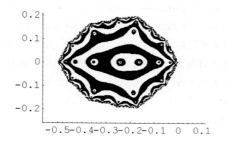

In the image below, times taken to converge to the fixed point -0.2416198487095662 are indicated alternately in black and white, using the function **Mod**, which was discussed in Chapter 1.

DensityPlot$\Big[$**Mod**$\Big[$**Length**$\Big[$**FixedPointList**$\Big[\dfrac{0.5\text{\textasciigrave }\,\#+\#^2}{0.2\text{\textasciigrave }+\#^2}$ **&, x + I y, 15,**

SameTest $\to$ **(Abs[#1 − (−0.2416198487095662)] < 10^{-3} &)**$\Big]\Big]$, **2**$\Big]$,

{x, −.55, .1}, {y, −.25, .2}, Mesh $\to$ **False, Frame** $\to$ **False, Axes** $\to$ **True,**

PlotPoints $\to$ **250, AspectRatio** $\to$ **Automatic,**

ColorFunction −> **(If[# ≥ 1, RGBColor[0, 0, 0], RGBColor[1, 1, 1]] &)**$\Big]$;

Example 2

A critical point at a pole.

$$b[z\_] := \frac{z + I}{(z - 1)^2} \, ;$$

In this case, 1 is a critical point, as b has a double pole at 1. Since b[1] = ∞ and b[∞] = 0, we must look at the orbit of 0 in order to see if it converges. We drop the first 35 iterates of the orbit of 0 and look at the next 6 iterates.

Drop[NestList[b, 0.0, 40], 35]

$\{-0.480791 + 0.408152\,i, \ -0.480792 + 0.408154\,i, \ -0.480792 + 0.408153\,i,$
$\quad -0.480791 + 0.408153\,i, \ -0.480792 + 0.408153\,i, \ -0.480792 + 0.408153\,i\}$

We find other finite critical point(s) and their orbits:

NSolve[D[b[z], z] == 0] $\{\{z \to -1. - 2.\,i\}\}$

NestList[b, −1.0 − 2.0 I, 10]

$\{-1. - 2.\,i, \ -0.125 + 0.125\,i, \ -0.287924 + 0.835217\,i,$
$\quad -0.760946 + 0.20613\,i, \ -0.324146 + 0.31742\,i, \ -0.477968 + 0.554051\,i,$
$\quad -0.554622 + 0.343943\,i, \ -0.421979 + 0.388365\,i,$
$\quad -0.492037 + 0.451537\,i, \ -0.499708 + 0.384822\,i, \ -0.460849 + 0.405939\,i\}$

It seems as though both orbits converge to the same fixed point. Using the techniques described in the previous example, we find that α = -0.480792+0.408153I is an attracting fixed point to which the orbits of the only 2 critical points converge.

We give a routine to generate the Julia set of b.

$$\text{DensityPlot}\Big[\text{Length}\Big[\text{FixedPointList}\Big[\frac{\# + I}{(\# - 1)^2} \ \&, \ x + I\,y, \ 50, \ \text{SameTest} \to$$

$$(\text{Abs}[\#1 - (-0.4807918484305007` + 0.40815313345915116`\,i)] < 10^{-3} \ \&)\Big]\Big],$$

$\{x, \ -1., \ 5.\}, \ \{y, \ -2., \ 4\}, \ \text{Mesh} \to \text{False}, \ \text{Frame} \to \text{False}, \ \text{Axes} \to \text{True},$
$\text{PlotPoints} \to 150, \ \text{AspectRatio} \to \text{Automatic},$
$\text{ColorFunction} -> (\text{If}[\# \geq 1, \ \text{RGBColor}[0, 0, 0], \ \text{Hue}[\#]] \ \&)\Big];$

Exercise:

Generate the Julia sets for the following functions:

$$j[z] := \frac{2z}{z^3 + 1} \, ;$$

$$c[z\_] := \frac{2z^2 + I}{(z - 1)^2} \, ;$$

$$f[z\_] := \frac{2\,z^3}{(z-1)^2\,(z+1)}.$$

■ Case 2: The Attracting Cycles of R are all of the Same Period p >1

In this case we construct the Julia set of R^p by the method just discussed as R and R^p have the same Julia set.

Alternatively, we can plot the set of points whose orbits, under the action of R come close to one of the attracting cycles after a sufficient number of iterations.

Example 1

$$k[z\_] := \frac{z^2 - 1}{z^2 + 1};$$

We find the finite critical points:

NSolve[D[k[z], z] = 0, z]   {{z → 0.}}

Orbit of 0:

NestList[k, 0, 10]   {0, −1, 0, −1, 0, −1, 0, −1, 0, −1, 0}

We check the orbit of k [∞] = 1 in case infinity is a critical point which converges to a different attracting cycle to that of the critical point 0.

NestList[k, 1, 10]   {1, 0, −1, 0, −1, 0, −1, 0, −1, 0, −1}

In this case, the finite critical point, 0, is part of a 2-cycle and the orbit of infinity converges to this 2-cycle, so infinity is eventually periodic. We find that $|\,\mathbf{D}[k^2[z]]\,| = 0$ at $z = 0$ and conclude that the above cycle is (super) attracting. As there is only one finite critical point, all points not in the Julia set of k^2 will converge to 0 or -1 under the action of k^2. We find and simplify $k^2[z]$:

Together[k[k[z]]] $-\frac{2\,z^2}{1+z^4}$

We use the following program to color white the points which, under the action of k^2, converge to 0 or to 1, and other points, which are in the Julia set of k^2, black.

As Julia sets are boundaries of other sets, they are very 'sparse'. In portraying a Julia set, there is a delicate balance between accuracy and visibility. If we use too many iterations, it may happen that most of the points are missed, and we may get a very poor or no image. If this is the case, we may use a smaller number of iterations, which gives us a not too accurate, but visible picture, giving us an idea of the Julia set. However, if a 'thick' Julia set is obtained, the number of iterations should be increased until the Julia set is still visible, but 'thin'. Increasing the number of plot-points sometimes improves the printed image but not the screen image.

$$\mathbf{DensityPlot}\Big[\mathbf{Length}\Big[\mathbf{FixedPointList}\Big[-\frac{2\,\#^2}{1+\#^4}\ \&,\ x+I\,y,$$

$$13,\ \mathbf{SameTest} \to (\mathbf{Abs[\#2]} < 10^{-3}\ \|\ \mathbf{Abs[\#2+1]} < 10^{-3}\ \&)\Big]\Big],$$

$$\{x,\ -2,\ 2\},\ \{y,\ -2,\ 2\},\ \mathbf{Mesh} \to \mathbf{False},\ \mathbf{Frame} \to \mathbf{False},\ \mathbf{Axes} \to \mathbf{True},$$
$$\mathbf{PlotPoints} \to 550,\ \mathbf{AspectRatio} \to \mathbf{Automatic},$$
$$\mathbf{ColorFunction} \to (\mathbf{If[\# \geq 1,\ RGBColor[0,\ 0,\ 0],\ RGBColor[1,\ 1,\ 1]]}\ \&)\Big];$$

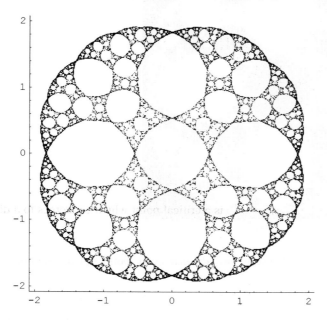

We can also plot the Julia set of k itself, using the fact that the orbit of each point not in the Julia set of k will get close to 0 and 1 alternately.

$$\mathbf{DensityPlot}\Big[\mathbf{Length}\Big[\mathbf{FixedPointList}\Big[\frac{\#^2-1}{\#^2+1}\ \&,\ x+I\,y,$$

$$25,\ \mathbf{SameTest} \to (\mathbf{Abs[\#2]} < 10^{-3}\ \|\ \mathbf{Abs[\#2+1]} < 10^{-3}\ \&)\Big]\Big],$$

$$\{x,\ -2,\ 2\},\ \{y,\ -2,\ 2\},\ \mathbf{Mesh} \to \mathbf{False},\ \mathbf{Frame} \to \mathbf{False},\ \mathbf{Axes} \to \mathbf{True},$$
$$\mathbf{PlotPoints} \to 250,\ \mathbf{AspectRatio} \to \mathbf{Automatic},$$
$$\mathbf{ColorFunction} \to (\mathbf{If[\# \geq 1,\ RGBColor[0,\ 0,\ 0],\ RGBColor[1,\ 1,\ 1]]}\ \&)\Big];$$

In the diagram below, the basin of attraction of the fixed point, 0, of k^2 is colored black:

DensityPlot$\Big[$

Length$\Big[$FixedPointList$\Big[-\dfrac{2\,\#^2}{1+\#^4}$ &, x + I y, 13, SameTest $\rightarrow$ (Abs[#2] < 10^{-3} &)$\Big]\Big]$,

{x, -2, 2}, {y, -2, 2}, Mesh $\rightarrow$ False, Frame $\rightarrow$ False,

Axes $\rightarrow$ True, PlotPoints $\rightarrow$ 250, AspectRatio $\rightarrow$ Automatic,

ColorFunction $\rightarrow$ (If[# $\geq$ 1, RGBColor[1, 1, 1], RGBColor[0, 0, 0]] &)$\Big]$;

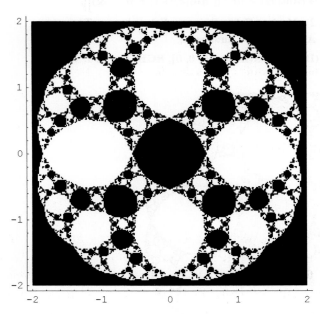

We can also plot the Julia set of k^2 using the fact that every point not in its Julia set will converge:

DensityPlot$\Big[$

Length$\Big[$FixedPointList$\Big[-\dfrac{2\,\#^2}{1+\#^4}$ &, x + I y, 15, SameTest $\rightarrow$ (Abs[#1 $-$ #2] < 10^{-3} &)$\Big]\Big]$,

{x, -2, 2}, {y, -2, 2}, Mesh $\rightarrow$ False, Frame $\rightarrow$ False,

Axes $\rightarrow$ True, PlotPoints $\rightarrow$ 250, AspectRatio $\rightarrow$ Automatic,

ColorFunction $->$ (If[# $\geq$ 1, RGBColor[0, 0, 0], Hue[1 $-$ #]] &)$\Big]$;

Exercise:

Generate the Julia sets for the following functions:

$$h[z\_] := \frac{z+1}{z^2 + 0.2}$$

$$j[z\_] := \frac{z+1}{z^3 - 1}$$

$$d[z\_] := \frac{1.5}{z^4 + 1}$$

$$f[z\_] := \frac{z^2 - 1}{2z^2 + 1}$$

$$e[z\_] := \frac{2z}{z^4 + z^2 + 1}$$

$$t[z\_] := \frac{2z - 1}{z^4 + z^2 + 1}$$

Example 2

In the following example, the only attracting cycle is $\{\infty,\ 0,\ -1,\ \infty,\ ...\}$:

$$f[z\_] := \frac{z - 1}{z^3 + 1};$$

and the finite critical points are obtained by using:

NSolve[D[f[z], z] = 0]

$\{\{z \to 1.67765\}, \{z \to -0.0888253 + 0.538652\, i\}, \{z \to -0.0888253 - 0.538652\, i\}\}$

We found that the orbits of all 3 finite critical points seemed to converge to the orbit of infinity. Also the derivative of f^3 at 0 is 0, so the orbit of infinity is an attracting 3-cycle. We did not calculate an escape criterion for $f^3[z]$, but first plotted the set of points whose orbits do not tend to 0 or -1, and obtained an estimate of the size of the Julia set and chose $|z| > 3$ as an escape criterion.

DensityPlot$\Big[$Length$\Big[$FixedPointList$\Big[\dfrac{\#-1}{1+\#^3}$ &, x + I y, 20,

  SameTest → (Abs[#2 + 1] < 10^{-3} || Abs[#2] < 10^{-3} || Abs[#2] > 3 &)$\Big]\Big]$,

  {x, −3, 2}, {y, −2, 2}, Mesh → False, Frame → False, Axes → True,

  PlotPoints → 650, AspectRatio → Automatic,

  ColorFunction → (If[# ≥ 1, RGBColor[0, 0, 0], RGBColor[1, 1, 1]] &)$\Big]$;

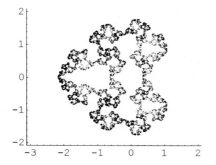

Exercise:
Construct the Julia set for the function:

$$d[z\_] := \frac{1.5}{z^4 + 1}.$$

■ **Case 3: The Attracting Cycles of R do not all have the Same Period**

Suppose that the attracting cycles have periods 1 and p, $p \geq 1$. In this case, the fixed points of R are also fixed points of R^p, and so we can use the methods of the previous section. If the attracting cycles have periods p and q, with $p > 1$, $q > 1$, $p \neq q$, then the points in these cycles are fixed points of R^s where s = LCM[p, q], and we may find the Julia set of R^s, or the set of points whose orbits under the action of R, do not tend to any of the attracting cycles. Similarly for the case of more than 2 different period cycles.

■ **Case 4: Some, but not all Attracting Cycles of R can be Found, or Some Have too High a Degree**

In this case, we plot the basin of attraction of one attracting cycle, the boundary of which is the Julia set of R, or we can choose a subset, S, of the attracting cycles and plot the union of the basins of attraction of the elements of S.

Example 1

$$e[z\_] := \frac{2z}{z^4 + z^2 + 1};$$

We find finite critical points:

NSolve[D[e[z], z] = 0, z] $\{\{z \to 0.87612\,i\}, \{z \to -0.87612\,i\}, \{z \to 0.658983\}, \{z \to -0.65898\}\}$

We show that ∞ is not a critical by showing that $\mathbf{D}\left[\frac{1}{e[\frac{1}{z}]}\right] \neq 0$ at $z = 0$.

We find part of orbit of one critical point:

NestList[e, 0.6589829632931833`, 10]

$\{0.658983, 0.812136, 0.77546, 0.790098, 0.784626,$
$\quad 0.786731, 0.785929, 0.786236, 0.786119, 0.786164, 0.786147\}$

It seems as though the orbit is converging to a fixed point. Similarly, the critical point -0.658983 seems to converge. We check, by finding fixed points, and if they are attracting:

Replace[z, NSolve[e[z] = z, z]] $\{1.27202\,i, -1.27202\,i, 0.786151, -0.786151, 0.\}$

Abs[D[e[z], z]] /. z → %   {2.61803, 2.61803, 0.381966, 0.381966, 2.}

So the 2 attracting fixed points are 0.786151 and -0.786151.

Testing the orbits of the other two critical points, we found after 10000 iterations, that the orbits were apparently converging to a cycle of length 12. As e^{12} is of very high degree, we decided to plot the basin of attraction of one of the attracting fixed points. We plot in black, the set of points which converge to the first attracting fixed point as follows:

$$\textbf{DensityPlot}\left[\textbf{Length}\left[\textbf{FixedPointList}\left[\frac{2\,\#}{\#^4 + \#^2 + 1}\ \&,\ x + I\,y,\right.\right.\right.$$

30, SameTest → (Abs[#2 − 0.7861513777574232`] < 10^{-4} &)]],

{x, −5, 8}, {y, −7, 7}, Mesh → False, Frame → False, Axes → True,
PlotPoints → 650, AspectRatio → Automatic,
ColorFunction –> (If[# ≥ 1, RGBColor[1, 1, 1], RGBColor[0, 0, 0]] &)];

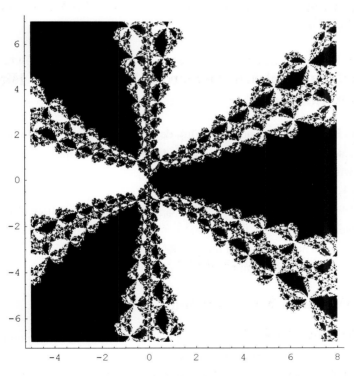

The following is a zoom-in on a plot, in black, of the set of points which do not tend to either of the attracting fixed points of e. The remaining points are colored white.

$$\text{DensityPlot}\Bigg[\text{Length}\Bigg[\text{FixedPointList}\Bigg[\frac{2\,\#}{\#^4 + \#^2 + 1}\,\&,$$

$$x + I\,y,\ 30,\ \text{SameTest} \to (\text{Abs}[\#2 - 0.7861513777574232`] < 10^{-4}\ ||$$

$$\text{Abs}[\#2 + 0.7861513777574232`] < 10^{-4}\ \&)\Bigg]\Bigg],$$

$$\{x,\ 5,\ 7\},\ \{y,\ 2,\ 4\},\ \text{Mesh} \to \text{False},\ \text{Frame} \to \text{False},\ \text{Axes} \to \text{True},$$

$$\text{PlotPoints} \to 650,\ \text{AspectRatio} \to \text{Automatic},$$

$$\text{ColorFunction} \to (\text{If}[\# \geq 1,\ \text{RGBColor}[0,\,0,\,0],\ \text{RGBColor}[1,\,1,\,1]]\ \&)\Bigg];$$

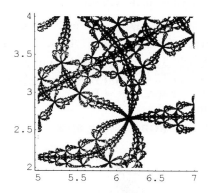

7.1.4 Julia Sets of Rational Functions with Numerator of Higher Degree than Denominator

Let $R[z] = \frac{P[z]}{Q[z]}$ where P, Q are polynomials with degree P > degree Q. In this case, $|R[z]| \to \infty$ as $z \to \infty$, so infinity is regarded as a fixed point. Also, infinity is an attracting fixed point, in the following 2 cases:

1) degree P > degree Q + 1;

2) degree P = degree Q + 1 and $\alpha = \left| \frac{q_{n-1}}{p_n} \right| < 1$, where q_{n-1} is the leading coefficient of Q, and p_n is the leading coefficient of P.

If $\alpha > 1$, then infinity is not an attracting fixed point of R.

If $\alpha = 1$, then infinity is a neutral fixed point.

■ Case 1: Infinity is an Attracting Fixed Point

One method of obtaining an idea of the Julia set of R is to plot in white, the set of points which tend to infinity as z tends to infinity and the remaining set black, as we did for quadratic functions. The boundary of the black set is the Julia set of R. To do this an escape criterion is needed. If all attracting cycles can be found, then the Julia set of R can be plotted by coloring in black those points whose orbits do not converge to any of the attracting cycles, nor tend to infinity, and the remainder in white.

▪ Finding an Escape Criterion for R

Recall from 7.1.1 that in order to find an escape criterion we need to find $S > 0$ such that if $|z| > S$, then $|R[z]| > |z|$. One method of finding such an S is as follows: use inequalities for absolute values of complex numbers such as $|z + w| \leq |z| + |w|$ and $|z - w| \geq |z| - |w|$ to find a real function f such that $|R[z]| \geq f[z] \geq 0$, for all sufficiently large values of $|z|$. Plot the graphs of the equations $y = x$ and $y = f[x]$, for x positive and f [x] defined. Suppose you find that there exists S such that the graph of f lies above that of the graph of the equation $y = x$ for $x > S$, you can use calculus methods to prove that $f[x] - x > 0$ for all $x > S$. It will then follow that $|R[z]| \geq f[|z|] \geq |z|$ for $|z| > S$ and so $|R^n[z]|$ tends to infinity as n tends to infinity. One should try to find an S which is as small as possible to avoid unnecessary calculations. The first example below illustrates this method.

Example 1

In this example degree $P >$ degree $Q + 1$.

$$p[z\_] := \frac{z^3 + I}{2(z-1)};$$

We follow the procedure for an escape criterion discussed above:

$$|z^3 + I| \geq |z|^3 - 1, \text{ and } \left| \frac{1}{2(z-1)} \right| \geq \left| \frac{1}{2(|z|+1)} \right|, \text{ if } |z| > 2.$$

It follows that

$$\left| p[z] \right| \geq \frac{|z|^3 - 1}{2(|z|+1)}, \text{ if } |z| > 2.$$

Define f:

$$f[x\_] := \frac{x^3 - 1}{2(x+1)};$$

Plot[{f[x], x}, {x, 2, 5}]

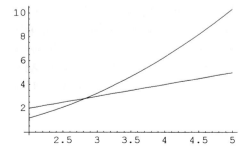

It seems that $f[x] > x$ if x is greater than the fixed point of f near 3. We find the fixed point of f near 3, obtaining 2.83118 and choose S = 2.84.

Now $f[x] - x = \frac{x^3 - 1 - 2x(1+x)}{2(1+x)}$, and using calculus it is easy to show that the function g defined by $g[x] = x^3 - 1 - 2x(1+x)$ is a strictly increasing function if $x > 2.84$. Also, $g[2.84] > 0$. It follows that $f[x] > x$ if $x \geq 2.84$, and so $|p[z]| > |z|$ if $|z| \geq 2.84$. So, we have found an escape criterion. We now plot, in white, the basin of attraction of infinity, and its complement in black, which is a 'filled Julia set' of p, using the escape-time algorithm. The common boundary of the 2 sets is the Julia set of p.

DensityPlot$\Big[$

      Length$\Big[$**FixedPointList**$\Big[\dfrac{\#^3 + I}{2(\# - 1)}$ **&, x + I y, 15, SameTest** $\rightarrow$ **(Abs[#1] > 2.84 &)**$\Big]\Big]$,

      {x, −2.5, 1.5}, {y, −2.4, 2.4}, Mesh $\rightarrow$ **False, Frame** $\rightarrow$ **False,**

      Axes $\rightarrow$ **True, PlotPoints** $\rightarrow$ **250, AspectRatio** $\rightarrow$ **Automatic,**

      ColorFunction $->$ **(If[#** $\geq$ **1, RGBColor[0, 0, 0], RGBColor[1, 1, 1]] &)**$\Big]$;

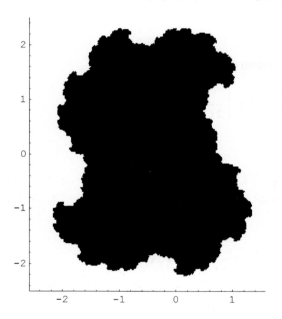

We shall now plot the Julia set of p. We find that there is one finite attracting fixed point, b = $-0.160437 - 0.389523\,I$, and 3 finite critical points, with orbits tending to infinity or to b, so there are no other attracting, and no neutral cycles. The Julia set of p is the set of points whose orbits do not tend to infinity and do not tend to b.

$$\text{DensityPlot}\!\left[\text{Length}\!\left[\text{FixedPointList}\!\left[\frac{\#^3 + I}{2\,(\# - 1)}\ \&,\ x + I\,y,\ 15,\right.\right.\right.$$

$$\left.\text{SameTest} \to (\text{Abs}[\#1 - (-0.16043748842164668\text{`} - 0.38952302579775655\text{`}\ i)] <\right.$$

$$\left.10^{-3}\ ||\ \text{Abs}[\#1] > 3\ \&)\right],\ \{x,\ -2.5,\ 1.5\},$$

$$\{y,\ -2.4,\ 2.4\},\ \text{Mesh} \to \text{False},\ \text{Frame} \to \text{False},\ \text{Axes} \to \text{True},$$

$$\text{PlotPoints} \to 250,\ \text{AspectRatio} \to \text{Automatic},$$

$$\text{ColorFunction} -\!> (\text{If}[\# \geq 1,\ \text{RGBColor}[0,\ 0,\ 0],\ \text{RGBColor}[1,\ 1,\ 1]]\ \&)\Big];$$

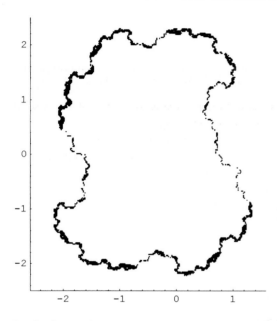

In the image below, we use **ContourPlot**, and color the contours and also color the contour lines with a set of contrasting colors. This technique is described in 2.8.2. The Julia set is colored white. (Color Fig 7.9)

$$\text{ContourPlot}\!\left[\text{Length}\!\left[\text{FixedPointList}\!\left[\frac{\#^3 + I}{2\,(\# - 1)}\ \&,\ x + I\,y,\ 15,\right.\right.\right.$$

$$\left.\text{SameTest} \to (\text{Abs}[\#1 - (-0.16043748842164668\text{`} - 0.38952302579775655\text{`}\ i)] < 10^{-3}\ ||\right.$$

$$\left.\text{Abs}[\#1] > 3\ \&)\right],\ \{x,\ -2.9,\ 2.3\},$$

$$\{y,\ -2.4,\ 2.4\},\ \text{Frame} \to \text{False},\ \text{Axes} \to \text{False},\ \text{PlotPoints} \to 250,$$

$$\text{AspectRatio} \to \text{Automatic},$$

$$\text{ColorFunction} -\!>$$

$$(\text{If}[\# \geq 1,\ \text{CMYKColor}[0,\ 0,\ 0,\ 0],\ \text{RGBColor}[\text{Abs}[2\,\# - 1],\ \text{Abs}[\text{Sin}[3\,\pi\,\#]],\ \#]]\ \&),$$

$$\text{ContourStyle} \to \text{Table}\!\left[\left\{\text{CMYKColor}\!\left[1 - \frac{n}{5} + \frac{n^2}{100},\ \frac{n}{20},\ 1 - \text{Sin}\!\left[n\,\frac{\pi}{20}\right],\ 0\right]\right\},\ \{n,\ 0,\ 15\}\right],$$

$$\text{Contours} \to 15,\ \text{PlotRegion} \to \{\{0,\ 1\},\ \{0.05,\ 1\}\},$$

$$\text{Epilog} \to \text{Text}["\text{Fig 7.7}",\ \{-0.25,\ -2.55\}]\Big];$$

Exercise:

1) Construct the Julia set and filled Julia set of the function: $p[z] = \frac{-1-z+2z^2+z^3}{2+z}$.

2) Construct the filled Julia sets of the following polynomials, calculating an escape time by the above method: $(z^2 - 0.5)^2$, $(z^2 + 0.5)^2$.

Example 2

Again degree P > degree Q + 1, but in this example, z^2 is a factor of Q [z].

$$b[z\_] := z^2 - \frac{0.001}{z^2} ;$$

Using the above techniques, we find that the orbits of all the non-zero finite critical points tend to infinity. It can also be shown that if $|z| > 1.01$ then $|f[z]| > |z|$, so $|f^n[z]| \to \infty$ as $n \to \infty$. We plot the basin of attraction of infinity in white and its complement in black. Also, 0 is a critical point with critical value ∞, which is an attracting fixed point.

> **DensityPlot[Length[FixedPointList[b, x + I y, 30, SameTest → (Abs[#1] > 1.01 &)]],**
> **{x, −1., 1.}, {y, −1, 1}, Mesh → False, Frame → False,**
> **Axes → True, PlotPoints → 650, AspectRatio → Automatic,**
> **ColorFunction −> (If[# ≥ 1, RGBColor[0, 0, 0], RGBColor[1, 1, 1]] &)];**

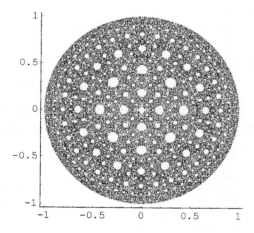

Example 3

In this example degree P = degree Q + 1, and infinity is an attracting fixed point:

$$\lambda[z\_] := \frac{1.2 z^2 + I}{(z - 1)} ;$$

We find that there are exactly 2 critical points, and the orbits of each tend to infinity. It can be proved that $|\lambda[z]| > 1.025|z|$, if $|z| > 7$. We use **ContourPlot** and the escape-time algorithm to generate the Julia set of λ. We color the contours and the contour lines. The Julia set is colored white. (Color Fig 7.10)

$$\text{ContourPlot}\Bigg[$$
$$\text{Length}\Bigg[\text{FixedPointList}\Bigg[\frac{1.2\,\#^2 + I}{(\# - 1)}\ \&,\ x + I\,y,\ 200,\ \text{SameTest} \to (\text{Abs}[\#1] > 7\ \&)\Bigg]\Bigg],$$
$$\{x,\ -6.,\ 2.\},\ \{y,\ -2.,\ 2\},\ \text{PlotPoints} \to 250,\ \text{ContourLines} \to \text{True},$$
$$\text{AspectRatio} \to \text{Automatic},\ \text{Frame} \to \text{False},\ \text{Contours} \to 26,$$
$$\text{ContourStyle} \to \text{Table}\Bigg[\Bigg\{\text{CMYKColor}\Bigg[\frac{n^2}{400},\ \Bigg(1 - \frac{n}{5} + \frac{n^2}{100}\Bigg),\ \text{Cos}\Bigg[\frac{n\pi}{40}\Bigg],\ 0\Bigg]\Bigg\},\ \{n,\ 1,\ 20\}\Bigg],$$
$$\text{Background} \to \text{RGBColor}[0.160159,\ 0.644541,\ 0.0781262],\ \text{ColorFunction} \to$$
$$(\text{If}[\# \geq 1,\ \text{RGBColor}[1,\ 1,\ 1],\ \text{RGBColor}[\text{Abs}[2\,\# - 1],\ 1 - \text{Abs}[\text{Sin}[3\,\pi\,\#]],\ \#]]\ \&)\Bigg]$$

Exercise:

Construct the Julia set of:

$$m[z\_] := \frac{1.2\,z^3 - z + I}{z^2 + z + 2\,I};$$

▪ A Rational Function with a Neutral Finite Fixed Point

We consider the Julia set of the quadratic function Q_c, with $c = -0.125 + 0.649519\,I$, known as the 'fat rabbit' (Keen (1989), Color Plate 9). Let $fr[z] = z^2 - 0.125 + 0.649519\,I$, then $p = -0.25 + 0.433013\,I$ is a neutral fixed point of fr. The sole critical point, 0, seems to converge to p, but the absolute value of the difference between p and the 100000th iteration is larger than 0.01. We can construct the filled Julia set (basin of attraction of infinity) as we did in 7.1.1.

▪ Case 2: Infinity is not an attracting Fixed Point

Any rational function of the form $R[z] = z - \frac{S[z]}{S'[z]} = \frac{P[z]}{Q[z]}$, where S is a polynomial of degree greater than or equal to 2 and any common factors of $S[z]$ and $S'[z]$ have been cancelled, is of this category. R is called the Newton iteration function of S.

It can be shown that the fixed points of R are zeros of S and that all fixed points of R are attracting. R may also have one or more attracting cycles.

To obtain the Julia set of R, plot, in black, the set, T, of points whose orbits do not converge, and the set U of remaining points whose orbits converge in a different coloring. The boundary of the set T or of the set U is the Julia set of R.

Example 1

$$S[z\_] := (z^2 + 1)\,(z^2 + I);$$

D[S[z], z] $2z(i + z^2) + 2z(1 + z^2)$

As the Newton iteration function of S is of degree 4, we first compile it, to save computation time:

$$NS = \text{Compile}\left[\{\{z, \_\text{Complex}\}\}, z - \frac{(z^2 + 1)(z^2 + I)}{2z(I + z^2) + 2z(1 + z^2)}\right];$$

We plot, in black, the set of points whose orbits do not converge. The Julia set will be the boundary of the set of white points.

```
DensityPlot[
    Length[FixedPointList[NS, x + I y, 35, SameTest → (Abs[#1 − #2] < 10⁻¹⁰ &)]],
    {x, −6, 6}, {y, −6, 6}, Mesh → False, Frame → False,
    Axes → True, PlotPoints → 450, AspectRatio → Automatic,
    ColorFunction –> (If[# ≥ 1, RGBColor[0, 0, 0], RGBColor[1, 1, 1]] &)];
```

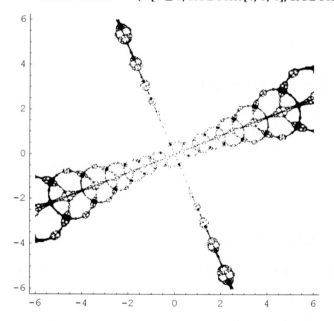

A zoom-in on a version of the above, in which the contours and contour lines are colored is shown in Color Fig 7.11. As can be seen in this image, the set of black points is not a boundary. This means that the black points themselves do not constitute the Julia set of NS. We have shown that if a rational function, R, has no neutral cycles, the Julia set of R is the set of points whose orbits do not converge to any attracting cycle. Since we have found all points whose orbits do not converge to attracting fixed points of NS, it follows that NS has neutral cycles, or NS has attracting cycles of order greater than 1. (It can be shown that NS has an attracting 3-cycle.)

ContourPlot[

    Length[FixedPointList[NS, x + I y, 35, SameTest → (Abs[#1 − #2] < 10^{-10} &)]],

    {x, 4.4, 4.9}, {y, 2.55, 3.05}, AspectRatio → Automatic,

    ContourLines → True, Contours → 25, PlotPoints → 250,

    Frame → False, Background → RGBColor[0.511727, 0.632822, 0.91017],

    ColorFunction −> $\Big($ If $\Big[$# ≥ 1, RGBColor[0, 0, 0], CMYKColor$\Big[$

        1 − Sin[π#], (4#^2 − 4# + 1), 1 − Cos$\Big[\pi\,\dfrac{\#}{3}\Big]$, 0$\Big]\Big]$ &$\Big)$, ContourStyle →

        Table$\Big[\Big\{$CMYKColor$\Big[\dfrac{n^2}{400}$, 1 − $\Big(1 − \dfrac{n}{5} + \dfrac{n^2}{100}\Big)$, 1 − Sin$\Big[\dfrac{n\,\pi}{20}\Big]$, 0$\Big]\Big\}$, {n, 1, 20}$\Big]\Big]$

Exercise:

Find the Julia sets for R where S[z] is given by:

    $S[z] = z^3 − 1;$
    $S[z] = z^3 − z − 1;$
    $S[z] = (z + 1)(z − 2)(z^2 + 1).$

■ A Rational Function with Infinity a Neutral Fixed Point

We define a function k:

    k[z_] := $\dfrac{z^3}{z^2 + 1}$;

In this case, infinity is a neutral fixed point. We find that 0 is a super attracting fixed point. The following command plots the basin of attraction of 0 in black, and points not attracted to 0 in white. The boundary of the set of black points represents the Julia set of k.

    DensityPlot[Length[FixedPointList[k, x + I y, 35, SameTest → (Abs[#1] < 10^{-4} &)]],

        {x, −8, 8}, {y, −1.2, 1.2}, Mesh → False, Frame → False,

        Axes → True, PlotPoints → 50, AspectRatio → Automatic,

        ColorFunction −> (If[# ≥ 1, RGBColor[1, 1, 1], RGBColor[0, 0, 0]] &)];

Exercise:

Represent as a boundary, the Julia set for the function a defined below:

    a[z_] := $\dfrac{z^4}{z^3 − 1}$;

■ **Locating Attracting Cycles**

It is not always possible to locate all attracting cycles. As the basin of attraction of each attracting cycle contains at least one critical point, we first find critical points or critical values of R as follows:

1) Check if infinity is an attracting fixed point (see 7.1.4).
2) Find finite critical points, by:
    a) checking if R has a pole of order greater than or equal to 2,
    b) solving the equation $\mathbf{D}[R[z]] = 0$ for z.
3) Check if infinity is a critical point (see 7.1.2). This step is not necessary if infinity is a fixed point, as we are interested in the orbits of the critical points, rather than the points themselves.

It may be possible to find attracting cycles by calculating parts of the orbits of critical points. If a critical point appears to be converging to an attracting cycle of period p, one may check this by solving the equation $R^p[z] = z$, and then checking if $|\mathbf{D}[R^p[z]]| < 1$ at a relevant solution of the afore-mentioned equation.

7.1.5 Julia Sets of Entire Transcendental Functions which are Critically Finite

■ **Critical and Asymptotic Values of Entire Transcendental Functions**

Let $f : \mathbb{C} \to \mathbb{C}$ be an entire transcendental function. If $f'[\alpha] = 0$, then $f[\alpha]$ is a critical value of f. A point $\beta \in \mathbb{C}$ is called an asymptotic value of f if there exists a continuous curve $g : \mathbb{R} \to \mathbb{C}$ satisfying $\lim_{t\to\infty} g[t] = \infty$ and $\lim_{t\to\infty} f[g[t]] = \beta$. The function f is said to be critically finite if it has at most finitely many critical and asymptotic values.
For example if $f[z] = \lambda E^z$ and $h[z] = \lambda Sin[z]$, then f has one aymptotic value, 0, and no critical values, while h has no asymptotic values, and 2 critical values, $-\lambda$ and λ.
We shall discuss the Julia sets of certain entire, critically finite transcendental functions.

The Julia set of a critically finite, entire transcendental function turns out to be the closure of the set of points whose orbits tend to infinity. So, in a sense, these Julia sets are the very opposite to those of polynomials. Although the Julia set of an entire transcendental function is the closure of the set of points whose orbits escape to infinity, and so contains much more than just the escaping orbits, our algorithmic process will simply test whether or not the orbit of a point is 'unbounded' and, if it is, will assume that the point lies in the Julia set.
We consider 2 types of transcendental functions separately, exponential functions and trigonometric functions.

▪ **Exponential Functions**

Determining an escape criterion is problematical. Devaney (1992) gives the following **Approximate Escape Criterion for Exponentials**: Suppose $E_\lambda[z] = \lambda E^z$. If the real part of $E_\lambda^n[z]$ exceeds 50, we say that the orbit of z 'escapes'.

The reason for this is that if $z = x + Iy$ and if the real part of z is large, say $x > 50$, then: $|E^z| = |E^{x+Iy}| = E^x > E^{50}$, which is very large.

Although it is not true that any orbit that reaches the half-plane $x > 50$ eventually escapes, it can be shown that there is often a nearby point whose orbit does escape, Devaney (1992).

As the value of λ changes, so the Julia sets vary. We examine a selection of these Julia sets.

For λ real, $\lambda > 0$, a bifurcation (change in dynamics) occurs at $\lambda = \frac{1}{E}$ as follows:

if $0 < \lambda \leq \frac{1}{E}$ the Julia set of E_λ lies in the right hand side of the plane $Re[z] \geq 1$, while for $\lambda > \frac{1}{E}$ the Julia set of E_λ is the entire complex plane. (In fact if $0 < \lambda < \frac{1}{E}$, E_λ has 2 real fixed points, q and p, with $q < 1 < p$, where q is attracting and p is repelling.) For these values of λ the Julia set of E_λ is a Cantor set of curves lying to the right of the vertical line $x = p$, Devaney (1989).

In the following examples we use a double escape time criterion:

if, for some N, $Re[z_n] > 50$ and if $Cos[Im[z_n]] \geq 0$ for $1 \leq n \leq N$, then we assume that the orbit of z_n escapes to infinity and color it accordingly.

The reason why we use the second part of this criterion is to ensure that the next iterate z_{n+1} will be large but will not lie in the left hand side of the plane where $Re[z] \leq 0$. If $z_n = x+Iy$, then $z_{n+1} = E_\lambda[z_n] = \lambda E^{x+Iy} = \lambda E^x(Cos[y] + ISin[y])$. So $Re[z_{n+1}] = \lambda E^x Cos[y] = \lambda E^x Cos[Im[z_n]]$.

Using this escape-time criterion to construct the Julia sets of λE^z for $\lambda = \frac{1}{E}$ and for $\lambda = \frac{1}{E} + 0.1$, we adapt the routine for generating the Julia set of a quadratic function, except that, in this case, the Julia set will be colored black and its complement white. As this escape criterion is more complicated than that for the previous Julia sets, we first compile the function **expJulia** before applying **DensityPlot** to it. We use a large number of plot points, in order to obtain a good printed image. The generation of the image may take a long time.

```
expJulia = Compile[{{c, _Complex}, {k, _Complex}}, Length[FixedPointList[
    k E# &, c, 40, SameTest → (Re[#2] > 50. && Cos[Im[#2]] ≥ 0 &)]]];
```

$$\text{DensityPlot}\Big[\text{expJulia}\Big[x + \mathrm{I}\, y, \ \frac{1}{E}\ \Big], \{x, 1, 7\}, \{y, -2, 2\}, \text{Mesh} \to \text{False}, \text{Frame} \to \text{False},$$

$$\text{Axes} \to \text{True}, \text{PlotPoints} \to \{1200, 800\}, \text{AspectRatio} \to \text{Automatic},$$

$$\text{ColorFunction} \to (\text{If}[\# \geq 1, \text{RGBColor}[1, 1, 1], \text{RGBColor}[0, 0, 0]] \ \&)\Big];$$

CompiledFunction::cfn : Numerical error encountered at instruction 18; proceeding with uncompiled evaluation.

CompiledFunction::cfn : Numerical error encountered at instruction 33; proceeding with uncompiled evaluation.

General::unfl : Underflow occurred in computation.

CompiledFunction::cfn : Numerical error encountered at instruction 33; proceeding with uncompiled evaluation.

General::stop : Further output of CompiledFunction::cfn will be suppressed during this calculation.

General::unfl : Underflow occurred in computation.

General::unfl : Underflow occurred in computation.

General::stop : Further output of General::unfl will be suppressed during this calculation.

General::ovfl : Overflow occurred in computation.

General::ovfl : Overflow occurred in computation.

General::ovfl : Overflow occurred in computation.

General::stop : Further output of General::ovfl will be suppressed during this calculation.

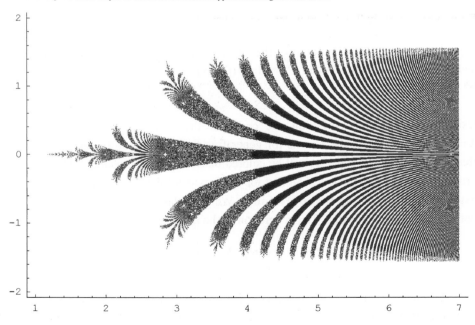

We received error messages from *Mathematica* when implementing the above program. Consider the following:

$$\text{FixedPointList}\left[\frac{1}{E}\,E^{\#}\,\&,\,5+1.2\,I,\,6,\,\text{SameTest}\rightarrow(\text{Re}[\#2]>50\,\&\&\,\text{Cos}[\text{Im}[\#2]]\geq0\,\&)\right]$$

General::unfl : Underflow occurred in computation.

$$\{5.+1.2\,i,\,19.7841+50.8876\,i,\,1.16873\times10^{8}+8.38125\times10^{7}\,i,$$
$$-5.6452782\times10^{50757205}-2.6241376\times10^{50757206}\,i,$$
$$\text{Underflow}[],\,\frac{\text{Underflow}[]+1}{e},\,e^{-1+\frac{\text{Underflow}[]+1}{e}}\}$$

In the above orbit of $5+1.2\,I$, the fourth iterate is enormous, and although the real part of the iterate is greater than 50, the cosine of the imaginary part must be negative, so the 'stopping test' has not been satisfied, however *Mathematica* cannot continue with the calculations, as the numbers involved are too large. It returns the value of **expJulia**$[5+I]$ as 41, the maximum number of iterations. There may be inaccuracy in this case, as the relevant point may be in the Julia set. The terms *underflow* and *overflow* are used by *Mathematica* to refer to numbers outside the range of Machine precision.

From Help - The *Mathematica* Book A.1.5:
In any implementation of Mathematica, the magnitudes of numbers (except 0) must lie between $MinNumber and $MaxNumber. Numbers with magnitudes outside this range are represented by Underflow[ ] and Overflow[ ].

From now on we shall not display error messages of the above type.

$$\text{DensityPlot}\left[\text{expJulia}\left[x+I\,y,\,\frac{1}{E}+0.1\right],\,\{x,\,-1,\,8\},\,\{y,\,-2,\,2\},\,\text{Mesh}\rightarrow\text{False},\right.$$
$$\text{Frame}\rightarrow\text{False},\,\text{Axes}\rightarrow\text{True},\,\text{PlotPoints}\rightarrow250,\,\text{AspectRatio}\rightarrow\text{Automatic},$$
$$\left.\text{ColorFunction}\rightarrow(\text{If}[\#\geq1,\,\text{RGBColor}[0,0,0],\,\text{RGBColor}[1,1,1]]\,\&)\right];$$

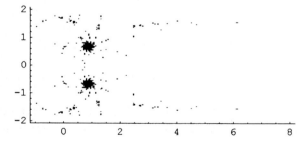

In the above example $\lambda>\frac{1}{E}$ and so the Julia set is the whole of the complex plane. We should therefore, theoretically, see a white rectangle. However we see bits of black due to the fact that our algorithm only computes at maximum the first 40 points along an orbit. If we were to increase the number of points being computed before coloring the non-escaping points black, we would see a concomitant decrease in the amount of black showing on the diagram.

For complex values of λ the Julia set of E_λ is the whole of the complex plane if $\lambda = k\pi I$, $k \in \mathbb{Z}$. However, for other complex values intricate diagrams are obtained. In the following examples of complex values of λ, we use the escape-time algorithm with the single criterion $\text{Re}[E_\lambda^n[z]]$ > 50 for $1 \leq n \leq 25$ and color the points which do not escape white and the other points according to the escape-time. The colored points represent the Julia set of the function concerned. (Color Fig 7.12)

$$\text{expJulia2} = \text{Compile}[\{\{z, \_\text{Complex}\}, \{c, \_\text{Complex}\}\},$$
$$\text{Length}[\text{FixedPointList}[c\, E^{\#}\, \&,\, z,\, 25,\, \text{SameTest} \rightarrow (\text{Re}[\#2] > 50.\, \&)]]];$$

$$\text{DensityPlot}\Big[-\text{expJulia2}[x + I\, y,\, 2\, I\,],\, \{x,\, -7,\, 5\},$$
$$\{y,\, -10,\, 1\},\, \text{Frame} \text{--}> \text{True},\, \text{Mesh} \rightarrow \text{False},\, \text{Frame} \rightarrow \text{False},$$
$$\text{Axes} \rightarrow \text{False},\, \text{PlotPoints} \rightarrow 250,\, \text{AspectRatio} \rightarrow \text{Automatic},$$
$$\text{ColorFunction} \rightarrow \Big(\text{If}\Big[\# <= 0,\, \text{RGBColor}[1,\, 1,\, 1],\, \text{RGBColor}\Big[\#,\, 1 - \frac{\#}{2},\, \#\Big]\Big]\, \&\Big)\Big];$$

Here is a zoom-in on the above: (Color Fig 7.13)

$$\text{DensityPlot}\Big[-\text{expJulia2}[x + I\, y,\, 2\, I\,],\, \{x,\, -6.2,\, -4\},\, \{y,\, -10,\, -7.9\},\, \text{Mesh} \rightarrow \text{False},$$
$$\text{Frame} \rightarrow \text{False},\, \text{Axes} \rightarrow \text{False},\, \text{PlotPoints} \rightarrow 250,\, \text{AspectRatio} \rightarrow \text{Automatic},$$
$$\text{ColorFunction} \rightarrow \Big(\text{If}\Big[\# <= 0,\, \text{RGBColor}[1,\, 1,\, 1],\, \text{RGBColor}\Big[\#,\, 1 - \frac{\#}{2},\, \#\Big]\Big]\, \&\Big)\Big];$$

In the following example, the orbit of every point tends to infinity, so the Julia set is the whole of the complex plane. (Color Fig 7.14)

$$\text{DensityPlot}[\text{expJulia2}[x + I\, y,\, 2\,],\, \{x,\, -4,\, 5\},\, \{y,\, -9,\, 3\},\, \text{Frame} \text{--}> \text{True},\, \text{Mesh} \rightarrow \text{False},$$
$$\text{Frame} \rightarrow \text{False},\, \text{Axes} \rightarrow \text{False},\, \text{PlotPoints} \rightarrow 250,\, \text{AspectRatio} \rightarrow \text{Automatic},$$
$$\text{ColorFunction} \rightarrow (\text{If}[\# \geq 1,\, \text{RGBColor}[1,\, 1,\, 1],\, \text{Hue}[\#]]\, \&)];$$

Exercise:
Construct the Julia sets of λE^z for $\lambda =$ 3, 4, 3I, -3I, -2.5+1.5I, 5+1.5I.

■ Trigonometric Functions

In the case of the trigonometric functions $S_\lambda[z] = \lambda \text{Sin}[z]$ and $C_\lambda[z] = \lambda \text{Cos}[z]$, orbits which escape do so with increasing absolute imaginary part. So if $S_\lambda^n[z]$ tends to infinity as n tends to infinity then $|\text{Im}[S_\lambda^n[z]]|$ tends to infinity as n tends to infinity. (Similarly for $C_\lambda^n[z]$.)

In the following examples we use the escape-time algorithm with the criterion:
if $|\text{Im}[S_\lambda^n[z]]| > 50$ for $1 \leq n \leq 25$ ($|\text{Im}[C_\lambda^n[z]]| > 50$ for $1 \leq n \leq 25$) then we assume that the orbit of z escapes to infinity and color the points which do not escape a chosen color and the other points according to the escape-time.

```
cosJulia = Compile[{{z, _Complex}, {c, _Complex}},
        Length[FixedPointList[c Cos[#] &, z, 25, SameTest → (Abs[Im[#2]] > 50 &)]]];
```

We choose the following palette for the escaping points: (Color Fig 7.15)

$$ContourPlot\left[x, \{x, 0, 20\}, \{y, 0, 2\},\right.$$

$$ColorFunction → \left(CMYKColor\left[1 - \#, 4\#^2 - 4\# + 1, Cos\left[\pi\,\frac{\#}{2}\right], 0\right] \&\right),$$

$$\left. AspectRatio → Automatic, Contours → 20\right]$$

The color **RGBColor[0, 0, 1]** (royal blue) does not occur in the above sequence of colors, and it contrasts well with them. (Color Fig 7.16)

$$ContourPlot\left[cosJulia[x + I\,y, 2.2], \{x, -2, 2\}, \{y, -2, 2\}, ContourLines \text{ -> } False,\right.$$

$$PlotPoints → 250, AspectRatio → Automatic, ColorFunction →$$

$$\left.\left(If\left[\# ≥ 1, RGBColor[0, 0, 1], CMYKColor\left[1 - \#, 4\#^2 - 4\# + 1, Cos\left[\pi\,\frac{\#}{2}\right], 0\right]\right] \&\right)\right];$$

We adapt the programme for members of the family S_λ.

```
sinJulia = Compile[{{z, _Complex}, {c, _Complex}},
        Length[FixedPointList[c Sin[#] &, z, 25, SameTest → (Abs[Im[#2]] > 50. &)]]];
```

In the following example, we use a coloring method described in the section on contour plots in Chapter 2. In this case we color the contours and the contour lines. The Fatou set is colored black. (Color Fig 7.17)

$$ContourPlot\left[sinJulia[x + I\,y, 1 + 0.4\,I], \{x, -4.3, 5\}, \{y, 4.5, -4.5\},\right.$$

$$Frame → False, Axes → False, PlotPoints → 250, AspectRatio → Automatic,$$

$$ContourStyle → Table\left[\left\{Hue\left[1 - \frac{n}{25}\right]\right\}, \{n, 1, 25\}\right], Contours → 25,$$

$$\left. ColorFunction \text{ -> } (If[\# ≥ 1, CMYKColor[0, 0, 0, 1], Hue[0.15 - 3\#^2]] \&)\right];$$

In constructing a Julia set image of a trigonometric function, using the command **ContourPlot**, if we color the contour lines and not the contours, the Fatou set will not be distinguished. However, if we color the contour lines and use the command **ColorFunction->(If [#>=1, c1, c2]&)** or **ColorFunction->(If #<=0, c1, c2]&)**, (where c1 and c2 are colors which do not occur in the palette for the contour line colors) then the Fatou set will be distinguished. Here is an example: (Color Fig 7.18)

ContourPlot[− cosJulia[x + I y, 2.95],

{x, −1.78, 1.78}, {y, −1.78, 1.78}, Frame → False, ContourStyle →

Table[{CMYKColor[0.0025 n^2 , 0.2 n − 0.01 n^2, 1 − Sin[0.05 n π], 0]}, {n, 1, 20}],

Background → RGBColor[0.737255, 0.415686, 0.858824],

PlotPoints → 250, AspectRatio → Automatic, Contours → 20,

ColorFunction → (If[# ≤ 0, RGBColor[0, 0, 0], RGBColor[1, 1, 1]] &)];

In the following example, we use a coloring technique described in 2.8.2. (Color Fig 7.19)

DensityPlot[sinJulia[x + I y, 0.81 + 0.8 I], {x, −3, 3},

{y, −4, 4}, Mesh → False, Frame → False, Axes → False,

PlotPoints → {300, 400}, AspectRatio → Automatic, ColorFunction →

(If[# ≥ 1, CMYKColor[0, 0, 0, 0], CMYKColor[1 − (2 #1 − UnitStep[−0.5` + #1])^2,

2 #1 − UnitStep[−0.5` + #1], Sin[$\pi$ (2 #1 − UnitStep[−0.5` + #1])], 0]] &)];

Exercise:

Experiment with **cosJulia[z, c]** and **sinJulia[z, c]** for the following values for c:

cosJulia[z,c]: 1.5; 2; 2.3; 2.94; I; 1+2I; 0.6; 0.7I; 3-0.6I;

sinJulia[z,c]: 1.3; 3.5; -2I; -3I; 2I; 1+I; 1-0.2I; 1-0.3I; 0.81+0.8I.

Here is a routine for generating the Julia set of a member of the family:

$\{\lambda$Cos + μ | λ, μ ∈ ℂ}, defined by $(\lambda$Cos + $\mu)[z]$ = λCos[z + μ]. The Fatou set is orange.

ContourPlot[Length[FixedPointList[(0.8 − 0.9 I) Cos[(0.8 − 0.7 I) + #] &,

x + I y, 25, SameTest → (Abs[Im[#2]] > 50 &)]], {x, −1, 1},

{y, −1, 1}, AspectRatio → Automatic, Frame → False, Axes → False,

PlotPoints → 250, Background → RGBColor[0.238285, 0.750011, 0.32813],

ContourStyle → Table[{CMYKColor[1 − Cos[$\frac{n\pi}{50}$], $\frac{4n}{25}$ − $\frac{4n^2}{625}$, 1 − Sin[$\frac{n\pi}{25}$], 0]},

{n, 1, 25}], Contours → 25,

ColorFunction −> (If[# ≥ 1, CMYKColor[0.2, 0.5, 1, 0], Hue[0.7 − $\frac{\#}{2}$]] &)];

Here is the Julia set of a member of the family {Sin ∗ λ | λ ∈ ℂ} defined by $(\text{Sin} ∗ \lambda)[z]$ = Sin[$z ∗ \lambda$]. The Fatou set is white. (Color Fig 7.20)

ContourPlot[Length[

FixedPointList[Sin[(0.5 I − 1) #] &, x + I y, 25, SameTest → (Abs[Im[#2]] > 50 &)]],

{x, −2, 2}, {y, −2, 2}, AspectRatio → Automatic, Frame → False, Axes → False,

PlotPoints → 350, ContourStyle → Table[{Hue[1 − $\frac{n}{25}$]}, {n, 1, 25}],

Contours → 25, ColorFunction → (If[# ≥ 1, CMYKColor[0, 0, 0, 0],

CMYKColor[Sin[π #], 1 − # ^ 2, 1 − (4 # ^ 2 − 4 # + 1), 0]] &)];

The following program gives a zoom-in on the Julia set of a member of the family $\{\lambda SinCos \mid \lambda \in \mathbb{C}\}$ defined by $(\lambda SinCos)[z] = \lambda Sin[Cos[z]]$. The Fatou set is dark blue.

> **sinJuliaCos = Compile[{{z, _Complex}, {c, _Complex}}, Length[**
> **FixedPointList[c Sin[Cos[#]] &, z, 25, SameTest → (Abs[Im[#2]] > 50. &)]]];**

> **ContourPlot[sinJuliaCos[x + I y, 1 − I], {x, 1.4, 2.4}, {y, .5, 1.5},**
> **ContourLines –> False, PlotPoints → 250, AspectRatio → Automatic,**
> **ColorFunction → (If[# ≥ 1, CMYKColor[1, 0.72, 0.372, 0],**
> **CMYKColor[1 − Sin[π #], #^2, 4 #^2 − 4 # + 1, 0]] &)];**

The following program gives the Julia set of a member of the family $\{\lambda Sin\mu \mid \lambda \in \mathbb{C}\}$ defined by $(\lambda Sin\mu)[z] = \lambda Sin[\mu z]]$.

> **DensityPlot[Length[FixedPointList[(0.2 − 0.5 I) Sin[2 #] &, x + I y, 25,**
> **SameTest → (Abs[Im[#2]] > 50 &)]], {x, −1.5, 1.5}, {y, −1.5, 1.5},**
> **Mesh → False, AspectRatio → Automatic, Frame → False, Axes → True,**
> **PlotPoints → 250, ColorFunction → (If[# ≥ 1, RGBColor[0, 0, 0], Hue[#]] &)];**

Exercise:
Experiment with members of the above families.

7.2 Parameter Sets

7.2.1 The Mandelbrot Set

The Julia set J_c of the function Q_c, where $Q_c[z] = z^2 + c$, is either totally disconnected or connected. It can be proved that J_c is connected if and only if the orbit of zero under iteration by Q_c is bounded.

The Mandelbrot set for the family $\{Q_c \mid c \in \mathbb{C}\}$ is defined to be the set:
$M = \{c \in \mathbb{C} \mid$ the orbit of 0 under iteration by Q_c is bounded$\}$.

It can be shown that if $|c| > 2$ then the orbit of 0 escapes to infinity, so we need only consider values of c for which $|c| \leq 2$. Further, if for any n, $|Q_c^n[0]| > 2$, then the orbit of 0 tends to infinity. Thus the Mandelbrot set lies inside the square with sides of length 4 parallel to the coordinate axes and center the origin.

If we consider part of the orbit of 0 under Q_c as follows:

> **NestList[#² + c &, 0, 3]** $\{0, c, c + c^2, c + (c + c^2)^2\}$

we see that the orbit of c under the action of Q_c converges to the same limit as the orbit of 0.

In the following adaptation of *Mathematica*'s program, we apply the pure function

$\#^2 + x + \mathrm{I}\,y$ repeatedly to each $x + \mathrm{I}\,y$ in the rectangle indicated below. The iteration stops if the absolute value of an iterate is greater than or equal to 2 or if the number of iterations exceeds 50. The length of the resulting orbit is then plotted with **DensityPlot**. The Mandelbrot set is colored black, and points outside the Mandelbrot set are colored according to their escape-time. (Color Fig 7.21)

> **DensityPlot[**
> **Length[FixedPointList[$\#^2$ + x + I y &, x + I y, 50, SameTest → (Abs[#2] > 2.0 &)]],**
> **{x, −2, 0.5}, {y, −1.2, 1.2}, Mesh → False, AspectRatio → Automatic,**
> **Frame → False, Axes → True, PlotPoints → 300,**
> **ColorFunction → (If[# ≥ 1, RGBColor[0, 0, 0], Hue[#]] &)];**

Color Fig 7.22 generated by the program below shows the rapid change of escape time near the border of the Mandelbrot set. For example, points in the broad band of vermilion and yellow escape after 1-10 iterations, points in the narrow band of green and turquoise escape after 11-25 iterations, while the purple and red band, constituting points which escape after 40-50 iterations, is so narrow that it can hardly be seen.

> **DensityPlot[x, {x, 1, 50}, {y, 0, 1}, PlotPoints → {50, 2}, AspectRatio → 0.1,**
> **ColorFunction → (If[# ≥ 1, RGBColor[0, 0, 0], Hue[#]] &)]**

The above-mentioned rapid change can also be illustrated by applying the command **Plot3D** to the escape-time function.

> **Plot3D[**
> **Length[FixedPointList[$\#^2$ + x + I y &, x + I y, 50, SameTest → (Abs[#2] > 2.0 &)]],**
> **{x, −2, 0.5}, {y, −1.3, 1.3}, Mesh → False,**
> **PlotPoints → 200, ViewPoint −> {−1.544, 1.640, 2.526}];**

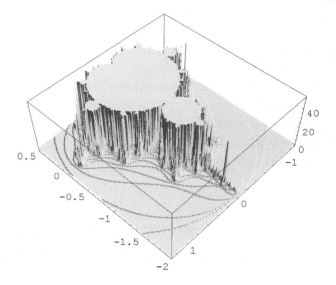

Interesting images can be obtained by zooming in.

We can zoom-in on parts of the Mandelbrot set especially the boundary by limiting the range of the density plot. It is important to increase the number of iterations according to the size of the zoom-in, as rapid changes in escape times occur near the boundary of the Mandelbrot set. (Color Fig 7.23)

> **DensityPlot[**
>   **Length[FixedPointList[#^{2} + x + I y &, x + I y, 100, SameTest → (Abs[#2] > 2.0 &)]],**
>   **{x, .347, 0.367}, {y, 0.636, 0.656}, Mesh → False, AspectRatio → Automatic,**
>   **Frame → False, Axes → False, PlotPoints → 350,**
>   **ColorFunction → (If[# ≥ 1, RGBColor[0, 0, 0], Hue[1 − 2 #]] &)];**

Here is another zoom-in: (Color Fig 7.24)

> **DensityPlot[**
>   **Length[FixedPointList[#^{2} + x + I y &, x + I y, 400, SameTest → (Abs[#2] > 2.0 &)]],**
>   **{x, −0.7454315, −0.7454255}, {y, 0.1130059, 0.1130119}, Mesh → False,**
>   **AspectRatio → Automatic, Frame → False, Axes → False, PlotPoints → 300,**
>   **ColorFunction → (If[# ≥ 1, RGBColor[0, 0, 0], Hue[#]] &)];**

We now construct a different version of a zoom-in, by indicating escape times alternately in black and white.

> **DensityPlot[Mod[**
>   **Length[FixedPointList[#^{2} + x + I y &, x + I y, 201, SameTest → (Abs[#2] > 2.0 &)]],**
>   **2], {x, −0.747, −0.745}, {y, 0.0977, 0.0997}, Mesh → False,**
>   **AspectRatio → Automatic, Frame → False, Axes → True, PlotPoints → 450];**

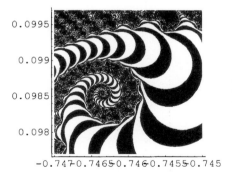

Exercise:

Let $f_\lambda[z] = z^n + \lambda$. In 7.1.1 we generated a Julia set for f_λ in the case n = 5. Adapt the Mandelbrot set program to generate a parameter set for f_λ in the case n = 5, using the escape criterion $|z| > 2$. Repeat for other values of n.

7.2.2 Parameter Sets for Entire Transcendental Functions

We consider the families $\{E_\lambda \mid \lambda \in \mathbb{C}\}, \{S_\lambda \mid \lambda \in \mathbb{C}\}$ and $\{C_\lambda \mid \lambda \in \mathbb{C}\}$ where $E_\lambda[z] = \lambda E^z$, $S_\lambda[z] = \lambda \mathrm{Sin}[z]$ and $C_\lambda[z] = \lambda \mathrm{Cos}[z]$.

The number 0 is an asymptotic value for each member of the family $\{E_\lambda \mid \lambda \in \mathbb{C}\}$ (Devaney (1994)), and the orbit of 0 is called a critical orbit. The members of the trigonometric families $\{S_\lambda \mid \lambda \in \mathbb{C}\}$ and $\{C_\lambda \mid \lambda \in \mathbb{C}\}$ do not have asymptotic values but have many critical points. However, they each have exactly 2 critical values, and so they are critically finite, (see section 7.1.5). It can be proved that the orbit of at least one critical point or asymptotic value lies in the basin of attraction of any attracting cycle, Devaney (1994).

■ **The Family** $\{E_\lambda \mid \lambda \in \mathbb{C}\}$

As each member of the family $\{E_\lambda \mid \lambda \in \mathbb{C}\}$ has no critical point, the role of the critical orbit is played by the orbit of 0. The parameter set for the family is the set of values of λ for which the critical orbit is bounded, Devaney (1994).

If $|E_\lambda^n[0]| \to \infty$ as $n \to \infty$, then J_λ, the Julia set of E_λ, is the whole complex plane. The diagram below shows part of the parameter set for $\{E_\lambda \mid \lambda \in \mathbb{C}\}$. Black points indicate the regions where the critical orbit is bounded, white points where it is not bounded i.e. where $J_\lambda = \mathbb{C}$.

```
DensityPlot[
    Length[FixedPointList[(x + I y) E# &, x + I y, 10, SameTest → (Re[#2] > 50 &)]],
    {x, −4, 6}, {y, −6, 6}, Mesh → False, AspectRatio → Automatic,
    Frame → False, Axes → True, PlotPoints → 450,
    ColorFunction → (If[# ≥ 1, RGBColor[1, 1, 1], RGBColor[0, 0, 0]] &)];
```

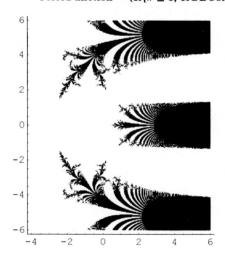

■ **The Families** $\{S_\lambda \mid \lambda \in \mathbb{C}\}$ **and** $\{C_\lambda \mid \lambda \in \mathbb{C}\}$

Each member of the family $\{S_\lambda \mid \lambda \in \mathbb{C}\}$ has infinitely many critical points since $\mathbf{D}[S_\lambda[z]] = 0$ if $z = \frac{\pm\pi}{2} + 2\,n\pi$. However, there are only 2 critical values, λ and $-\lambda$. It is easily shown that the orbits of λ and $-\lambda$ both tend to infinity or are both bounded. The parameter set for the family is taken to be the set $\{\lambda \in \mathbb{C} \mid$ orbit of λ is bounded$\}$. (Color Fig 7.25)

```
DensityPlot[Length[
    FixedPointList[(x + I y) Sin[#] &, x + I y, 30, SameTest → (Abs[Im[#2]] > 50 &)]],
    {x, −7.5, 7.5}, {y, −2.5, 2.5}, Mesh → False, AspectRatio → Automatic,
    Frame → False, Axes → False, PlotPoints → 350,
    ColorFunction → (If[# ≥ 1,  RGBColor[0, 0, 0], Hue[0.5 − 3 #]] &)];
```

Exercise:

1) Construct the parameter set for the family $\{C_\lambda \mid \lambda \in \mathbb{C}\}$ noting that the critical values for this family are also λ and $-\lambda$.

Take as the parameter set for the family, the set $\{\lambda \in \mathbb{C} \mid$ orbit of λ is bounded$\}$.

2) Zoom in on some of the Mandelbrot-set-like sections of the above images.

3) Construct the parameter sets for the families $\{\lambda C_\lambda \mid \lambda \in \mathbb{C}\}$ and $\{\lambda S_\lambda \mid \lambda \in \mathbb{C}\}$ defined by $\lambda C_\lambda[z] = \lambda \mathrm{Cos}[\lambda z]$ and $\lambda S_\lambda[z] = \lambda \mathrm{Sin}[\lambda z]$, first checking that the critical values are λ and $-\lambda$, and that their orbits are both bounded, or both not bounded.

7.3 Illustrations of Newton's Method

7.3.1 Classifying Starting Points for Newton's Method

In Chapter 1 we discussed Newton's iterative method of finding solutions to an equation of the form $f[z] = 0$. This method amounts to iterating the function F defined by $F[z] = z - \frac{f[z]}{f'[z]}$ applied to a suitable starting point.

When using Newton's iterative method to solve a complex equation of the form $f[z] = 0$, with starting point t, it is of interest to know which points in the complex plane converge to each of the different roots of the equation. We give a method of displaying the regions of the complex plane corresponding to each root in the case that f is a polynomial. If f is of degree 4 or less, exact solutions of the equation can be found, using the command **Solve**. For f of degree 5 or more, approximate solutions can be found using the command **NSolve**.

We start with the polynomial $S[z] = (z^2 + 1)(z^2 + I)$ which was considered in section 7.1.4.

$$S[z\_] := (z^2 + 1)(z^2 + I);$$

We find the zeros of S:

Solve[S[z] = 0, z] $\{\{z \to -i\}, \{z \to i\}, \{z \to -(-1)^{3/4}\}, \{z \to (-1)^{3/4}\}\}$

and the derivative of S[z] with respect to z:

Together[D[S[z], z]] $(2 + 2\,i)\,(z + (1 - i)\,z^3)$

So Newton's iteration function for the above polynomial is: **iter**[z] = $z - \frac{(z^2+1)\,(z^2+I)}{(2+2\,I)\,(z+(1-I)\,z^3)}$;

Given a point z in the plane, we wish to find out if the orbit of z under the action of **iter** does or does not converge to one of the roots of the equation, and if so, which one. Consider the following definition:

$$\text{roots1}[z\_] := \text{FixedPoint}\left[N\left[\# - \frac{(\#^2 + 1)\,(\#^2 + I)}{2\,\#\,(i + \#^2) + 2\,\#\,(1 + \#^2)}\right]\,\&,\right.$$

$$\left. z, 25, \text{SameTest} \to (\text{Abs}[\#1 - \#2] < 10^{-10}\,\&)\right];$$

roots1[2 + I] $-1.54074 \times 10^{-32} + 1.\,i$

When **roots1** is applied to z, the orbit of z under the action of **iter** is calculated until the absolute value of the last 2 iterations differs by an amount less than 10^{-10}, or until 25 iteration have been carried out. The output is the last orbit value calculated. We construct a function which assigns one of 5 numbers to each point in the plane, according to the outcome of **roots1**. We then associate these numbers with colors. We use the command **Which** (discussed in Chapter1). The function **which1** below assigns the numbers 0 (red), 0.15 (yellow), 0.4 (green) and 0.6 (blue) to z if its orbit converges to $-I$, I, $-(-1)^{\frac{3}{4}}$, $(-1)^{\frac{3}{4}}$ respectively. It assigns the number 1 (black) if the orbit does not converge to any of the points. (Color Fig 7.26)

which1[z_] := Which[Abs[z − (−I)] < 10^{-10}, 0, Abs[z − I] < 10^{-10}, 0.15,
Abs[z + (−1)³/⁴] < 10^{-10}, 0.4, Abs[z − (−1)³/⁴] < 10^{-10}, 0.6, True, 1];

DensityPlot[which1[roots1[x + I y]], {x, −6, 6}, {y, −6, 6}, Mesh → False,
PlotPoints → 400, ColorFunction → (If[# ≥ 1, RGBColor[0, 0, 0], Hue[#]] &)]

Color Fig 7.27, generated by the program below shows a zoom-in of the above. The fact that the set of black points is not a boundary, means that the black points do not constitute the Julia set of NS. (It can be shown that the orbit of one of the critical points converges to an attracting 3-cycle).

DensityPlot[which1[roots1[x + I y]], {x, 3, 6}, {y, 1, 4}, Mesh → False,
PlotPoints → 400, ColorFunction → (If[# ≥ 1, RGBColor[0, 0, 0], Hue[#]] &)]

Exercise:
Apply the above procedure to Newton's iteration functions of various polynomials.

7.3.2 Choosing a Starting Point for Using Newton's Method to Solve Transcendental Equations

When applying Newton's method to find a solution of an equation of the form f [z] = 0, where f is transcendental, one has to choose a starting point, which may be problematical. Here is a method which may work:

Use the basin scanning algorithim applied to the Newton iteration function of f and then use a program similar to that used for generating Julia sets of rational functions and you may be able to pick out approximate roots of the equation f [z] = 0 (which are fixed points of the Newton iteration function), from the resulting image.

For example, suppose you wish to find a solution, near the origin, with positive real and imaginary parts, to the equation $f [z] = Cos[z] + \frac{z^2}{2} = 0$

In the previous example, we could find the roots of the relative equation and we were interested in the behaviour of points in the plane. In this case, it is not so easy to calculate roots unless we know their approximate location. The following program plots in black the points whose orbits do not converge, and colors the points whose orbits do converge according to the rate of convergence. (Color Fig 7.28)

$$\mathbf{DensityPlot}\left[\mathbf{Length}\left[\mathbf{FixedPointList}\left[\left(\# - \frac{\left(Cos[\#] + \frac{\#^2}{2}\right)}{-Sin[\#] + \#}\right) \&,\right.\right.\right.$$

$$\left.\left.x + I\,y, 30, \mathbf{SameTest} \to (Abs[\#1 - \#2] < 10^{-10} \&)\right]\right],$$

$$\{x, -3\,\pi, 3\,\pi\}, \{y, -3\,\pi, 3\,\pi\}, \mathbf{Mesh} \to \mathbf{False}, \mathbf{Frame} \to \mathbf{False},$$

$$\mathbf{Axes} \to \mathbf{True}, \mathbf{PlotPoints} \to 250, \mathbf{AspectRatio} \to \mathbf{Automatic},$$

$$\left.\mathbf{ColorFunction} \to (If[\# \geq 1, \mathbf{RGBColor}[0, 0, 0], \mathbf{Hue}[1 - 2\,\#]] \&)\right];$$

Note that the function we iterated is neither rational nor entire. We are not investigating the Julia set of the function, but only the set of points whose orbits converge.

Attractive fixed points of a function have a basin of attraction, which may be disconnected. The component which contains the fixed point is called the *immediate basin of attraction*. It seems reasonable to suppose that the immediate basin of attraction is larger than the other components and that the closer a point in the basin to the fixed point the faster it converges to that point, so it seems reasonable to expect that fixed points in the above diagram will be inside the larger blue spots. It looks as though there are 4 fixed points near the origin. We find co-ordinates of a point in the upper right blue spot with the mouse, obtaining the point representing the complex number 1.46 + 1.71 I. We now use *Mathematica*'s **FindRoot** command to find a better solution of the equation near this point. Having found a solution of the equation, we shall also investigate which points in the plane converge to this point. For this reason, we use the options **AccuracyGoal** and **WorkingPrecision** (discussed in Chapter 1) in order to obtain a better approximation to the fixed point.

$$\textbf{FindRoot}\left[\textbf{Cos[z]} + \frac{z^2}{2} = 0, \{z, 1.46 + 1.71\ I\},\right.$$

$$\left.\textbf{AccuracyGoal} \rightarrow 20, \textbf{WorkingPrecision} \rightarrow 30\right]$$

$\{z \rightarrow 1.49584251021296429561738129479 + 1.62220903968193976974838129835i\}$

We check this result to confirm our guess about the location of solutions of the given equation:

$$\textbf{Cos[z]} + \frac{z^2}{2}\ /.\ z \rightarrow 1.49584251021296429561738129478862118463752211`29.9817 +$$

$$1.62220903968193976974838129835142498050293`30.0169i$$

$-0. \times 10^{-29} + 0. \times 10^{-29}\ i$

We construct a zoom-in on the above figure: (Color Fig 7.29)

$$\textbf{DensityPlot}\left[\textbf{Length}\left[\textbf{FixedPointList}\left[\left(\# - \frac{\left(\textbf{Cos[\#]} + \frac{\#^2}{2}\right)}{-\textbf{Sin[\#]} + \#}\right)\ \&, x + I\,y, 30,\right.\right.\right.$$

$$\left.\left.\textbf{SameTest} \rightarrow (\textbf{Abs[\#1} - \#2] < 10^{-10}\ \&)\right]\right], \{x, 5, 7\}, \{y, -4, -3\}, \textbf{Mesh} \rightarrow \textbf{False},$$

$$\textbf{Frame} \rightarrow \textbf{False}, \textbf{Axes} \rightarrow \textbf{True}, \textbf{PlotPoints} \rightarrow 250, \textbf{AspectRatio} \rightarrow \textbf{Automatic},$$

$$\textbf{ColorFunction} \rightarrow (\textbf{If[\# } \geq 1, \textbf{RGBColor[0, 0, 0]}, \textbf{Hue[1} - 2\#]]\ \&)\right]$$

We now investigate the basin of attraction of the fixed point we found above. The program below colors in black, points whose orbits do not converge to the chosen fixed point, and colors the other points according to their rate of convergence. The image illustrates how complicated a basin of attraction may be. (Color Fig 7.30)

$$\textbf{DensityPlot}\left[\textbf{Length}\left[\textbf{FixedPointList}\left[\left(\# - \frac{\left(\textbf{Cos[\#]} + \frac{\#^2}{2}\right)}{-\textbf{Sin[\#]} + \#}\right)\ \&, x + I\,y, 30, \textbf{SameTest} \rightarrow\right.\right.\right.$$

$$(\textbf{Abs[\#1} - (1.49584251021296429561738129478862118463752211`29.9817 +$$

$$1.62220903968193976974838129835142498050293`30.0169i)] <$$

$$10^{-10}\ \&)\Big]\Big], \{x, -3\pi, 3\pi\}, \{y, -3\pi, 3\pi\}, \textbf{Mesh} \rightarrow \textbf{False},$$

$$\textbf{Frame} \rightarrow \textbf{False}, \textbf{Axes} \rightarrow \textbf{True}, \textbf{PlotPoints} \rightarrow 250, \textbf{AspectRatio} \rightarrow$$

$$\textbf{Automatic},$$

$$\textbf{ColorFunction} \rightarrow (\textbf{If[\# } \geq 1, \textbf{RGBColor[0, 0, 0]}, \textbf{Hue[0.7} - 2\#]]\ \&)\Big];$$

In the following example, the contours and contour lines are colored. (Color Fig 7.31)

$$\text{ContourPlot}\left[-\text{Length}\left[\text{FixedPointList}\left[\left(\# - \frac{(\text{Cosh}[\#] - \text{Sin}[\#])}{\text{Sinh}[\#] - \text{Cos}[\#]}\right) \&, x + I\,y, 40,\right.\right.\right.$$

$$\left.\left.\text{SameTest} \rightarrow (\text{Abs}[\#1 - \#2] < 10^{-10} \&)\right]\right], \{x, -8, 8\}, \{y, -8, 8\}, \text{Frame} \rightarrow \text{False},$$

$$\text{Axes} \rightarrow \text{False}, \text{PlotPoints} \rightarrow 250, \text{Contours} \rightarrow 40, \text{ColorFunction} \mathrel{-\!>}$$
$$(\text{If}[\# \geq 1, \text{RGBColor}[1, 1, 1], \text{RGBColor}[\text{Abs}[2\# - 1], \text{Abs}[\text{Sin}[2\,\pi\#]], \#]] \&),$$
$$\text{AspectRatio} \rightarrow \text{Automatic}, \text{ContourStyle} \rightarrow$$

$$\text{Table}\left[\left\{\text{CMYKColor}\left[\left(1 - \frac{n}{5} + \frac{n^2}{100}\right), \frac{n}{20}, 1 - \text{Sin}\left[\frac{n\,\pi}{20}\right], 0\right]\right\}, \{n, 1, 20\}\right]\right];$$

A zoom-in on the above: (Color Fig 7.32)

$$\text{ContourPlot}\left[-\text{Length}\left[\text{FixedPointList}\left[\left(\# - \frac{(\text{Cosh}[\#] - \text{Sin}[\#])}{\text{Sinh}[\#] - \text{Cos}[\#]}\right) \&, x + I\,y, 40,\right.\right.\right.$$

$$\left.\left.\text{SameTest} \rightarrow (\text{Abs}[\#1 - \#2] < 10^{-10} \&)\right]\right], \{x, 2.5, 4\}, \{y, 2.5, 4\}, \text{Frame} \rightarrow \text{False},$$

$$\text{Axes} \rightarrow \text{False}, \text{PlotPoints} \rightarrow 250, \text{Contours} \rightarrow 40, \text{ColorFunction} \mathrel{-\!>}$$
$$(\text{If}[\# \geq 1, \text{RGBColor}[1, 1, 1], \text{RGBColor}[\text{Abs}[2\# - 1], \text{Abs}[\text{Sin}[2\,\pi\#]], \#]] \&),$$
$$\text{AspectRatio} \rightarrow \text{Automatic}, \text{ContourStyle} \rightarrow$$

$$\text{Table}\left[\left\{\text{CMYKColor}\left[\left(1 - \frac{n}{5} + \frac{n^2}{100}\right), \frac{n}{20}, 1 - \text{Sin}\left[\frac{n\,\pi}{20}\right], 0\right]\right\}, \{n, 1, 20\}\right]\right];$$

Another example: (Color Fig 7.33)

$$\text{ContourPlot}\left[\text{Length}\left[\text{FixedPointList}\left[\left(\# - \frac{\text{Cos}[\#] + 1 + \frac{\#^2}{2}}{-\text{Sin}[\#] + \#}\right) \&, x + I\,y,\right.\right.\right.$$

$$\left.\left.30, \text{SameTest} \rightarrow (\text{Abs}[\#1 - \#2] < 10^{-10} \&)\right]\right], \{x, 5., 6.5\}, \{y, -3, -1.5\},$$

$$\text{ContourLines} \rightarrow \text{False}, \text{Axes} \rightarrow \text{False}, \text{PlotPoints} \rightarrow 250, \text{AspectRatio} \rightarrow \text{Automatic},$$

$$\text{Frame} \rightarrow \text{False}, \text{ColorFunction} \rightarrow \left(\text{If}\left[\# \geq 1, \text{RGBColor}[0, 0, 0],\right.\right.$$

$$\left.\left.\text{CMYKColor}\left[1 - \text{Sin}[\pi\#], 4\#^2 - 4\# + 1, 1 - \text{Cos}\left[\pi\,\frac{\#}{3}\right], 0\right]\right] \&\right)\right];$$

Exercise:
Experiment with applying the above construction to Newton iteration functions of polynomials and transcendental functions, for example:
$\text{Cosh}[z] - \text{Sin}[z], \quad \text{Sin}[z^2] + 4, \quad \text{Cos}[z] - 2 + z^2, \quad \text{Cos}[2\,z] + z^2, \quad \text{Cosh}[z] - z, \quad \text{Sinh}[z] + z^2,$
$\text{Cos}[z] + 2 + z^2, \text{Cosh}[z] + \text{Cos}[2\,z] + z^2$.

Chapter 8

Miscellaneous Design Ideas

Introduction

In this Chapter we describe more complicated constructions of fractals, patterns and natural forms. We shall construct some more Julia sets, describe some tiling techniques and show how to write programs which generate 3D shell-like structures.

8.1 Sierpinski Relatives as Julia Sets

In Chapter 5 we used the following routine to construct a fractal called a Sierpinski relative:

```
sr1[x_, n_] :=
    Show[Graphics[Nest[IFS[{AffineMap[180°, 180°, 0.5, −0.5, −1, 1], AffineMap[90°,
            90°, 0.5, 0.5, −1, −1], AffineMap[90°, 90°, 0.5, −0.5, 1, −1]}], x, n]],
        Axes → False, AspectRatio → Automatic, AxesOrigin → {0, 0}];

sr1[Point[{0, 0}], 8]
```

Recall that all Sierpinski relatives were constructed by applying an IFS consisting of 3 affine contraction mappings each of which mapped a square, S, into itself. The 3 images, S_1, S_2 and S_3 of the square were disjoint except for boundary points. Also, the 3 affine maps map the attractor of the IFS onto subsets of itself which are disjoint except for points on the boundaries of S_1, S_2 and S_3. Such an IFS is called 'just touching'. In the case we are considering, the affine maps are one-to-one, and so have inverses. It can be proved that each Sierpinski relative is the Julia set of a certain complex function. We shall illustrate this by constructing a function whose Julia set is the attractor of the IFS defined above.

The programs we have used to generate Julia sets involve the iteration of complex functions. Every map from $\mathbb{R}^2$ into $\mathbb{R}^2$ can be represented as a map from $\mathbb{C}$ into $\mathbb{C}$. The affine maps used in construction of Sierpinski relatives were of the forms $w_1 = $ **AffineMap[θ, θ, r, r, p, q]** and $w_2 = $ **AffineMap[θ, θ, r, −r, p, q]** with r = 0.5 or r = −0.5, θ a multiple of $\frac{\pi}{2}$. We shall express these maps as $\mathbb{C}$ to $\mathbb{C}$ maps, W_1 and W_2. We first load the following package of Roman Maeder:

<< ProgrammingInMathematica`AffineMaps`

AffineMap[θ, θ, r, r, p, q] [{Re[z], Im[z]}]

$\{p + r \operatorname{Cos}[\theta] \operatorname{Re}[z] - r \operatorname{Im}[z] \operatorname{Sin}[\theta], q + r \operatorname{Cos}[\theta] \operatorname{Im}[z] + r \operatorname{Re}[z] \operatorname{Sin}[\theta]\}$

$W_1[z] = p + r \operatorname{Cos}[\theta] \operatorname{Re}[z] - r \operatorname{Im}[z] \operatorname{Sin}[\theta] + I (q + r \operatorname{Cos}[\theta] \operatorname{Im}[z] + r \operatorname{Re}[z] \operatorname{Sin}[\theta]) = p + I q + r E^{I\theta} z.$

Similarly,

$W_2[z] = p + I q + r E^{I\theta} \operatorname{Conjugate}[z].$

As $r \neq 0$, the above maps are invertible. We find their inverses:

$$W_1^{-1}[z] = -\frac{E^{-I\theta} (p + I q - z)}{r};$$
$$W_2^{-1}[z] = -\frac{E^{I\theta} (p - Iq - \operatorname{Conjugate}[z])}{r} = \operatorname{Conjugate}[W_1^{-1}[z]]$$

We now use the above formulae to find the inverses of the affine maps, in complex form, used in **sr1**.

The first affine map, u_1, its complex form, U_1, and the inverse thereof:

$u_1 = $ AffineMap[180°, 180°, 0.5, −0.5, −1, 1]

$U_1[z] = p + I q + r E^{I\theta} \operatorname{Conjugate}[z]$ /. $\{p \to -1, q \to 1, \theta \to \pi, r \to 0.5\}$

$(-1 + i) - 0.5 \operatorname{Conjugate}[z]$

$-\dfrac{E^{I\theta} (p - I q - \operatorname{Conjugate}[z])}{r}$ /. $\{p \to -1, q \to 1, \theta \to \pi, r \to 0.5\}$

$2. ((-1 - i) - \operatorname{Conjugate}[z])$

So **$U_1^{-1}[z] = 2. ((-1 - i) - \operatorname{Conjugate}[z])$**

Similarly,
$$U_2^{-1}[z] = 2.\,i\,((-1-i)-z) \text{ and } U_3^{-1}[z] = -2.\,i\,((1+i) - \text{Conjugate}[z])$$

Now u_1, u_2 and u_3 map the large square in the diagram onto the squares S_1, S_2 and S_3 respectively.

If $z \in S_1$, then $\text{Re}[z] < 0$ and $\text{Im}[z] > 0$, while if $z \in S_2$, $\text{Re}[z] < 0$ and $\text{Im}[z] < 0$ and if $z \in S_3$, $\text{Re}[z] > 0$. (We have excluded the boundaries of S_1, S_2 and S_3.)

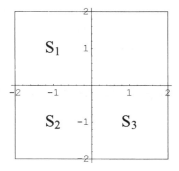

Now consider the function F defined below:

$$\mathbf{F[z\_] := 2\,((-1-I) - \text{Conjugate}[z]) \;/; \text{Re}[z] < 0 \;\&\&\; \text{Im}[z] \geq 0;}$$
$$\mathbf{F[z\_] := 2\,I\,((-1-I) - z) \;/; \text{Re}[z] < 0 \;\&\&\; \text{Im}[z] < 0;}$$
$$\mathbf{F[z\_] := -2\,I\,((1+I) - \text{Conjugate}[z]) \;/; \text{Re}[z] \geq 0;}$$

$F[z]$ is defined for all $z \in \mathbb{C}$.

If we let $T_1 = \{z \in \mathbb{C} \mid \text{Re}[z] < 0 \text{ and } \text{Im}[z] \geq 0\}$, let $T_2 = \{z \in \mathbb{C} \mid \text{Re}[z] < 0 \text{ and } \text{Im}[z] < 0\}$ and let $T_3 = \{z \in \mathbb{C} \mid \text{Re}[z] \geq 0\}$, then the inverses of the function F restricted to T_1, T_2, T_3 are u_1, u_2, u_3 respectively, expressed as maps from $\mathbb{C}$ to $\mathbb{C}$. It can be proved that the Julia set of F is the set of points in $\mathbb{C}$ whose orbits are bounded under F, and is also the attractor of the IFS defined above in **sr1**, a Sierpinski relative. Similar results are true for the other Sierpinski relatives. We know that the attractor is contained within the square of side-length 2 with axes parallel to the co-ordinate axes and centre the origin. This fact determines the escape criterion we use in the following application of the escape-time algorithm to F. (Color Fig 8.1)

$$\mathbf{ContourPlot[Length[FixedPointList[F, x + I\,y, 50, SameTest \to (Abs[\#2] > 4\,\&)]],}$$
$$\mathbf{\{x, -2, 2\}, \{y, -2, 2\}, ContourLines \to False, Frame \to False,}$$
$$\mathbf{Axes \to True, PlotPoints \to 250, AspectRatio \to Automatic,}$$
$$\mathbf{ColorFunction \to (If[\# \geq 1, CMYKColor[0, 0, 0, 0],}$$
$$\mathbf{CMYKColor[Random[], Random[], Random[], 0]]\,\&)];}$$

The next image (Color Fig 8.2) shows a 3D plot of the above function **Length[FixedPointList...]**. We color the image using the command below which was defined in Chapter 2.

cf[RGBColor[r1_, g1_, b1_], RGBColor[r2_, g2_, b2_]] :=
    (RGBColor[r1 + (r2 − r1) #, g1 + (g2 − g1) #, b1 + (b2 − b1) #]) &;

Plot3D[Length[FixedPointList[F, x + I y, 100, SameTest → (Abs[#2] > 4 &)]], {x, −2, 2},
    {y, −2, 2}, Mesh → False, Axes → False, PlotPoints → 80, BoxRatios → {2, 2, 0.5},
    Boxed → False, ViewPoint −> {−3.000, 0.966, 1.231}, ColorFunction →
    cf[RGBColor[0.925795, 0.902358, 0.453132], RGBColor[0.0, 0.0, 0.0]]]];

Exercise:

Construct other Sierpinski relatives as Julia sets. One way to do this directly is to define a function F with $F[z] = W_1^{-1}[z]$ or its conjugate, with $\{p, q\} = \{-1, 1\}$, $\{-1, -1\}$, $\{1, -1\}$ respectively for $z \in S_1, S_2, S_3$, and choosing r as 0.5 or -0.5 and θ as a multiple of $\frac{\pi}{2}$ in each case.

8.2 Patterns Formed from Randomly Selected Circular Arcs

We shall construct a random pattern consisting of arcs of circles.

    ? Circle

    Circle[{x, y}, r] is a two−dimensional graphics primitive that represents
        a circle of radius r centered at the point x, y. Circle[{x, y}, {rx, ry}] yields an ellipse
        with semi−axes rx and ry. Circle[{x, y}, r, {theta1, theta2}] represents a circular arc.

We wish to tile part of the x-y plane with a random arrangement of the following pair of square tiles, each with side-length 1:

We shall call the tiles $tile_0$ and $tile_1$. If these tiles are chosen at random to form a rectangular array, the curves formed are always continuous and smooth. At each lattice point $\{m, n\}$ in a section of the x-y plane, we shall plot $tile_r$ with r randomly chosen by *Mathematica*, and lower left-hand corner the point $\{m, n\}$. We can express each tile as the union of a pair of circular arcs. Here are commands defining the tiles:

$tile_0[\{m\_, n\_\}]$:=

    Union$\left[\left\{\text{Circle}\left[\{m + 1, n\}, 0.5, \left\{\frac{\pi}{2}, \pi\right\}\right], \text{Circle}\left[\{m, n + 1\}, 0.5, \left\{\frac{3\pi}{2}, 2\pi\right\}\right]\right\}\right]$;

$tile_1[\{m\_, n\_\}] :=$

$$Union\left[\left\{Circle\left[\{m, n\}, 0.5, \left\{0, \frac{\pi}{2}\right\}\right], Circle\left[\{m + 1, n + 1\}, 0.5, \left\{\pi, \frac{3\pi}{2}\right\}\right]\right\}\right];$$

We display $tile_0[\{-1, 0\}]$ (dark gray) and $tile_1[\{0, 0\}]$ (light gray):

Show[Graphics[{{GrayLevel[0.9], Rectangle[{0, 0}, {1, 1}]},
{GrayLevel[0.7], Rectangle[{-1, 0}, {0, 1}]}, tile$_0$[{-1, 0}], tile$_1$[{0, 0}]},
AspectRatio → Automatic, Axes → {True, False}]]

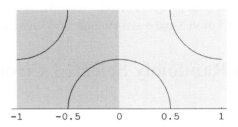

The command $tile_{Random[Integer]}$ returns $tile_0$ or $tile_1$. Here is a program for generating the random pattern:

Show[Graphics[Table[tile$_{Random[Integer]}$[{m, n}], {m, 1, 10}, {n, 1, 10}]],
AspectRatio → Automatic]

Adding random coloring and using sectors of circles instead of arcs of circles (Color Fig 8.3) we get:

$$\text{tileS}_0[\{m\_, n\_\}] := \text{Union}\Big[\Big\{\Big\{\text{Hue[Random[]]}, \text{Disk}\Big[\{m+1, n\}, 0.5, \Big\{\frac{\pi}{2}, \pi\Big\}\Big]\Big\},$$

$$\Big\{\text{Hue[Random[]]}, \text{Disk}\Big[\{m, n+1\}, 0.5, \Big\{\frac{3\pi}{2}, 2\pi\Big\}\Big]\Big\}\Big\}\Big];$$

$$\text{tileS}_1[\{m\_, n\_\}] := \text{Union}\Big[\Big\{\Big\{\text{Hue[Random[]]}, \text{Disk}\Big[\{m, n\}, 0.5, \Big\{0, \frac{\pi}{2}\Big\}\Big]\Big\},$$

$$\Big\{\text{Hue[Random[]]}, \text{Disk}\Big[\{m+1, n+1\}, 0.5, \Big\{\pi, 3\frac{\pi}{2}\Big\}\Big]\Big\}\Big\}\Big];$$

Show[Graphics[Table[tileS$_{\text{Random[Integer]}}$[{m, n}], {m, 1, 10}, {n, 1, 10}]],
    AspectRatio → Automatic]

Exercise:

1) Add colors to the circular arcs of the first construction. Add a background color or a thickness directive so that the circular arcs show up well.

2) Try tiling with tiles formed in a different way with circular arcs or sectors or line segments.

3) Try a random tiling with a pair of images such as a white and a black square or a red triangle and a blue circle.

4) In the case of a random tiling with a pair of the same shape, but different coloring, such as black and white squares, write a single command program which generates the tiling. Generalise your program to a procedure with 2 parameters r and s, say, which determine the size of the array of tiles.

5) Examine the constituent tiles in the image below, and then write a program to construct a similar image.

8.3 Constructing Images of Coiled Shells

8.3.1 Shell Anatomy

Many mollusc shells consist of an ever-widening, spirally coiled tube, which may have one of a variety of cross-sectional shapes. We shall show how to construct such tubes which have circular or elliptical cross-section.

The logarithmic spiral has polar equation $r = E^{a\theta}$, and parametric equations given by $\{E^{a\theta} \, Sin[\theta], \, E^{a\theta} \, Cos[\theta]\}$ where a is a positive parameter. We shall be considering only parametric functions of θ with θ having maximum value of 0. Here is an example, with $a = \frac{1}{4}$. Notice that, as $\theta \to -\infty$, the point on the curve with parameter θ tends to the origin. The initial point of the spiral is $\{0, 1\}$.

$$ParametricPlot\left[\left\{E^{\frac{\theta}{4}} \, Sin[\theta], \, E^{\frac{\theta}{4}} \, Cos[\theta]\right\}, \, \{\theta, \, -6\pi, \, 0\}, \, AspectRatio \to Automatic,\right.$$

$$\left. PlotRange \to All, \, Ticks \to \{\{-0.6, \, -0.2, \, 0.2\}, \, Automatic\}\right]$$

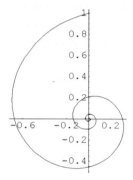

The larger the value of a, the more quickly the spiral expands. Here is an example with $a = \frac{1}{2}$.

$$ParametricPlot\left[\left\{E^{\frac{\theta}{2}} \, Sin[\theta], \, E^{\frac{\theta}{2}} \, Cos[\theta]\right\}, \, \{\theta, \, -6\pi, \, 0\},\right.$$

$$\left. AspectRatio \to Automatic, \, PlotRange \to All, \, Ticks \to \{False, \, True\}\right]$$

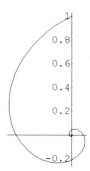

Dawkins (1996) defines the 'flare', f, of the logarithmic spiral to be the ratio of the distance from the origin of the point with parameter $\theta + 2\pi$ and the point with parameter θ. So $f = E^{2a\pi}$. Solving for a we get $a = \frac{\text{Log}[f]}{2\pi}$.

A 3D spiral can be constructed from the above by lowering each point on the spiral (regarded as a curve in 3D) by a distance proportional to its distance from the center of the 2D spiral $E^{a\theta}$. Thus the point on the spiral with parameter θ will have z-co-ordinate $-(s\,E)^{a\theta}$, where s is a constant. We call the constant s the 'spire', in agreement with Dawkins (1996).

$\{E^{a\theta}\,\text{Sin}[\theta],\ E^{a\theta}\,\text{Cos}[\theta],\ -s\,E^{a\theta}\}$ are the parametric equations for the resulting 3D spiral, Spiral[a, s]. The point with parameter θ on this curve tends to the point $\{0, 0, 0\}$ as θ tends to $-\infty$, so, as the maximum value of θ is 0, the vertical height of the complete curve is s.

ParametricPlot3D $\left[\left\{E^{\frac{\theta}{4}}\,\textbf{Sin}[\theta],\ E^{\frac{\theta}{4}}\,\textbf{Cos}[\theta],\ -1.3\,E^{\frac{\theta}{4}}\right\},\ \{\theta,\ -7\pi,\ 0\},\ \textbf{PlotRange} \rightarrow \textbf{All}\right]$

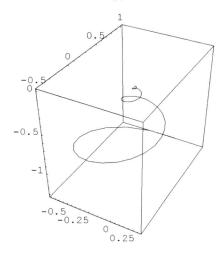

As Dawkins (1996) states: '*Because of their method of expanding at the margin, shells all have the same general form. It is a solid version of the so-called logarithmic or equiangular spiral*'.

Thus all coiled shells will have the same general form: a coiled, ever-widening tube whose outer edge is a spiral with parametric equation of the above form.

We start with a circle in the z-y plane with center $\{0, t, 0\}$ and radius r, such that $t + r = 1$ and $t \geq 0$. We call this the initial circle. We shall then rotate the circle successively through $n\,\phi$ radians where $0 \leq n \leq m$, say, and ϕ is a small angle, reducing the radius suitably. We then reduce the z-co-ordinate of the center of the circle by $s\,E^{a\theta}$ and plot all these circles, possibly coloring them.

ParametricPlot3D$\left[\left\{\left\{E^{\frac{\theta}{4}} Sin[\theta], E^{\frac{\theta}{4}} Cos[\theta], 0\right\}, \{0, 0.2 Sin[\theta] + 0.8, 0.2 Cos[\theta]\}\right\},\right.$

$\left.\{\theta, -6\pi, 0\}, \textbf{AspectRatio} \rightarrow \textbf{Automatic, PlotRange} \rightarrow \textbf{All}\right]$

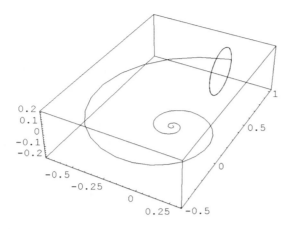

Our problem is to choose our parameters in such a way that the resulting coiled tube looks 'shell-like'.

We first show how to construct coiled shells whose vertical cross-section resembles the image below. The outer margin of the coiled tube is the spiral, Spiral[a, s]. We call this a 'just touching' shell with maximum radius.

We need to find the relationship between the parameters a and s in order that the coils of the tube forming the shell fit together. (Note that this can also happen if the cross-sectional circles do not meet the vertical central axis of the shell. We shall not discuss this case.)

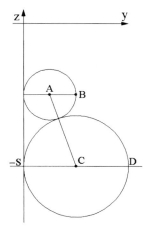

The above diagram shows a vertical cross section in the z-y plane of part of the coiled tube we wish to construct. The points B and D lie on the spiral with parametric equations given by $\{E^{a\theta}\mathrm{Sin}[\theta],\ E^{a\theta}\mathrm{Cos}[\theta],\ -sE^{a\theta}\}$. The point D has parameter 0 and co-ordinates $\{0,\ 1,\ -s\}$. The point B has parameter -2π and co-ordinates $\{0,\ E^{-2a\pi},\ -sE^{-2a\pi}\}$. The points A and C have co-ordinates $\{0,\ 0.5E^{-2a\pi},\ -sE^{-2a\pi}\}$ and $\{0, 0.5, -s\}$ respectively. The radii of the circles in the diagram with centres A and C are $0.5E^{-2a\pi}$ and 0.5 respectively. We use the distance formula for the length of AC and the fact that the length of AC is the sum of the radii of the 2 circles in the diagram above to obtain the following equation, giving the relationship between s and a:

$$s = \frac{\mathrm{Cosech}[a\pi]}{2}$$

See appendix to 8.3.1 for details.

Note that if $s > \frac{\mathrm{Cosech}[a\pi]}{2}$, the coils of the tube will not touch each other, while, if $s < \frac{\mathrm{Cosech}[a\pi]}{2}$, the coils of the tube will overlap.

8.3.2 Shell Construction

We now wish to find parametric equations for the circular cross-section, $C[\theta]$, of the tube which meets Spiral[a, s] at the point with parameter θ. We project the spiral and $C[\theta]$ onto the x-y plane. In the diagram below, $B\hat{O}A = \theta$, OA is the diameter of $C[\theta]$, A is the point $\{E^{a\theta}\mathrm{Sin}[\theta],\ E^{a\theta}\mathrm{Cos}[\theta]\}$, so the radius of $C[\theta]$ is $r[\theta] = 0.5E^{a\theta}$. The center of $C[\theta]$ is center$[\theta] = \{0.5E^{a\theta}\mathrm{Sin}[\theta],\ 0.5E^{a\theta}\mathrm{Cos}[\theta],\ -sE^{a\theta}\}$.

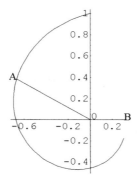

We start with the circle C '[θ] in the y-z plane, center the origin, radius r[θ], and rotate it through the angle θ about the z-axis. We shall then translate it to center[θ]. We load the necessary package:

<< Geometry`Rotations`

The parametric equations for C '[θ] are given by $\{0, 0.5\, E^{a\theta}\, \mathrm{Sin}[t], 0.5\, E^{a\theta}\, \mathrm{Cos}[t]\}$. We rotate each point with these co-ordinates about the z-axis through the angle θ:

Rotate3D[{0, y, z}, θ, 0, 0] /. {y $\to$ 0.5 E$^{a\theta}$ Sin[t], z $\to$ 0.5 E$^{a\theta}$ Cos[t]}

$\{0.5\, e^{a\theta}\, \mathrm{Sin}[t]\, \mathrm{Sin}[\theta], 0.5\, e^{a\theta}\, \mathrm{Cos}[\theta]\, \mathrm{Sin}[t], 0.5\, e^{a\theta}\, \mathrm{Cos}[t]\}$

We now translate to center[θ], to obtain the following parametric expression for the circle C[θ]:

$\{0.5`\ e^{a\theta}\, \mathrm{Sin}[t]\, \mathrm{Sin}[\theta] + 0.5\, E^{a\theta}\, \mathrm{Sin}[\theta],$
$\quad 0.5`\ e^{a\theta}\, \mathrm{Cos}[\theta]\, \mathrm{Sin}[t] + 0.5\, E^{a\theta}\, \mathrm{Cos}[\theta], 0.5`\ e^{a\theta}\, \mathrm{Cos}[t] - s\, E^{a\theta}\}$

We rename C[θ]:

circle[a_, s_, θ_] := {0.5` $e^{a\theta}$ Sin[t] Sin[θ] + 0.5 E$^{a\theta}$ Sin[θ],
0.5` $e^{a\theta}$ Cos[$\theta$] Sin[t] + 0.5 E$^{a\theta}$ Cos[$\theta$], 0.5` $e^{a\theta}$ Cos[t] − s E$^{a\theta}$};

We now choose parameters:
let f = 2, so that $a = \frac{\mathrm{Log}[f]}{2\pi} = 0.110318$ and $s = \frac{\mathrm{cosech}[a\pi]}{2} = 1.414213$. We shall construct a table of circles: C [n d], for d a small angle and $-m \le n \le 0$. We choose $d = \frac{\pi}{30}$, and m = 360.

circle$\left[0.110318, 1.414213, \dfrac{n\pi}{30}\right]$

$\Big\{0.5\, e^{0.0115525n}\, \mathrm{Sin}\Big[\dfrac{n\pi}{30}\Big] + 0.5\, e^{0.0115525n}\, \mathrm{Sin}\Big[\dfrac{n\pi}{30}\Big]\, \mathrm{Sin}[t],$
$\quad 0.5\, e^{0.0115525n}\, \mathrm{Cos}\Big[\dfrac{n\pi}{30}\Big] + 0.5\, e^{0.0115525n}\, \mathrm{Cos}\Big[\dfrac{n\pi}{30}\Big]\, \mathrm{Sin}[t],$
$\quad -1.41421\, e^{0.0115525n} + 0.5\, e^{0.0115525n}\, \mathrm{Cos}[t]\Big\}$

$$\textbf{ParametricPlot3D}\Big[\textbf{Evaluate}\Big[$$

$$\textbf{Table}\Big[\Big\{0.5\text{`}\; e^{0.011552473945290626\text{`} n}\; \textbf{Sin}\Big[\frac{n\,\pi}{30}\Big]+0.5\text{`}\; e^{0.011552473945290626\text{`} n}\; \textbf{Sin}\Big[\frac{n\,\pi}{30}\Big]\textbf{Sin}[t],$$

$$0.5\text{`}\; e^{0.011552473945290626\text{`} n}\; \textbf{Cos}\Big[\frac{n\,\pi}{30}\Big]+0.5\text{`}\; e^{0.011552473945290626\text{`} n}\; \textbf{Cos}\Big[\frac{n\,\pi}{30}\Big]\textbf{Sin}[t],$$

$$-1.414213\text{`}\; e^{0.011552473945290626\text{`} n}+0.5\text{`}\; e^{0.011552473945290626\text{`} n}\; \textbf{Cos}[t]\Big\},\{n,-360,0\}\Big],$$

$$\{t,0,2\,\pi\}\Big],\; \textbf{Axes}\rightarrow\textbf{False},\; \textbf{PlotRange}\rightarrow\textbf{All},\; \textbf{Boxed}\rightarrow\textbf{False},$$

$$\textbf{ViewPoint}->\{3.383,\,-0.040,\,-0.026\}\Big]$$

8.3.3 Coloring Methods

We now demonstrate a coloring method. Recall that in the command **Hue[h, s, b]**, h represents hue, s saturation and b brightness. Consider the color directive:

$$\textbf{Hue}\Big[0.05+0.04\,\textbf{Random}[\,],$$

$$\textbf{If}\Big[\textbf{Mod}[t+n,3]<\textbf{Random}[\textbf{Integer},\{0,1\}]\,||\,\textbf{Mod}\Big[\frac{t}{4},2\Big]<1,0.4,0.1\Big],$$

$$\textbf{If}[\textbf{Abs}[t]<40,0.9,1]\Big]$$

In the above, h varies at random between 0.05 and 0.09. Note that **Hue[h, 1, 1]**, with h between 0.05 and 0.09 produces shades of bright orange. Also, the command for s ensures an irregularly striped pattern, depending on n and t, with saturaton varying between 0.4 and 0.1, producing a creamy color or a medium orange. The specification for b ensures that if **Abs[t]< 40**, then the coloring is slightly darker, giving a slight shadow effect where the coils of the shell meet. This helps to define the coils, which may not be clear in a complicated color pattern. (Note that we allowed t to vary between -180 and 180, instead of between 0 and 360. Note also that we have introduced a thickness directive.) (Color Fig 8.4)

$$
\text{ParametricPlot3D}\Big[\text{Evaluate}\Big[\text{Table}\Big[
$$
$$
\Big\{0.5` \, e^{0.011552473945290626` \, n}\, \text{Sin}\Big[\frac{n\pi}{30}\Big] + 0.5` \, e^{0.011552473945290626` \, n}\, \text{Sin}\Big[\frac{n\pi}{30}\Big]\, \text{Sin}\Big[\frac{\pi}{180}\, t\Big],
$$
$$
0.5` \, e^{0.011552473945290626` \, n}\, \text{Cos}\Big[\frac{n\pi}{30}\Big] + 0.5` \, e^{0.011552473945290626` \, n}\, \text{Cos}\Big[\frac{n\pi}{30}\Big]\, \text{Sin}\Big[\frac{\pi}{180}\, t\Big],
$$
$$
-1.414213` \, e^{0.011552473945290626` \, n} + 0.5` \, e^{0.011552473945290626` \, n}\, \text{Cos}\Big[\frac{\pi}{180}\, t\Big],
$$
$$
\Big\{\text{Hue}\Big[.05 + 0.04\,\text{Random}[], \text{If}\Big[\text{Mod}[t+n,3] < \text{Random}[\text{Integer},\{0,1\}]\Big]\,\|\!
$$
$$
\text{Mod}\Big[\frac{t}{4},2\Big] < 1, 0.4, 0.1\Big], \text{If}[\text{Abs}[t] < 40, 0.9, 1]\Big], \text{Thickness}[0.035]\Big\}\Big\},
$$
$$
\{n,-360,0\}\Big], \{t,-180,180\}\Big], \text{Axes} \rightarrow \text{False}, \text{PlotRange} \rightarrow \text{All},
$$
$$
\text{Boxed} \rightarrow \text{False}, \text{PlotRange} \rightarrow \text{All}, \text{ViewPoint} -> \{-2.764, 0.160, 1.945\}\Big]
$$

In the following example, we increase the value of s, while retaining the value for a. This means that the coils do not 'fit together'. Also, stripes along the length of the tube have been introduced by a color directive which depends on t. (Color Fig 8.5)

$$
\text{ParametricPlot3D}\Big[\text{Evaluate}\Big[
$$
$$
\text{Table}\Big[\Big\{0.5` \, e^{0.011552473945290626` \, n}\, \text{Sin}\Big[\frac{n\pi}{30}\Big] + 0.5` \, e^{0.011552473945290626` \, n}\, \text{Sin}\Big[\frac{n\pi}{30}\Big]\, \text{Sin}[\pi\, t],
$$
$$
0.5` \, e^{0.011552473945290626` \, n}\, \text{Cos}\Big[\frac{n\pi}{30}\Big] + 0.5` \, e^{0.011552473945290626` \, n}\, \text{Cos}\Big[\frac{n\pi}{30}\Big]\, \text{Sin}[\pi\, t],
$$
$$
-2.5\, e^{0.011552473945290626` \, n} + 0.5` \, e^{0.011552473945290626` \, n}\, \text{Cos}[\pi\, t],
$$
$$
\{\text{Thickness}[0.035], \text{Hue}[0.08, 0.3, \text{If}[\text{Mod}[\text{Floor}[25\, t], 5] < 1\,\|\, n > -1, 1, 0.7]]\}\Big\},
$$
$$
\{n,-360,0\}\Big], \{t,0,2\}\Big], \text{Axes} \rightarrow \text{False}, \text{PlotRange} \rightarrow \text{All},
$$
$$
\text{Boxed} \rightarrow \text{False}, \text{ViewPoint} -> \{3.383, -0.040, -0.026\},
$$
$$
\text{Background} -> \text{RGBColor}[0.652354, 0.917983, 0.859388]\Big]
$$

In the following example, the coiled tube has been given an elliptical cross-section simply by changing the box-ratios. The color varies with the t parameter from white to medium pink. (Color Fig 8.6)

ParametricPlot3D[Evaluate[Table[

$$\{0.5` \, e^{0.011552473945290626` \, n} \, Sin\left[\frac{n\pi}{30}\right] + 0.5` \, e^{0.011552473945290626` \, n} \, Sin\left[\frac{n\pi}{30}\right] \, Sin\left[\frac{\pi}{180} \, t\right],$$

$$0.5` \, e^{0.011552473945290626` \, n} \, Cos\left[\frac{n\pi}{30}\right] + 0.5` \, e^{0.011552473945290626` \, n} \, Cos\left[\frac{n\pi}{30}\right] \, Sin\left[\frac{\pi}{180} \, t\right],$$

$$-1.414213` \, e^{0.011552473945290626` \, n} + 0.5` \, e^{0.011552473945290626` \, n} \, Cos\left[\frac{\pi}{180} \, t\right],$$

$$\left\{Hue\left[1, \, Abs\left[0.5 - \frac{Abs[t]}{360}\right], \, 1\right], \, Thickness[0.04]\right\}\}, \, \{n, \, -360, \, 0\}\right], \, \{t, \, -180, \, 180\}\right],$$

Axes → **False, BoxRatios** → {1, 1, 2}, **PlotRange** → **All,**
Boxed → **False, PlotRange** → **All,**
Background –> **RGBColor[0.753918, 0.843763, 0.906264],**
ViewPoint –> {3.276, −0.039, 0.845}]

We now construct a rainbow-colored, transparent image, showing some of the internal anatomy of the shell. We replace θ by $\frac{n\pi}{60}$, and use the default thickness. We also choose s and a such that $s < \frac{Cosech[a\pi]}{2}$, so that the coils of the shell are overlapping. (Color Fig 8.7)

$$circle\left[0.110318, \, 1.2, \, \frac{n\pi}{60}\right]$$

$$\left\{0.5 \, e^{0.00577624n} \, Sin\left[\frac{n\pi}{60}\right] + 0.5 \, e^{0.00577624n} \, Sin\left[\frac{n\pi}{60}\right] \, Sin[t],\right.$$

$$0.5 \, e^{0.00577624n} \, Cos\left[\frac{n\pi}{60}\right] + 0.5 \, e^{0.00577624n} \, Cos\left[\frac{n\pi}{60}\right] \, Sin[t],$$

$$\left. -1.2 \, e^{0.00577624n} + 0.5 \, e^{0.00577624n} \, Cos[t]\right\}$$

ParametricPlot3D[Evaluate[

$$Table\left[\left\{0.5` \, e^{0.005776236972645313` \, n} \, Sin\left[\frac{n\pi}{60}\right] + 0.5` \, e^{0.005776236972645313` \, n} \, Sin\left[\frac{n\pi}{60}\right] \, Sin[t],\right.\right.$$

$$0.5` \, e^{0.005776236972645313` \, n} \, Cos\left[\frac{n\pi}{60}\right] + 0.5` \, e^{0.005776236972645313` \, n} \, Cos\left[\frac{n\pi}{60}\right] \, Sin[t],$$

$$\left. -1.2` \, e^{0.005776236972645313` \, n} + 0.5` \, e^{0.005776236972645313` \, n} \, Cos[t], \, Hue\left[\frac{t}{2\pi}\right]\right\},$$

$$\{n, \, -540, \, 0\}\right], \, \{t, \, 0, \, 2\pi\}\right], \, \textbf{Axes} \rightarrow \textbf{False, PlotRange} \rightarrow \textbf{All,}$$

Boxed → **False, ViewPoint** –> {3.383, −0.040, −0.026}]

8.3.4 Constructing Shell Images as 3D Surface Plots

A coiled shell can also be represented as a 3D surface plot, instead of as a sequence of 3D curves. In the function **circle[a, s, θ]**, we replace a and θ by our chosen values, and retain θ as the second parameter for **ParametricPlot3D**.

Here is an example with f = 3, a = 0.1748495762830299, s = 0.8660254037844385`

$$circle[a\_, s\_, \theta\_] := \{0.5` e^{a\theta} Sin[t] Sin[\theta] + 0.5 E^{a\theta} Sin[\theta],$$
$$0.5` e^{a\theta} Cos[\theta] Sin[t] + 0.5 E^{a\theta} Cos[\theta], 0.5` e^{a\theta} Cos[t] - s E^{a\theta}\} /.$$

$$circle[0.1748495762830299, 0.8660254037844385, \theta]$$

$$\{0.5 e^{0.17485\theta} Sin[\theta] + 0.5 e^{0.17485\theta} Sin[t] Sin[\theta],$$
$$0.5 e^{0.17485\theta} Cos[\theta] + 0.5 e^{0.17485\theta} Cos[\theta] Sin[t], -0.866025 e^{0.17485\theta} + 0.5 e^{0.17485\theta} Cos[t]\}$$

If a color directive is used, the option **Lighting→False** must be included. The stripes in this case encircle the cross-sections of the coil. The color directive depends on θ. (Color Fig 8.8)

$$\textbf{ParametricPlot3D}\Big[\Big\{0.5` e^{0.1748495762830299` \theta} Sin[\theta] + 0.5` e^{0.1748495762830299` \theta} Sin[t] Sin[\theta],$$
$$0.5` e^{0.1748495762830299` \theta} Cos[\theta] + 0.5` e^{0.1748495762830299` \theta} Cos[\theta] Sin[t],$$
$$-0.8660254037844385` e^{0.1748495762830299` \theta} + 0.5 e^{0.1748495762830299` \theta} Cos[t],$$
$$\textbf{Hue}\Big[0.05, .2 \textbf{ Random}[] + \frac{1}{4} \textbf{ Mod}[7\,\theta, 2], 1\Big]\Big\}, \{\theta, -6\,\pi, 0\}, \{t, -\pi, \pi\},$$
$$\textbf{PlotRange} \to \textbf{All}, \textbf{PlotPoints} \to \{80, 40\}, \textbf{Boxed} \to \textbf{False}, \textbf{Axes} \to \textbf{False},$$
$$\textbf{Lighting} \to \textbf{False}, \textbf{ViewPoint} -> \{2.070, -0.025, 2.676\},$$
$$\textbf{Background} -> \textbf{RGBColor}[0.597665, 0.843763, 0.871107]\Big]$$

In the case of a 3D parametric surface construction, omitting the polygon edges when using a color directive does not always give a good image. The shell sometimes looks too smooth, or sometimes not well-defined. One loses *Mathematica*'s clever shading and lighting properties that one obtains with the default coloring.

In the following example, we use *Mathematica*'s default coloring:

We wish to make a 2-image stereogram of the shell. The commands j and k below are used to construct a shell image from 2 different view-points. The command **DisplayFunction→ Identity** is used so that the images are not displayed. The commands c and d are used to remove the mesh from j and k, as was explained in 2.7. Then the resulting images are displayed next to each other using the command **GraphicsArray**. (Color Fig 8.9)

$$j = \textbf{ParametricPlot3D}[\{0.5` e^{0.1748495762830299` \theta} Sin[\theta] + 0.5` e^{0.1748495762830299` \theta} Sin[t] Sin[\theta],$$
$$0.5` e^{0.1748495762830299` \theta} Cos[\theta] + 0.5` e^{0.1748495762830299` \theta} Cos[\theta] Sin[t],$$
$$-0.8660254037844385` e^{0.1748495762830299` \theta} + 0.5 e^{0.1748495762830299` \theta} Cos[t]\},$$
$$\{\theta, -8\,\pi, 0\}, \{t, -\pi, \pi\}, \textbf{PlotRange} \to \textbf{All}, \textbf{PlotPoints} \to \{80, 40\}, \textbf{Boxed} \to \textbf{False},$$
$$\textbf{Axes} \to \textbf{False}, \textbf{ViewPoint} -> \{2.070, -0.025, 2.676\}, \textbf{DisplayFunction} \to \textbf{Identity}]$$

k = **ParametricPlot3D[**{**0.5** $e^{0.1748495762830299`\,\theta}$ **Sin[**$\theta$**]** + **0.5** $e^{0.1748495762830299`\,\theta}$ **Sin[**t**] Sin[**θ**]**,
0.5 $e^{0.1748495762830299`\,\theta}$ **Cos[**$\theta$**]** + **0.5** $e^{0.1748495762830299`\,\theta}$ **Cos[**θ**] Sin[**t**]**,
$-$**0.8660254037844385** $e^{0.1748495762830299`\,\theta}$ + **0.5** $e^{0.1748495762830299`\,\theta}$ **Cos[**t**]**},
{θ, -8π, **0**}, {t, $-\pi$, π}, **PlotRange** $\rightarrow$ **All, PlotPoints** $\rightarrow$ {**80, 40**}, **Boxed** $\rightarrow$ **False,**
Axes $\rightarrow$ **False, ViewPoint** $->$ {**2.170,** $-$**0.075, 2.676**}, **DisplayFunction** $\rightarrow$ **Identity]**

c = j /. **Polygon[p_]** $->$ {**EdgeForm[ ], Polygon[p]**}

d = k /. **Polygon[p_]** $->$ {**EdgeForm[ ], Polygon[p]**}

Show[GraphicsArray[{c, d}, **GraphicsSpacing** $\rightarrow$ $-$**0.1],**
Background $->$ **RGBColor[0.847669, 0.792981, 0.480476]]**

Exercise:
Construct other shell images by varying the parameters, the thickness and color directives.

Appendices

Appendix to 5.4.2

Theorem

Let τ be a contraction mapping on $\mathcal{H}[\mathbb{R}^2]$. Let $B \in \mathcal{H}[\mathbb{R}^2]$ and let I be the identity map on $\mathbb{R}^2$.
Let $w = I \bigcup \tau$.
a) The sequence $(w^n[B])$ converges in the Hausdorff metric to an element, A, of $\mathcal{H}[\mathbb{R}^2]$ where A is the fixed point of a contraction mapping on $\mathcal{H}[\mathbb{R}^2]$.
b) If $D \in \mathcal{H}[\mathbb{R}^2]$, and $D \neq B$, then $(w^n[D])$ may converge to a different limit.

Proof:

a) Let C_B be the constant mapping on $\mathcal{H}[\mathbb{R}^2]$ satisfying $C_B[V] = B$ for all $V \in \mathcal{H}[\mathbb{R}^2]$ and let $v = C_B \bigcup \tau$. Since C_B and τ are contraction maps on $\mathcal{H}[\mathbb{R}^2]$, so is v, hence, by the Contraction Mapping Theorem, v has a unique fixed point $A \in \mathcal{H}[\mathbb{R}^2]$ such that for all $V \in \mathcal{H}[\mathbb{R}^2]$, the sequence $(v^n[V])$ converges to A.

We shall prove that, for all $n \in \mathbb{N}$, $w^n[B] = v^n[B]$. It will then follow that $(w^n[B])$ converges to A.

We prove by induction that $w^n[B] = B \bigcup \tau[B] \bigcup ..... \bigcup \tau^n[B]$.

Firstly, $w^1[B] = I[B] \bigcup \tau[B] = B \bigcup \tau[B]$.

Suppose that, for some $n \in \mathbb{N}$, $w^n[B] = B \bigcup \tau[B] \bigcup ..... \bigcup \tau^n[B]$..........................(I)

Then $w^{n+1}[B] = w[w^n[B]]$

$\qquad\qquad = I[w^n[B]] \bigcup \tau[w^n[B]]$

$\qquad\qquad = w^n[B] \bigcup \tau[B \bigcup \tau[B] \bigcup ...... \bigcup \tau^n[B]] \qquad$ (by (I))

$\qquad\qquad = w^n[B] \bigcup \tau[B] \bigcup \tau^2[B] \bigcup ... \bigcup \tau^{n+1}[B]$

$\qquad\qquad = B \bigcup \tau[B] \bigcup ..... \bigcup \tau^{n+1}[B]$.

Similarly it can be proved by induction that $v^n[B]$ has the same value:

Firstly, $v^1[B] = C_B[B] \bigcup \tau[B] = B \bigcup \tau[B]$.

Suppose $v^n[B] = B \bigcup \tau[B] \bigcup .... \bigcup \tau^n[A]$ for some $n \in \mathbb{N}$. Then

$\qquad v^{n+1}[B] = C_B[v^n[B]] \bigcup \tau[v^n[B]] = B \bigcup \tau[B] \bigcup ... \bigcup \tau^{n+1}[B]$.

The result now follows.

The above proof also applies to $\mathbb{R}$, $\mathbb{C}$, and $\mathbb{R}^3$ (and indeed to any complete metric space).

b) We prove this result by means of an example:

In $\mathcal{H}[\mathbb{R}]$, let $B = \{1\}$, $D = \{2\}$ and $\tau : x \to \frac{x}{2}$.

Then $w^n[B] = \{1, \frac{1}{2}, \frac{1}{4}, ..., \frac{1}{2^n}\} \to \{1, \frac{1}{2}, \frac{1}{4}, ...\} \bigcup \{0\}$ as $n \to \infty$, while

$\qquad w^n[D] = \{2, 1, \frac{1}{2}, \frac{1}{4}, ..., \frac{1}{2^{n-1}}\} \to \{2, 1, \frac{1}{2}, \frac{1}{4}, ...\} \bigcup \{0\}$ as $n \to \infty$.

Result b) is well illustrated by the last 2 images in 5.4.2.

It now follows that if an IFS, $\{\mathbb{R}^2 \,|\, w_0, w_1, w_2, ... w_n\}$ consists of n contraction mappings $w_1, w_2, ... w_n$ together with the identity map, w_0, then the sequence $(w^n[B])$ converges in the Hausdorff metric to an element of $\mathcal{H}[\mathbb{R}^2]$ where $w[B] = w_0[B] \bigcup w_1[B] \bigcup ... \bigcup w_n[B]$.

Appendix to 7.1.1

■ Conjugate Mappings

A rational map of the form $z \to \frac{az+b}{cz+d}$, with $ad - bc \neq 0$, is called a Möbius map. Two rational maps R and S are said to be conjugate if and only if there is a Möbius map g with $S = g \circ R \circ g^{-1}$, Beardon (1991). If g is linear, then it is a similarity (shape-preserving) transformation, and so is g^{-1}.

Any quadratic polynomial, $P[z] = \alpha z^2 + \beta z + \gamma$ ($\alpha, \beta, \gamma \in \mathbb{C}$) is conjugate to a quadratic polynomial of the form $Q_c[z] = z^2 + c$ ($c \in \mathbb{C}$) via a conjugation of the form $h[z] = az + b$ (a, $b \in \mathbb{C}$). This follows since $h^{-1}[Q_c[h[z]]] = (a^2 z^2 + 2abz + b^2 + c - b)/a$. So by choosing a, b and c to satisfy the equations: $\alpha = a$, $\beta = b$ and $\gamma = (b^2 + c - b)/a$, we can ensure that $h^{-1} \circ Q_c \circ h = P$.

Now, since $h^{-1} \circ Q_c \circ h = P$ we have that $h^{-1} \circ Q^k_c \circ h = P^k$ for all k. This means that the mapping h transforms the dynamical picture of P to that of Q_c. In particular, z is a periodic point of period k of P if and only if $h[z]$ is a periodic point of period k of Q_c.

So the (filled) Julia set of P is the image of the (filled) Julia set of Q_c under the transformation h. Further, since h is a similarity transformation, the (filled) Julia set of any quadratic polynomial is geometrically similar to the (filled) Julia set of Q_c for some $c \in \mathbb{C}$.

For this reason we examine only the filled Julia sets of Q_c for polynomials of degree 2.

Here is an example of a polynomial of degree 2 which is conjugate to Q_c, with $c = 0.377 - 0.248 I$, whose Julia set was constructed in 7.1.1. Let $h[z] = 2z - I$, then $h^{-1}[z] = \frac{z+I}{2}$, and $h \circ Q_c \circ h^{-1}[z] = 0.5 z^2 + Iz - 2.5 - 0.254 - 1.496 I$. We construct the Julia set of $h \circ Q_c \circ h^{-1}$. (Notice that a different escape criterion is required.)

$$\textbf{DensityPlot}\left[\textbf{Length}\left[\textbf{FixedPointList}\left[(0.254 - 1.496\,i) + i\,\#1 + \frac{\#1^2}{2}\,\&,\,x + I\,y,\,101,\right.\right.\right.$$

$$\textbf{SameTest} \to (\textbf{Abs}[\#2] > 5.0\ \&)\bigg]\bigg],\,\{x,\,-2,\,2.\},\,\{y,\,-3.5,\,1.5\},\,\textbf{Mesh} \to \textbf{False},$$

$$\textbf{Frame} \to \textbf{False},\,\textbf{PlotPoints} \to 250,\,\textbf{AspectRatio} \to \textbf{Automatic},\,\textbf{Axes} -> \textbf{True},$$

$$\textbf{ColorFunction} \to (\textbf{If}[\# \geq 1,\,\textbf{RGBColor}[0,\,0,\,0],\,\textbf{RGBColor}[1,\,1,\,1]]\ \&)\bigg];$$

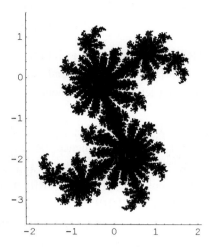

Appendix to 7.1.2

Theorem 1

Let $R[z] = \frac{P[z]}{Q[z]}$ be a rational function of degree $d \geq 2$, such that infinity is a an attracting fixed point of R.

There exists $S > 0$, such that:

1) if $|z| > S$, then $|R[z]| > |z|$;

2) if $|z| > S$ then $|R^n[z]| \to \infty$ as $n \to \infty$.

Proof

1) Let $P[z] = p_0 + p_1 z + \ldots + p_n z^n$, $Q[z] = q_0 + q_1 z + \ldots + q_m z^m$, with $p_n\,q_m \neq 0$ and $n > m$.

Suppose that $n > m + 1$.

Then $|\frac{R[z]}{z}| \to \infty$ as $z \to \infty$, and so there exists $S > 0$ such that:

$|z| > S \Rightarrow |\frac{R[z]}{z}| > 1 \Rightarrow |R[z]| > |z|$.

Suppose that n = m + 1.

Then $\left|\frac{R[z]}{z}\right| \to \left|\frac{p_n}{q_m}\right|$ as $z \to \infty$.

Also, since infinity is an attracting fixed point of R, we have that $\left|\frac{p_n}{q_m}\right| > 1$. It follows that there exists S > 0 such that $|z| > S \Rightarrow |\frac{R[z]}{z}| > 1 \Rightarrow |R[z]| > |z|$.

Since, R has at most finitely many poles we can choose S so that R is continuous at z if $|z| \geq S$.

2) Suppose there exists a with $|a| \geq S$ such that $|R^n[a]|$ does not tend to infinity as n tends to infinity. Now $|R[a]| > |a| > S$, so $|R^2[a] > |R[a]| > |a| > S$, and so forth. So, $(|R^n[a]|)$ is a strictly increasing sequence which does not tend to infinity, hence it converges to a finite limit, b, say.

Then $S \leq |R^n[a]| < b$ for all $n \in \mathbb{N}$. Thus, for all $n \in \mathbb{N}$, $R^n[a] \in D(0, b) = \{z \in \mathbb{C} \mid |z| \leq b\}$, which is (sequentially) compact. It follows that $(R^n[a])$ has a subsequence, $(R^{b_n}[a])$, which converges to t, say with $S \leq |t| \leq b$. So $|R^{b_n}[a]|$ converges to $|t|$ and so $|R^{b_n}[a]| < |t| (= b)$, for all $n \in \mathbb{N}$. Since R is continuous at t, $R[R^{b_n}[a]] \to R[t]$ as $n \to \infty$. But $(R[R^{b_n}[a]])$ is a subsequence of the sequence $(|R^n[a]|)$, so $|R[R^{b_n}[a]]| \to t$ as $n \to \infty$. So $|R[t]| = |t|$. But $|R[t]| > |t|$. We have a contradiction, so $|R^n[a]| \to \infty$ as $n \to \infty$ for all a such that $|a| \geq S$.

Theorem 2

In the following theorem, 'attracting' shall include 'super-attracting'.

If R is a rational function of degree greater than 1, with no neutral cycles, then the Julia set of R, J_R, is the set of points whose orbits do not converge to any attracting cycle.

Proof

In this proof, theorem numbers refer to those in the book by Beardon (1991).

Firstly, if $z \in J_R$, then, as z is in the boundary of the basin of attraction of every attracting cycle (each of which is open), the orbit of z does not converge to any attracting cycle.

Now suppose that z is in the Fatou set, F_R of R. We shall prove that the orbit of z converges to some attracting cycle, $\{\alpha, R[\alpha], ..., R^{k-1}[\alpha]\}$, of R. To do this, it is sufficient to show that $R^{nk}[z] \to R^t[\alpha]$ as $n \to \infty$ for some t.

Now z is in a component, C, of F_R, since F_R is the disjoint union of its components. Also, R[C] is connected, as R is continuous. So R[C] is a subset of a component, D, of F_R. By Theorem 5.5.4, R[C]=D. It follows that for each $n \in \mathbb{N}$, $R^n[C]$ is a component of F_R. Also C is said to be periodic if for some positive integer n, $R^n[C] = C$ and eventually periodic if for some positive integer m, $R^m[C]$ is periodic. Every component of the Fatou set of a rational map is eventually periodic (Theorem 8.1.2).

It follows that there exist positive integers k and m such that $R^k[R^m[C]] = R^m[C]$, so $R^m[C]$ is forward invariant under R^k. As R and hence R^k has no neutral fixed point, and 'Siegel disks' and 'Hermann rings' are only present if there are neutral attracting cycles, we conclude that $R^m[C]$ contains an attracting fixed point, α, of R^k (Theorem 7.1.2). Hence, the sequence of functions (R^{nk}) converges to the constant function α (defined by $\alpha[z]=\alpha$) on $R^m[C]$ (Theorem 6.3.1). It follows that $(R^{nk}[R^m[z]]) = (R^{nk+m}[z])$ converges to α, and thus $(R^{nk}[z])$ converges to α. So the orbit of z converges to an attracting cycle of R.

We have shown that $z \in J_R$ if and only if the orbit of z does not converge to any attracting cycle of R.

Appendix to 8.3.1

With reference to Fig 1, 8.3.1:

$$(r_1 + r_2)^2 = (0.5 + 0.5 \operatorname{Exp}[-2\, a\pi])^2 = \operatorname{Exp}[-2a\pi]\, \operatorname{Cosh}^2[a\pi].$$

Also, $AC^2 = \left(\frac{\operatorname{Exp}[-2\pi a]-1}{2}\right)^2 + 4\, s^2\left(\frac{\operatorname{Exp}[-2\pi a]-1}{2}\right)^2 = \operatorname{Exp}[-2\pi a]\, \operatorname{Sinh}^2[\pi a]\, (1 + 4\, s^2).$

For the 'just-touching' case, we equate $(r_1 + r_2)^2$ and AC^2, and solve for s in terms of a, to obtain $s = 0.5\, \operatorname{Cosech}[a\,\pi]$. (We used the fact that $a\,\pi > 0$.)
Also:
1) if $(r_1 + r_2)^2 > AC^2$ then $s > 0.5\, \operatorname{Cosech}[a\,\pi]$;
2) if $(r_1 + r_2)^2 < AC^2$, then $s < 0.5\, \operatorname{Cosech}[a\,\pi]$.

Bibliography

Barnsley, M. (1988), *Fractals Everywhere*, Second Edition, Academic Press.

Beardon, A. F. (1991), *Iteration of Rational Functions*, First Edition, Springer-Verlag, New York.

Carleson, L. C. and Gamelin, T.W. (1993), *Complex Dynamics*, First Edition, Springer-Verlag, New York.

Chossat P. and Golubitsky M. (1998), *Symmetry-increasing Bifurcation of Chaotic Attractors*, p. 423, Physica D 32.

Dawkins, R. (1996), *Climbing Mount Improbable*, Penguin Group.

Devaney, R. L. (1988), *Fractal Patterns Arising in Chaotic Dynamical Systems*, The Science of Fractal Images, Edited by H. O. Peitgen and D. Saupe, Springer-Verlag, New York.

Devaney, R. L. (1989), *An Introduction to Chaotic Dynamical Systems*, Second Edition, Addison-Wesley.

Devaney, R. L. (1992), *A First Course in Chaotic Dynamical Systems*, First Edition, Addison-Wesley, Boston.

Devaney, R. L. (1994), *Complex Dynamics and Entire Functions*, p.161, Proceedings of Symposia in Applied Mathematics Volume 49.

Field M. and Golubitsky M. (1995), *Symmetry in Chaos*, Oxford University Press.

Holmgren, R. A. (1994), *A First Course in Discrete Dynamical Systems*, First Edition, Springer-Verlag, New York.

Keen, L. (1989), *Julia Sets*, p. 57, Proceedings of Symposia in Applied Mathematics Volume 39.

Keen, L. (1994), *Julia Sets of Rational Maps*, p. 181, Proceedings of Symposia in Applied Mathematics Volume 49.

Maeder, Roman E. (1996), Programming in *Mathematica*, Third Edition, Addison-Wesley.

Maurer, P.M. (1987), A Rose is a Rose, Amreican Mathematical Monthly, Volume 94 No. 7.

Milnor, J. (1999), *Dynamics in One Complex Variable*, Friedr: Vieweg & Sohn Verlagsgesellschaft mbH, Braunschweig/Wiesbaden.

Peitgen, H. O., Jurgens, H. and Saupe, D. (1992), *Chaos and Fractals*, New Frontiers of Science, Springer-verlag.

Raup, D. M. (1966), *Geometric Analysis of Shell Coiling: General Problems*, p.1178-1190, Journal of Paleontology, Volume 40, No. 5

Steinmetz, N. (1993), *Rational Iteration*. de Gruyter Studies in Mathematics 16, Edited by H. Bauer, J.L. Kazdan and E. Zehnder, Berlin - New York.

Index

If an entry has more than one reference, one of them is sometimes in boldface, to indicate the main reference. The other references are usually examples of use.
Mathematica commands are in boldface.

Fig 2.1

Fig 2.2

Fig 2.3

Fig 2.4

Fig 2.5

Fig 2.6

Fig 2.7

Fig 2.8

Fig 2.9

Fig 2.10

Fig 2.11

Fig 2.12

Fig 2.13

Fig 2.14

Fig 2.15

Fig 2.16

310

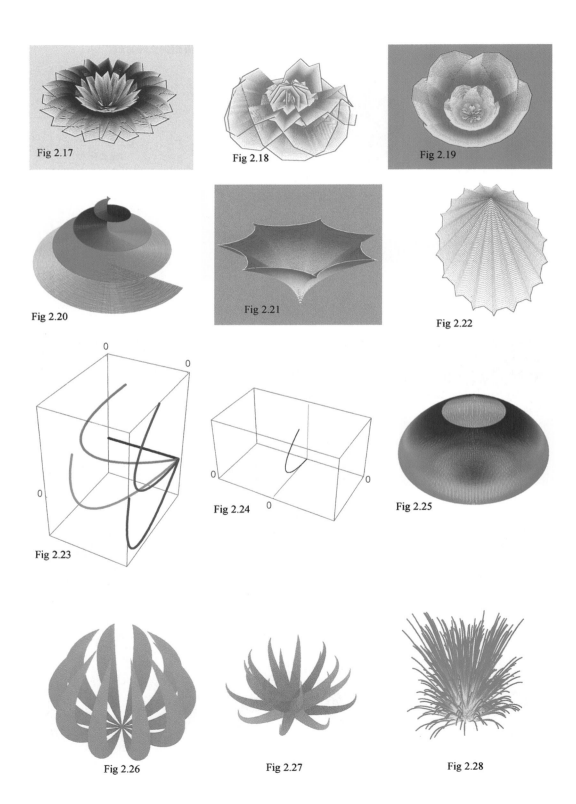

Fig 2.17

Fig 2.18

Fig 2.19

Fig 2.20

Fig 2.21

Fig 2.22

Fig 2.23

Fig 2.24

Fig 2.25

Fig 2.26

Fig 2.27

Fig 2.28

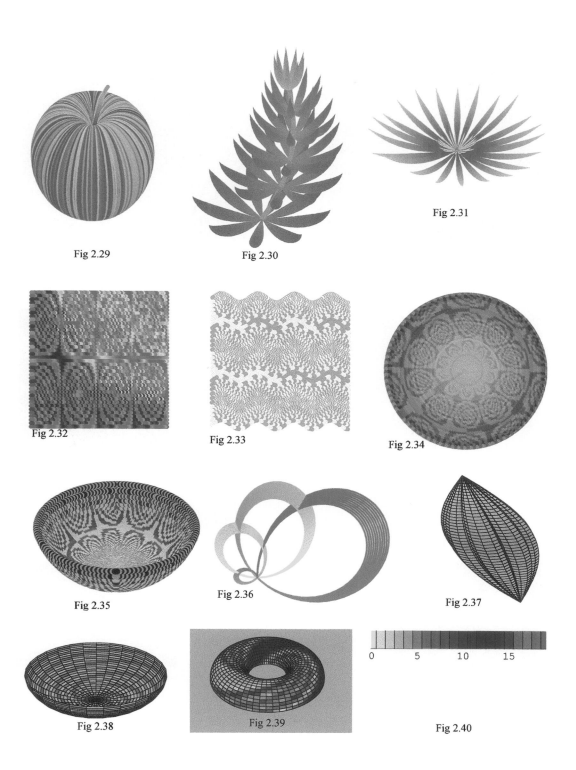

Fig 2.29

Fig 2.30

Fig 2.31

Fig 2.32

Fig 2.33

Fig 2.34

Fig 2.35

Fig 2.36

Fig 2.37

Fig 2.38

Fig 2.39

Fig 2.40

312

Fig 2.41

Fig 2.42

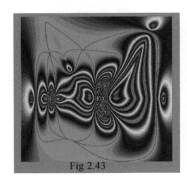

Fig 2.43

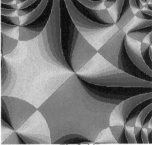

Fig 2.44

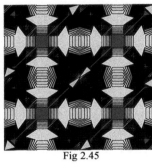

Fig 2.45

Fig 2.46

Fig 2.47

Fig 2.48

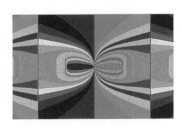

Fig 2.49

Fig 2.50

Fig 2.51

Fig 2.52

Fig 2.53

Fig 2.54

Fig 2.55

Fig 2.56

Fig 2.57

Fig 2.58

314

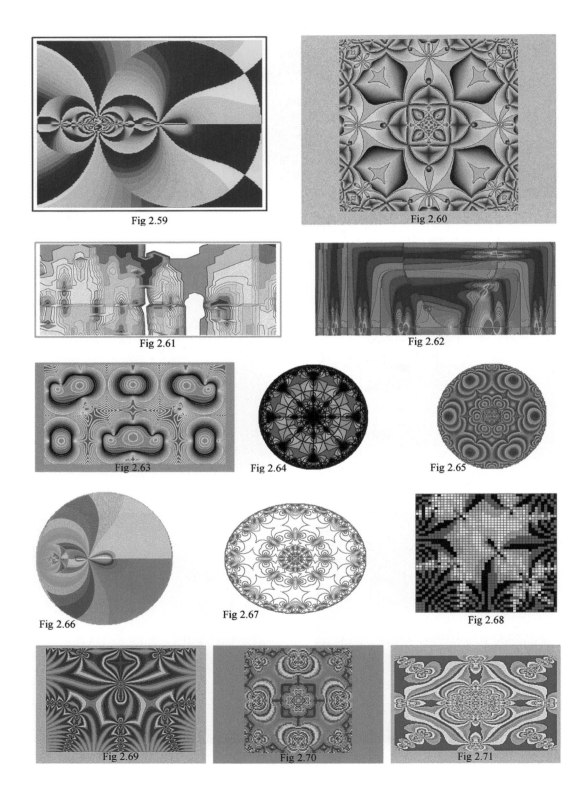

Fig 2.59

Fig 2.60

Fig 2.61

Fig 2.62

Fig 2.63

Fig 2.64

Fig 2.65

Fig 2.66

Fig 2.67

Fig 2.68

Fig 2.69

Fig 2.70

Fig 2.71

Fig 2.72

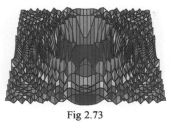

Fig 2.73

Fig 2.74

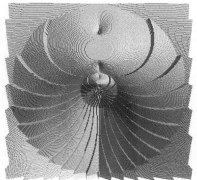

Fig 2.75

Fig 2.76

Fig 2.77

Fig 2.78

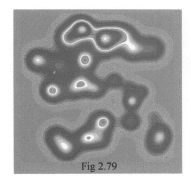

Fig 2.79

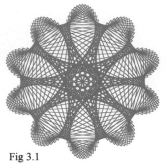

Fig 3.1

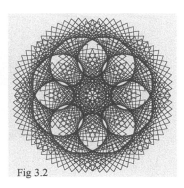

Fig 3.2

316

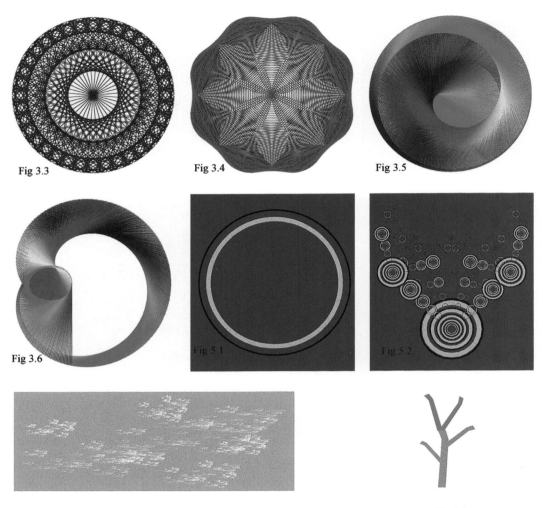

Fig 3.3

Fig 3.4

Fig 3.5

Fig 3.6

Fig 5.1

Fig 5.2

Fig 5.3

Fig 5.4

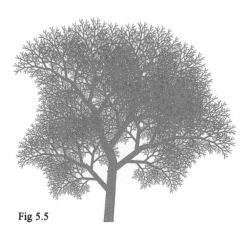

Fig 5.5

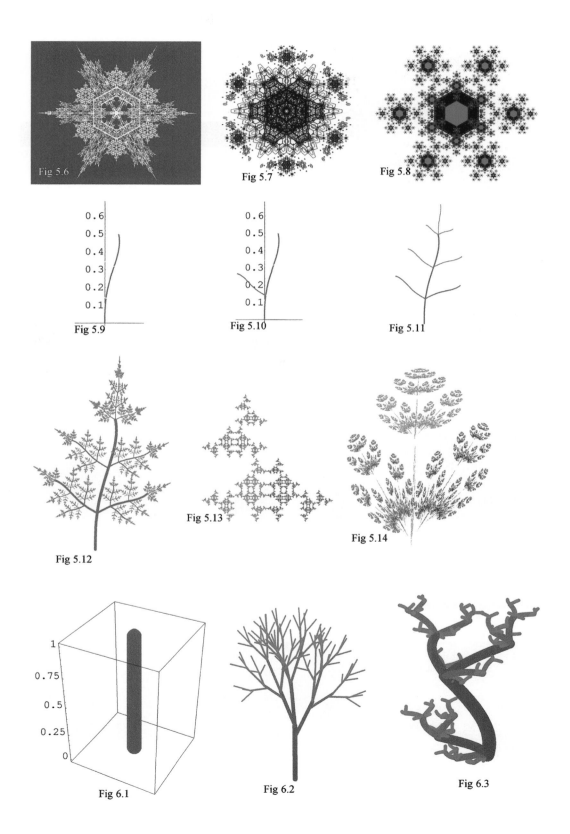

Fig 5.6

Fig 5.7

Fig 5.8

Fig 5.9

Fig 5.10

Fig 5.11

Fig 5.12

Fig 5.13

Fig 5.14

Fig 6.1

Fig 6.2

Fig 6.3

318

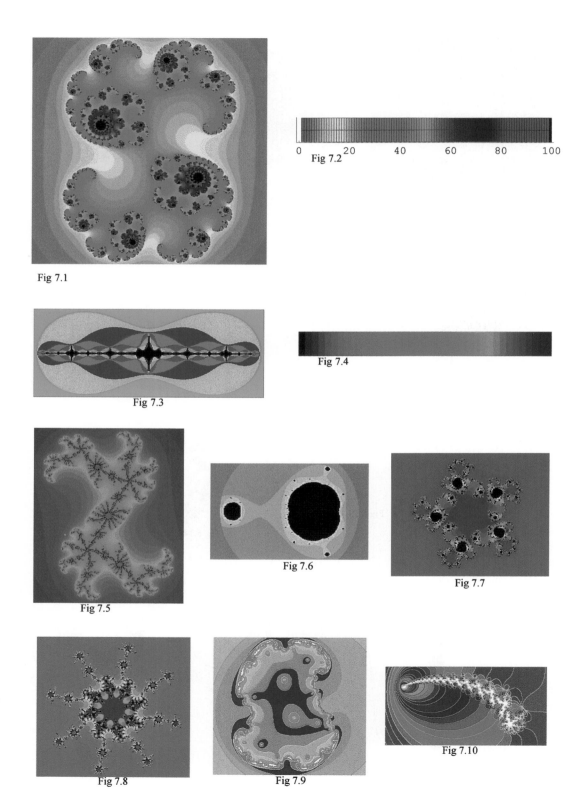

Fig 7.1

Fig 7.2

Fig 7.3

Fig 7.4

Fig 7.5

Fig 7.6

Fig 7.7

Fig 7.8

Fig 7.9

Fig 7.10

Fig 7.11

Fig 7.12

Fig 7.13

Fig 7.14

Fig 7.15

Fig 7.16

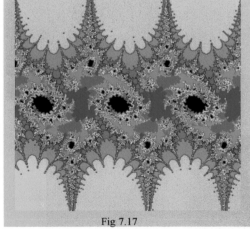

Fig 7.17

Fig 7.18

320

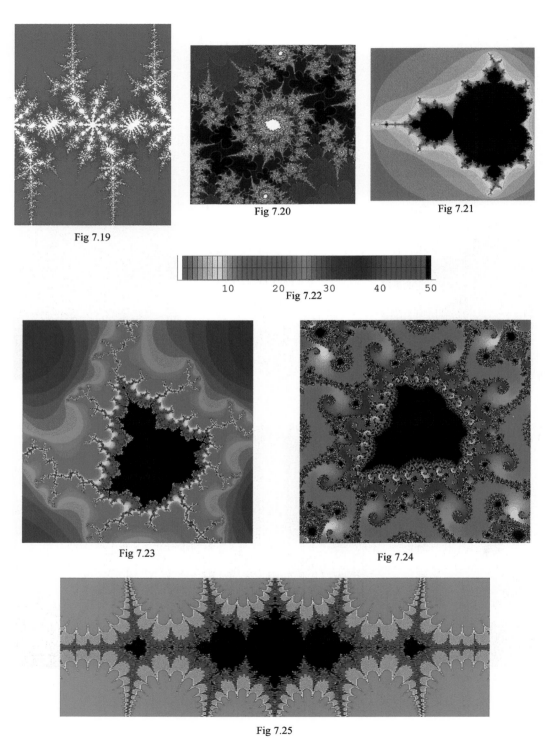

Fig 7.19

Fig 7.20

Fig 7.21

Fig 7.22

Fig 7.23

Fig 7.24

Fig 7.25

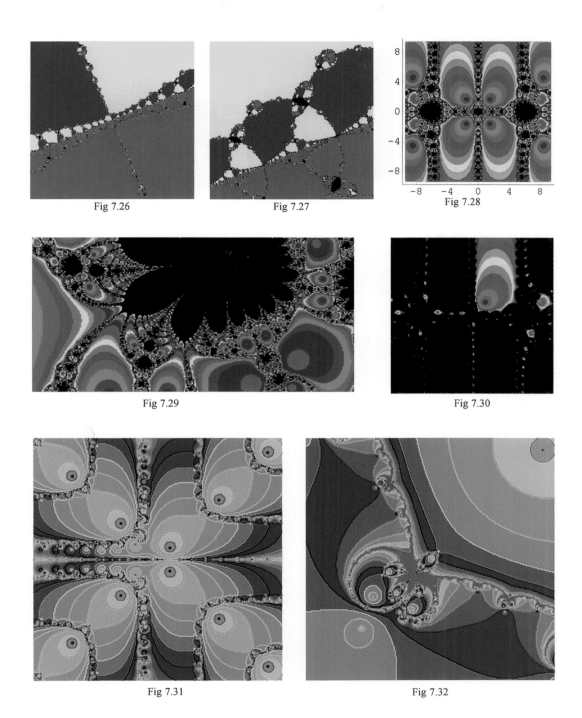

Fig 7.26

Fig 7.27

Fig 7.28

Fig 7.29

Fig 7.30

Fig 7.31

Fig 7.32

322

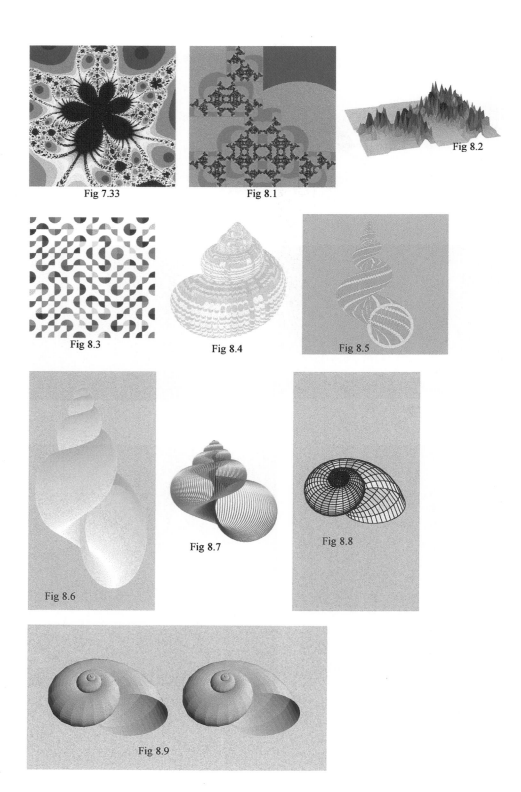

Fig 7.33

Fig 8.1

Fig 8.2

Fig 8.3

Fig 8.4

Fig 8.5

Fig 8.6

Fig 8.7

Fig 8.8

Fig 8.9